Order, Disorder and Criticality

Advanced Problems of Phase Transition Theory

Volume 3

Order, Disorder and Criticality

Advanced Problems of Phase Transition Theory

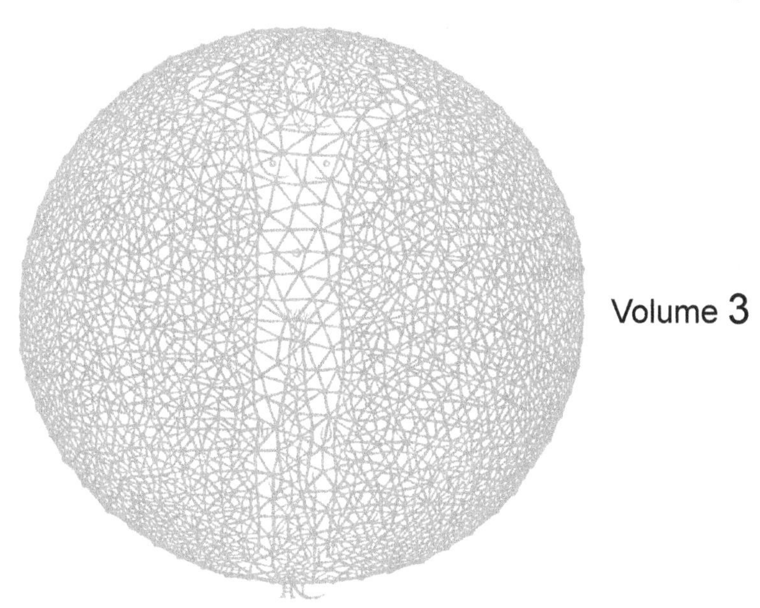

Volume 3

Editor

Yurij Holovatch

National Academy of Sciences, Ukraine

World Scientific

NEW JERSEY • LONDON • SINGAPORE • BEIJING • SHANGHAI • HONG KONG • TAIPEI • CHENNAI

Published by

World Scientific Publishing Co. Pte. Ltd.
5 Toh Tuck Link, Singapore 596224
USA office: 27 Warren Street, Suite 401-402, Hackensack, NJ 07601
UK office: 57 Shelton Street, Covent Garden, London WC2H 9HE

British Library Cataloguing-in-Publication Data
A catalogue record for this book is available from the British Library.

ORDER, DISORDER AND CRITICALITY
Advanced Problems of Phase Transition Theory
Volume 3

ISBN 978-981-4417-88-4

Printed in Singapore.

Preface

This book continues a series of review volumes on phase transitions and critical phenomena which has been initiated recently by World Scientific.[1] The principal goal of the series is to give pedagogical introductory reviews of problems that are currently discussed in the scientific literature and are related to phase transitions and criticality. This series aims to demonstrate that phase transition theory, which experienced its golden age during the 1970s and 1980s, is far from complete and there is still a good deal of work to be done, both at a fundamental level and in respect of applications.

This, the third volume of the series, deals with a broad spectrum of problems connected with criticality. It covers its theoretical backgrounds (such as the origin and form of scaling relations for logarithmic-correction exponents, discussed in the chapter by Ralph Kenna), analytical approaches to describe criticality in specific systems (such as criticality of ionic fluids, the chapter by Oksana Patsahan and Ihor Mryglog), numerical simulations of critical systems (the chapter by Wolfhard Janke) as well as well as phase transitions on complex networks (Krzysztof Suchecki and Janusz A. Hołyst) and in the minority game model (František Slanina).

The reader of this volume has the possibility to discover how the appearance of long-range correlations gives rise to common features in phenomena that occur in very different systems, descriptions of which sometimes belong to different fields of science. Indeed, the systems under consideration range from lattice magnets and soft matter (complex fluids and polymers) to models of econo- and sociophysics. Another unique feature of this book is that it allows the reader to get acquainted with the variety of methods used in theoretical description and modeling of these systems. Indeed, whereas the reviews presented in the former two volumes of this series were more concentrated on the renormalization group approaches (see their contents on pages vii and vii) the methods discussed in this volume include phenomenological thermodynamical approaches, Lee-Yang analysis

[1] *Order, Disorder and Criticality. Advanced Problems of Phase Transition Theory*, edited by Yu. Holovatch (World Scientific, Singapore), vol. 1 – 2004; vol. 2 – 2007.

of partition function zeros, functional integration, variants of mean-field based calculations, the replica trick to treat quenched disorder, Monte Carlo simulations, time series analyses and many others. The way these methods are introduced and the manner of presentation of results make the presentation interesting both for experts as well as for those who want to get simple but comprehensive reviews of the latest activities in the field.

One more common feature of the reviews collected in this book is that in one or another way all of them are related to so-called complex systems. The notion of a complex system is currently understood by the physics community as either a system who's behaviour crucially depends on the details of the system[2] or as a systems of many interacting entities of non-physical origin.[3] Papers presented in this volume deal with complexity in both above mentioned senses. Since all complex systems involve cooperative behaviour between many interconnected components, the field of phase transitions and critical phenomena gives a very natural conceptual and methodological framework for their study.

The review chapters of this book, as those of the previous volumes of this series stem from the lectures presented during the workshops called "The Ising Lectures".[4] These workshops occur annually in Lviv (Ukraine) and provide a possibility for participants to attend review lecture courses, given by leading experts working in the field. I am deeply indebted to my colleagues, the authors of this volume, for coming to Lviv, for lecturing at the workshops, and for writing these reviews. Special thanks are due to World Scientific for their interest in getting this book published.

Yurij Holovatch
Institute for Condensed Matter Physics,
National Academy of Sciences of Ukraine,
Lviv, 16.06.2012

[2]P.W. Anderson. *Science* **177** (1972) 293; G. Parisi. *Physica A* **263** (1999) 557.
[3]D. Stauffer. *Physica A* **285** (2000) 121.
[4]See: http://www.icmp.lviv.ua/ising/

Contents of Vol. 1

Contents of Vol. 2

Contents

Chapter 1

Universal Scaling Relations for Logarithmic-Correction Exponents

Ralph Kenna

Applied Mathematics Research Centre,
Coventry University, Coventry CV1 5FB, England
r.kenna@coventry.ac.uk

By the early 1960's advances in statistical physics had established the existence of universality classes for systems with second-order phase transitions and characterized these by critical exponents which are different to the classical ones. There followed the discovery of (now famous) scaling relations between the power-law critical exponents describing second-order criticality. These scaling relations are of fundamental importance and now form a cornerstone of statistical mechanics. In certain circumstances, such scaling behaviour is modified by multiplicative logarithmic corrections. These are also characterized by critical exponents, analogous to the standard ones. Recently scaling relations between these logarithmic exponents have been established. Here, the theories associated with these advances are presented and expanded and the status of investigations into logarithmic corrections in a variety of models is reviewed.

Contents

1. Introduction

Phase transitions are abundant in nature. They are involved in the evolution of the universe and in a multitude of phenomena in the physical, biological and socio-economic sciences. First-order, temperature-driven, phase transitions, such as melting and evaporation , exhibit discontinuous changes in the internal energy through emission or absorption of a latent heat as the transition point is traversed. Higher-order transitions, in contrast, involve a continuous change in the internal energy. Unlike first-order transitions, they can involve divergences at the transition temperature T_c. One of the major achievements of statistical physics is the fundamental explanation of such phenomena and, 150 years after their experimental discovery,[1] their ubiquity ensures that their study remains one of the most exciting areas of modern physics. This review concerns such higher-order phase transitions.

Second-order transitions are particularly common and examples include ferromagnets, superconductors and superfluids in three-dimensional condensed-matter physics, as well as the Higgs phenomenon in four-dimensional particle physics. Indeed, the edifice of lattice quantum gauge theory relies upon second-order phase transitions to achieve a continuum limit. The Kosterlitz-Thouless transitions of a Coulomb gas in two dimensions or in thin films of ^{4}He are examples of infinite-order phase transitions.

A cornerstone in the study of phase transitions is the principal of universality. This maintains that entire families of systems behave identically in the neighbourhood of a Curie or critical point. Nearby, thermodynamic observables and critical exponents, which characterise the transition, do not depend on the details of intermolecular interactions. Instead they depend only on the range of interactions, inherent symmetries of the Hamiltonian and dimensionality d of the system. Universality arises as the system develops thermal or quantum fluctuations of all sizes near the critical point, which wash out the details of interaction and render the system scale invariant. This remarkable lack of dependency on the details allows us to understand real materials through simplified mathematical models which incorporate the same dimensions, symmetries and interaction ranges. Furthermore, systems with disparate physics can be categorised into common universality classes. Renormalization group (RG) theory provides a satisfying, fundamental explanation of critical behaviour and universality. Indeed, this is one of the major achievements of statistical mechanics.

Each higher-order phase transition is characterised by a set of universal critical exponents. These exponents describe the strength of the phase transition in terms of power-laws. In the 1960's, before the discovery of the renormalization group, a set of scaling relations between these critical exponents was developed and this set is now well established and of foundational importance in the study of critical phenomena. Because of the importance of these power laws and the associated scaling relations, circumstances where they are modified must be scrutinised and understood. In certain situations, these power laws are modified by multiplicative logarithmic factors, which themselves are raised to certain powers. Since 2006 a set of scaling relations between the powers of these logarithms was discovered,[2] analogous to the conventional scaling relations between the leading critical exponents. This review focuses on these universal scaling relations for logarithmic-correction exponents.

For spin models on d-dimensional systems, logarithmic corrections of this type occur when the mean-field descriptions, valid in high dimensions, turn to non-trivial power laws for smaller dimensionalities due to the importance of thermal fluctuations there. The value of d which marks the onset of the importance of these fluctuations is known as the upper critical dimension d_c. Another prominent example where multiplicative logarithmic corrections are manifest is in the q-state Potts model in two dimensions. For $q \leq 4$ there is a second-order phase transition there, while for $q > 4$ this transition is first order. Logarithmic corrections arise when $q = 4$. A more subtle example is the $d = 2$ Ising model with or without non-magnetic impurities. In each of these cases, exponents characterise the logarithmic corrections in a manner analagous to the way in which the standard critical exponents characterise the leading power-law scaling behaviour.

The examples cited above concern well-defined Euclidean lattices. There, the notion of space dimensionality is crucial, the logarithmic corrections only arising at special values of d. In recent times, the study of critical phenomena has also focused on systems defined on networks or random graphs. There, instead of the system's dimensionality, a set of probabilistically-distributed coordination numbers characterised by a parameter λ is associated with the network. Logarithmic corrections arise at a critical value of λ.

Although logarithmic corrections to scaling are also encountered in the study of surface effects, tricritical points, the Casimir effect, and elsewhere in physics, the emphasis in this review is on bulk critical phenomena and second-order phase transitions in particular. In the following, the leading, power-law scaling and the associated scaling relations are discussed in Sec.2, where the logarithmic-correction counterparts are also summarised. The standard derivation of the scaling relations in terms of homogeneous functions and the block-spin renormalization group is recalled in Sec.3. The logarithmic corrections are presented in Sec.4, where relations between them are derived is a self-consistent manner. Fisher renormalisation for logarithmic-correction exponents is discussed in Sec.5. In Sec.6 the values of the correction exponents (together with the leading exponents) are given for various models exhibiting second-order phase transitions.

2. Scaling Relations at Second-Order Phase Transitions

Second-order phase transitions are characterised by a power-law divergence in the correlation length ξ (the length scale which describes coherent behaviour of the system). A consequence of this is the power-law behaviour of many other physical observables. Although second-order phase transitions are also manifest in fluids, partical physics and other arenas, in this exposition we adhere to the language of magnetism for definiteness, the translation to other fields being easily facilitated. We have in mind, then, a system of spins s_i located at the sites i of a d-dimensional lattice, whose partition function is of the form

$$Z = \sum_{\{s_i\}} e^{-\beta E - hM} . \tag{1}$$

Here E and M represent the energy and magnetization, respectively, of a given configuration and the summation is over all such configurations accessible by the system. The parameters $\beta = 1/k_B T$ and $h = \beta H$ are the inverse temperature (divided by the Boltzmann constant) and the reduced external field (H is the absolute strength of an applied external field).

The Helmholtz free energy is usually defined as

$$F_L = -k_B T \ln Z_L(T, H). \tag{2}$$

At $h = 0$, a system with $N = L^d$ sites has entropy given by

$$Ns_L = -\frac{\partial F_L}{\partial T} = k_B \ln Z_L + \frac{1}{T} N e_L, \tag{3}$$

where

$$Ne_L = -\frac{\partial \ln Z_L}{\partial \beta} = \langle E \rangle, \tag{4}$$

is the internal energy and $\langle \ldots \rangle$ refer to expectation values. Since the prefactor in Eq.(2) plays no essential role in what is to come, we drop it in the definition of the reduced free energy

$$f_L = -\frac{1}{N} \ln Z_L(T, H). \tag{5}$$

With this set-up, the reduced internal energy and reduced entropy, being the first derivative of the reduced free energy, become essentially the same.

According to a (modified) Ehrenfest classification scheme, the order of the transition is that of the first derivative of the free energy which displays a non-analycity in the form of a discontinuity or divergence. For magnetically-symmetric systems this occurs at $h = 0$ and at the Curie critical temperature T_c. We write the reduced temperature as

$$t = \frac{|T - T_c|}{T_c}, \tag{6}$$

to express the distance away from criticality in a dimensionless way. The internal energy e and specific heat c are now defined as the first and second derivatives of the free energy with respect to temperature, respectively, while the magnetization (of the entire system) m and the susceptibility χ are defined as the first and second derivatives with respect to the external field.

2.1. *Leading scaling behaviour*

Since we are interested in behaviour near the critical point $(t, h) = (0, 0)$, we write thermodynamic functions in terms of the reduced variables, *viz.* $e_L(t, h)$, $c_L(t, h)$, $m_L(t, h)$ and $\chi_L(t, h)$, respectively. Here the subscripts indicate the size of the system. Then the leading, power-law, scaling behaviour which describes the approach to criticality at a phase transition of second order is conventionally described as

$$e_\infty(t, 0) \sim t^{1-\alpha}, \qquad e_\infty(0, h) \sim h^{\epsilon}, \tag{7}$$

$$c_\infty(t, 0) \sim t^{-\alpha}, \qquad c_\infty(0, h) \sim h^{-\alpha_c}, \tag{8}$$

$$m_\infty(t, 0) \sim t^{\beta} \quad \text{for } T < T_c, \qquad m_\infty(0, h) \sim h^{\frac{1}{\delta}}, \tag{9}$$

$$\chi_\infty(t, 0) \sim t^{-\gamma}, \qquad \chi_\infty(0, h) \sim h^{\frac{1}{\delta}-1}. \tag{10}$$

The numbers α, β, γ, δ, ϵ and α_c introduced here are called critical exponents.

The above thermodynamic functions – derivable from the partition function – describe how the entire system responds to tuning the temperature and/or external field near the phase transition. To characterise local behaviour within the system we need to introduce the correlation function and the correlation length. The correlation function is given by

$$\mathcal{G}_\infty(x,t,h) \sim x^{-(d-2+\eta)}\tilde{\mathcal{G}}\left(xt^\nu, ht^{-\Delta}\right), \tag{11}$$

and the correlation length in the even (temperature) and odd (magnetic field) sectors is usually written

$$\xi_\infty(t,0) \sim t^{-\nu}, \qquad \xi_\infty(0,h) \sim h^{-\nu_c}. \tag{12}$$

Again, η, ν, Δ and ν_c are critical exponents, with η being called the anomalous dimension.

In addition to the above thermodynamic and correlation functions, one may consider the zeros of the partition function. These are the complex values of the temperature or magnetic-field parameters at which the partition function vanishes. For example, when $h = 0$, the partition function in Eq.(1) becomes essentially a polynomial in $u = \exp(-\beta)$ when L is finite. Moreover, this polynomial has real coefficients. As such, its zeros are strictly complex in the variable u and appear in complex-conjugate pairs. The zeros in the complex temperature plane at real h values are denoted by $t_j(L,h)$ and are called Fisher zeros.[4] The set of such zeros becomes more dense as the lattice size increases. In the infinite-volume limit, Fisher zeros most commonly also lie on curves in the complex temperature plane,[4] but may also be dispersed across 2-dimensional areas (typically this happens when anisotropy is present[5]). In the thermodynamic limit they pinch the real temperature axis at the point where the phase transition occurs (namely at $t = 0$).

Similarly, when β is real and fixed, the partition function in Eq.(1) becomes a polynomial in $\exp(-h)$. Zeros in the complex magnetic-field plane (for real values of the reduced temperature t) are denoted by $h_j(L,t)$, are both t- and L-dependent and called Lee-Yang zeros after their inventors.[3] In the infinite-volume limit the Lee-Yang zeros also usually form curves in the complex plane. In fact, in many circumstances the Lee-Yang theorem[3] ensures that these zeros are purely imaginary.

The impact of the Lee-Yang or Fisher zeros onto the real magnetic-field or temperature axis precipitates the phase transition and, in this sense, the zeros may be considered as "proto-critical" points[6] – they have the potential to become critical points. Above the actual critical point $T > T_c$, the linear locus of Lee-Yang zeros remains away from the real axis. In the thermodynamic ($L \to \infty$) limit, its lowest point is called the Lee-Yang edge[6] and denoted by $r_{\mathrm{YL}}(t)$. The edge approaches the real temperature axis as t reduces to its critical value $t = 0$ and that approach is also characterised by a power law. Similarly, the lowest Fisher zero is denoted $t_1(h)$. For an infinite-sized system, this approaches the real axis as h vanishes.[7] The leading scaling behaviour for the edge of the distribution of Lee-Yang zeros are[7]

$$r_{\mathrm{YL}}(t) \sim t^{\Delta}, \qquad t_1(h) \sim h^{\frac{1}{\Delta}}. \tag{13}$$

Besides the critical indices listed above, one is often interested in the so-called shift exponent λ_{shift}. This characterises how the pseudocritical point in a finite-sized system is shifted away from the critical point. The pseudocritical point is the size-dependent value of the temperature $t_{\mathrm{pseudo}}(L)$ which marks the specific heat peak or the real part of the lowest Fisher zero. For a system of linear extent L it also scales as a power-law

$$t_{\mathrm{pseudo}}(L) \sim L^{\lambda_{\mathrm{shift}}}, \tag{14}$$

to leading order. We next summarise how these critical exponents are linked through the scaling relations.

2.2. *Scaling relations for leading exponents*

Thus the leading power-law scaling associated with second-order phase transitions is fully described by 10 critical exponents α, β, γ, δ, ϵ, α_c, ν, ν_c, Δ and η (or 11 if the exponent λ_{shift} characterising the finite-size scaling (FSS) of the pseudo-critical point is included). The following 8 scaling relations (9 including one for the shift exponent) are well established for the leading critical exponents both in the (even) thermal and (odd) magnetic sectors (see, e.g., Ref.[8] and references therein):

$$\nu d = 2 - \alpha, \tag{15}$$

$$2\beta + \gamma = 2 - \alpha, \tag{16}$$

$$\beta(\delta - 1) = \gamma, \tag{17}$$

$$\nu(2 - \eta) = \gamma, \tag{18}$$

$$\epsilon = 2 - \frac{(\delta - 1)(\gamma + 1)}{\delta\gamma}, \tag{19}$$

$$\alpha_c = -2 + \frac{(\gamma + 2)(\delta - 1)}{\delta\gamma}, \tag{20}$$

$$\nu_c = \frac{\nu}{\Delta}, \tag{21}$$

$$\Delta = \frac{\delta\gamma}{\delta - 1} = \delta\beta = \beta + \gamma. \tag{22}$$

The relation (15) was developed by Widom[9,10] using dimensional considerations, with alternative arguments given by Kadanoff.[11] Widom[9] also showed how a logarithmic singularity can arise in the specific heat if $\alpha = 0$, but does not have to, leaving instead a finite discontinuity (see also Ref.[4,10]). Later Josephson[12] derived a related inequality on the basis of some plausible assumptions and Eq.(15) is sometimes called Josephson's relation.[13] It can also be derived from the hyperscaling hypothesis, namely that the free energy behaves near criticality as the inverse correlation volume: $f_\infty(t, 0) \sim \xi_\infty^{-d}(t)$. Twice differentiating this relation recovers formula (15). For this reason, Eq.(15) is also frequently called the hyperscaling relation. It is conspicuous in the set (15)–(22) in that it is the only scaling relation involving the dimensionality d.

The equality (16) was originally proposed by Essam and Fisher[14] and a related inequality rigorously proved by Rushbrooke.[15] The relation (17) was advanced by Widom,[16] with a related inequality being proved by Griffiths.[17] Equalities (16) and (17), sometimes called Rushbrooke's and Griggiths' laws, respectively,[13] were rederived by Abe[18] and Suzuki[19] using an alternative route involving Lee-Yang zeros. Eq.(18) was derived by Fisher,[20] with a related inequality proved in Ref.[21]. The relations (19) and (20) were also derived in Refs.[18,19,22] and (21) was derived in Ref.[23]. The reader is also referred to Ref.[24]. Finally, the relationship between the gap exponent and the other exponents was established in Refs.[18,19,22].

In addition to the above scaling relations, one usually finds that

$$\lambda_{\text{shift}} = \frac{1}{\nu}. \tag{23}$$

But this is not always true and in some cases, such as in the Ising model in two dimensions with special boundary conditions, it can deviate from this value.[25] A criterion for when this may happens is given[26] in Sec.3.4. It turns out that Eq.(23) may be violated when the specific-heat amplitude ratio is 1.

Because of the scaling relations, only two of the exponents listed are actually independent. The scaling relations (15)–(18) are often listed as standard in textbooks, being the most frequently used. Relations (19)–(21) involve exponents characterizing the scaling behaviour of the even thermodynamic functions e and c as well as of the correlation length, in field. Although less frequently encountered, their fundamental importance is similar to that of the other scaling relations.

The formulae (7)-(14) characterise the *leading* behaviour of thermodynamic and correlation functions in the vicinity of a second-order phase transition. There are additive corrections to these scaling forms coming from both confluent and analytic sources. Each scaling formula is also associated with amplitudes, so that a more complete description of the susceptibility (for example) is

$$\chi(t,0) = \Gamma_{\pm} t^{-\gamma} \left(1 + \mathcal{O}(t^{\theta}) + \mathcal{O}(t)\right), \tag{24}$$

where the amplitude Γ_+ refers to the $t > 0$ symmetric phase and Γ_- corresponds to the broken-symmetry $T < T_c$ sector. Amplitude terms such as these and additive-correction exponents are outside the remit of this review and the reader is referred to the literature.[13,27]

2.3. *Logarithmic scaling corrections*

Instead we focus on circumstances where the dominant corrections to the scaling forms (7)–(14) are powers of logarithms, which couple multiplicatively to the leading power laws as

$$e_\infty(t,0) \sim t^{1-\alpha}|\ln t|^{\hat{\alpha}}, \tag{25}$$
$$e_\infty(0,h) \sim h^{\epsilon}|\ln h|^{\hat{\epsilon}}, \tag{26}$$
$$c_\infty(t,0) \sim t^{-\alpha}|\ln t|^{\hat{\alpha}}, \tag{27}$$
$$c_\infty(0,h) \sim h^{-\alpha_c}|\ln h|^{\hat{\alpha}_c}, \tag{28}$$
$$m_\infty(t,0) \sim t^{\beta}|\ln t|^{\hat{\beta}} \quad \text{for } T < T_c, \tag{29}$$
$$m_\infty(0,h) \sim h^{\frac{1}{\delta}}|\ln h|^{\hat{\delta}}, \tag{30}$$
$$\chi_\infty(t,0) \sim t^{-\gamma}|\ln t|^{\hat{\gamma}}, \tag{31}$$
$$\chi_\infty(0,h) \sim h^{\frac{1}{\delta}-1}|\ln h|^{\hat{\delta}}, \tag{32}$$
$$\xi_\infty(t,0) \sim t^{-\nu}|\ln t|^{\hat{\nu}}, \tag{33}$$
$$\xi_\infty(0,h) \sim h^{-\nu_c}|\ln h|^{\hat{\nu}_c}, \tag{34}$$
$$r_{\mathrm{YL}}(t) \sim t^{\Delta}|\ln t|^{\hat{\Delta}} \quad \text{for } t > 0. \tag{35}$$

In addition to these, the scaling of the correlation function at $h = 0$ (which is the case most often considered) may be expressed as

$$\mathcal{G}_\infty(x,t,0) \sim x^{-(d-2+\eta)}(\ln x)^{\hat{\eta}} D\left(\frac{x}{\xi_\infty(t,0)}\right), \tag{36}$$

The above list of functions describe the salient features of a second-order phase transition, which is only manifest in the thermodynamic limit. The pseudocritical-point FSS is

$$t_{\rm pseudo}(L) \sim L^{\lambda_{\rm shift}} (\ln L)^{\hat{\lambda}_{\rm shift}}, \tag{37}$$

It will turn out that the correlation length of the finite-sized system will play a crucial role and may also take logarithmic corrections, and for this reason we write

$$\xi_L(0) \sim L(\ln L)^{\hat{q}}. \tag{38}$$

Note that this allows for the correlation length of the system to exceed its actual length. For a long time this was thought not to be possible in finite-size scaling theory.[28] However we shall see that this is an essential feature of systems at their upper critical dimensionality. Some implications of this phenomenon are discussed in Sec.7.

2.4. *Scaling relations for logarithmic exponents*

Over the past 5 years a set of universal scaling relations for the logarithmic-correction exponents has been developed,[2,29,30] which connects the hatted exponents in a manner analogous to the way the standard relations (15)–(22) relate the leading critical exponents. While the derivation of these relations is a theme of later sections, we gather them here for convenience. The scaling relations for logarithmic corrections are

$$\hat{\alpha} = \begin{cases} 1 + d(\hat{q} - \hat{\nu}) & \text{if} \quad \alpha = 0 \quad \text{and} \quad \phi \neq \pi/4 \\ d(\hat{q} - \hat{\nu}) & \text{otherwise,} \end{cases} \tag{39}$$

$$2\hat{\beta} - \hat{\gamma} = d(\hat{q} - \hat{\nu}), \tag{40}$$

$$\hat{\beta}(\delta - 1) = \delta\hat{\delta} - \hat{\gamma}, \tag{41}$$

$$\hat{\eta} = \hat{\gamma} - \hat{\nu}(2 - \eta), \tag{42}$$

$$\hat{\epsilon} = \frac{(\gamma + 1)(\hat{\beta} - \hat{\gamma})}{\beta + \gamma} + \hat{\gamma}, \tag{43}$$

$$\hat{\alpha}_c = \frac{(\gamma + 2)(\hat{\beta} - \hat{\gamma})}{\beta + \gamma} + \hat{\gamma}, \tag{44}$$

$$\hat{\delta} = d\hat{q} - d\hat{\nu}_c, \tag{45}$$

$$\hat{\Delta} = \hat{\beta} - \hat{\gamma}, \tag{46}$$

$$\hat{\lambda}_{\rm shift} = \frac{\hat{\nu} - \hat{q}}{\nu}. \tag{47}$$

In the first of these, ϕ refers to the angle at which the complex-temperature zeros impact onto the real axis. If $\alpha = 0$, and if this impact angle is any value other than $\pi/4$, an extra logarithm arises in the specific heat. This is expected to happen in $d = 2$ dimensions, but not in $d = 4$, where $\phi = \pi/4$.[2] The static scaling relations (40), (41), (43), (44), (46), and (47) can be deduced from the Widom scaling hypothesis that the Helmholtz free energy is a homogeneous function[9] but the others require more careful deliberations, as we shall see.

3. Standard Derivation of Leading Scaling Relations

In this section we show how the leading-power-law scaling relations (15)–(18) are derived using the Widom scaling hypothesis and Kadanoff's block-spin approach. The remaining leading-power-law relations (19)–(21) are derived in a similar manner. The presentation here is necessarily elementary and the reader is refered to the standard literature (e.g., Ref.[27]) for more in-depth treatments. The derivation of relation (22) is reserved for the subsequent section where the emphasis is on partition function zeros.

3.1. *Static scaling*

The Widom (or static) scaling hypothesis is that the free energy f (or at least its singular part, which is responsible for divergences in the thermodynamic functions at the critical point) is a generalized homogeneous function (see Appendix A). It is convenient to express Widom homogeneity as

$$f(t,h) = b^{-d} f(b^{y_t} t, b^{y_h} h), \tag{48}$$

where b is a dimensionless rescaling parameter. In the renormalization group (RG) context, t and h are called *linear scaling fields* and y_t and y_h are *RG eigenvalues.* Differentiating Eq.(48) with respect to h gives the magnetization and susceptibility as

$$m(t,h) = b^{y_h - d} m(b^{y_t} t, b^{y_h} h) \quad \text{and} \quad \chi(t,h) = b^{2y_h - d} m(b^{y_t} t, b^{y_h} h), \tag{49}$$

while appropriate differentiation with respect to t gives the internal energy (entropy) and specific heat,

$$e(t,h) = b^{y_t - d} f(b^{y_t} t, b^{y_h} h) \quad \text{and} \quad c(t,h) = b^{2y_t - d} f(b^{y_t} t, b^{y_h} h). \tag{50}$$

At $h = 0$, the choice $b = t^{-1/y_t}$ recovers the first expressions in each of Eqs.(7)–(10) provided

$$\alpha = \frac{2y_t - d}{y_t}, \quad \beta = \frac{d - y_h}{y_t} \quad \text{and} \quad \gamma = \frac{2y_h - d}{y_t}. \tag{51}$$

On the other hand, at $t = 0$, the choice $b = h^{-1/y_h}$ recovers the remaining parts of Eqs.(7)–(10) if

$$\delta = \frac{y_h}{d - y_h}, \quad \epsilon = \frac{d - y_t}{y_h} \quad \text{and} \quad \alpha_c = \frac{2y_t - d}{y_h}. \tag{52}$$

The scaling hypothesis therefore allows one to express the six static critical exponents α, β, γ, δ, ϵ and α_c in terms of y_t/d and y_h/d. In particular,

$$\frac{y_t}{d} = \frac{1}{\beta(\delta + 1)}, \qquad \frac{y_h}{d} = \frac{\delta}{\delta + 1}. \tag{53}$$

These can now be eliminated from the remaining equations (51) and (52) to give scaling relations (16), (17), (19) and (20).

The hyperscaling scaling relation (15) and Fisher's relation (18) are of rather a different status than the others in that they involve the exponents ν and η, which are associated with local rather than global properties of the system. To derive these, we need more than the Widom homogeneous scaling hypothesis.

3.2. *Renormalization group*

To illustrate the renormalization group, we have in mind $N = L^d$ Ising spins, for example, on a d-dimensional lattice with spacing a (Fig.1). The Hamiltonian for the system is

$$\mathcal{H} = -J \sum_{\langle i,j \rangle} s_i s_j - H \sum_i s_i, \tag{54}$$

where $s_i \in \{\pm 1\}$ is an Ising spin at the ith lattice site and where the interaction is between nearest neighbours. In anticipation of the renormalization group (RG), we generalise this by introducing other locally interacting terms such as sums over plaquettes, next-nearest-neighbour interactions, etc., and we consider a reduced Hamiltonian

$$\bar{\mathcal{H}} = \beta\mathcal{H} = -K_t \sum_{\langle i,j \rangle} s_i s_j - K_h \sum_i s_i - K_3 \sum_{\langle i,j,k,l \rangle} s_i s_j s_k s_l - K_4 \sum_{\langle\langle i,j \rangle\rangle} s_i s_j - \dots, \tag{55}$$

where $K_t = \beta J$, $K_h = h$, and the remaining K_n are similar coupling strengths. Here, all of the additional terms have to be symetrical under $s_i \to s_j$ and $\ll \cdots \gg$ indicates interactions between next-nearest neighbours.

As illustrated in Fig.1, we place the spins into blocks of length ba and rescale the $N' = N/b^d$ block spins so they each have magnitude 1,

$$s'_I = \mathcal{F}\left(\{s_i\}_{i \in I}\right) = \frac{c(b)}{b^d} \sum_{i \in I} s_i, \tag{56}$$

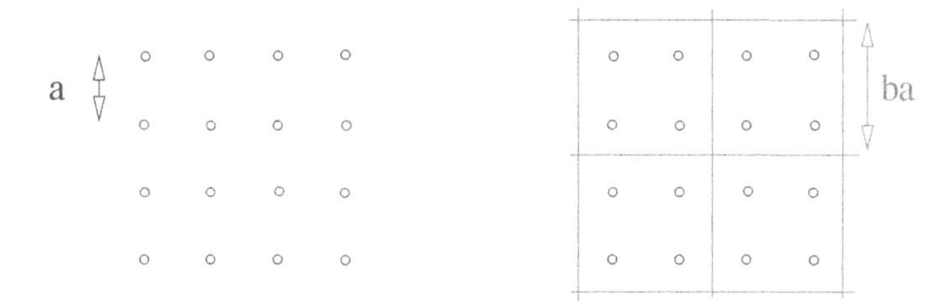

Fig. 1. In the block-spin process, the original lattice with spacing a and N spins becomes one of spacing ba with N/b^d block spins.

where s'_I labels an Ising block spin at site I of the new lattice. Here $c(b)$ is a spin dilatation factor. (If all spins in a block were fully aligned then $c(b)$ would be 1.) We require that the partition function of the blocked system is the same as the original one:

$$Z_{N'}(\bar{\mathcal{H}}') = Z_N(\bar{\mathcal{H}}). \tag{57}$$

This requirement induces a transformation from the point $\vec{\mu} = (K_1, K_2, \dots)$ in the parameter space to another point $\vec{\mu}' = (K'_1, K'_2, \dots)$ and we write $\vec{\mu}' = R_b\vec{\mu}$. The RG approach presumes the existence of a fixed point $\vec{\mu}^*$ which is invariant under R_b. If $\vec{\mu} = \vec{\mu}^* + \delta\vec{\mu}$ and $\vec{\mu}' = \vec{\mu}^* + \delta\vec{\mu}'$ are nearby this fixed point, we linearise the RG by a Taylor expansion,

$$\vec{\mu}^* + \delta\vec{\mu}' = R_b(\vec{\mu}^* + \delta\vec{\mu}) = \vec{\mu}^* + R'_b(\vec{\mu}^*)\delta\vec{\mu} + \dots, \tag{58}$$

so that

$$\delta\vec{\mu}' = R'_b(\vec{\mu}^*)\delta\vec{\mu}. \tag{59}$$

Let the eigenvectors and eigenvalues of $R'_b(\vec{\mu}^*)$ be given by

$$R'_b(\vec{\mu}^*)\vec{v}_i = \lambda_i(b)\vec{v}_i. \tag{60}$$

Demanding that two successive applications of the RG transformation using scale factors b_1 and b_2 are equivalent to a single transformation with scale factor b_1b_2, we have that $\lambda_i(b_1b_2) = \lambda_i(b_1)\lambda_i(b_2)$, so that λ_i is a homogeneous function and (see Appendix A)

$$\lambda_i(b) = b^{y_i}. \tag{61}$$

Vectors $\vec{\mu}$ and $\vec{\mu}^*$ may be expanded in terms of the eigenvectors $\vec{v}_i$:

$$\delta\vec{\mu} = \sum_i t_i\vec{v}_i, \qquad \delta\vec{\mu}' = \sum_i t'_i\vec{v}_i. \tag{62}$$

The t_i here are called linear scaling fields. Eq.(61) then gives the linear version of Eq.(59) to be

$$\sum_i t'_i \vec{v}_i = \sum_i t_i b^{y_i} \vec{v}_i, \tag{63}$$

and the linear scaling fields transform under the RG as

$$t'_i = b^{y_i} t_i. \tag{64}$$

There are three cases to distinguish:

- If $y_i > 0$, the associated linear scaling field t_i is called relevant – it increases away from the fixed point under successive applications of the RG.
- If $y_i < 0$, the scaling field decreases towards the fixed point under the RG and is called irrelevant.
- If $y_i = 0$, the field t_i is termed marginal. In this critical exponents may be continuously dependent on the parameters of the Hamiltonian or logarithmic corrections may arise.

Since the partition functions (57) of the original and renormalized systems are the same, the free energies near the fixed point are related by

$$N' f_{N'}(\bar{\mathcal{H}}') = N f_N(\bar{\mathcal{H}}), \tag{65}$$

or, since $N' = N/b^d$,

$$f(t_1, t_2, t_3, \dots) = b^{-d} f(t'_1, t'_2, t'_3, \dots) = b^{-d} f(b^{y_1} t_1, b^{y_2} t_2, b^{y_3} t_3, \dots). \tag{66}$$

This is precisely the Widom scaling form (48) leading to the static scaling relations.

To investigate local properties, consider the correlation function of the block spins,

$$G(r_{IJ}) = \langle s'_I s'_J \rangle = \frac{1}{Z} \sum_{\{s'\}} s'_I s'_J e^{-\bar{\mathcal{H}}'(s')}, \tag{67}$$

where r_{IJ} is the distance between block I and block J in units of the block-spin distance. The sum is over the block-spin configurations. Now, for a given s'_I value at a block-spin-site I, there is a multitude of sub-configurations for the set $\{s_i\}$ with $i \in I$. The probability that the Ith blocked spin takes a particular value (say $s'_I = 1$) is then $\sum_{\{s_i\}_{i \in I}} \delta(s'_I - 1) e^{-\mathcal{H}(s)}/Z$. Similarly, the probability that the blocked system is in any particular configuration given by Eq.(56) is $e^{-\mathcal{H}(s)}/Z$ where

$$e^{-\mathcal{H}'(s')} = \sum_{\{s\}} \prod_{I} \delta(s'_I - \mathcal{F}(\{s_i\}_{i\in I}))e^{-\mathcal{H}(s)}. \tag{68}$$

Putting this expression into Eq.(67), we obtain

$$G_{IJ} = \langle s'_I s'_J \rangle = \frac{1}{Z} \sum_{\{s\}} s'_I s'_J \prod_{K} \delta(s'_K - \mathcal{F}(\{s_k\}_{k\in K})\, e^{-\mathcal{H}(s)} \tag{69}$$

$$= \frac{1}{Z}\frac{c^2(b)}{b^{2d}} \sum_{\{s\}} \left(\sum_{i\in I} s_i\right)\left(\sum_{j\in J} s_j\right) e^{-\mathcal{H}(s)} \tag{70}$$

$$= \frac{c^2(b)}{b^{2d}} \sum_{i\in I}\sum_{j\in J} \langle s_i s_j \rangle \tag{71}$$

$$\approx c^2(b)\langle s_i s_j \rangle = c^2(b) G_{ij}, \tag{72}$$

provided the two blocks I and J are sufficiently far apart that $\langle s_{i'} s_{j'} \rangle \approx \langle s_i s_j \rangle$ for $i, i' \in I$ and $j, j' \in J$. If

$$G_{ij} \sim e^{-r_{ij}/\xi}, \tag{73}$$

then since $r_{IJ} = r_{ij}/b$, we have the property that

$$\xi' = \frac{\xi}{b}, \tag{74}$$

where ξ' is the block-spin correlation length, and that

$$G\left(\frac{r}{b}, \mu'\right) \approx c^2(b) G(r, \mu). \tag{75}$$

Demanding as usual that $c(b_1 b_2) = c(b_1)c(b_2)$ (homogeneity) means a power law, which we write as

$$c(b) = b^{d_\phi}. \tag{76}$$

Here d_ϕ is called the *anomalous dimension of the field.* The choice $b = r$ gives that $G(r) \sim c^{-2}(r) \sim r^{-2d_\phi}$. Comparing this to Eq.(11), we obtain

$$d_\phi = \frac{1}{2}(d - 2 + \eta). \tag{77}$$

Eq.(64) with $t_i = t$ and $y_i = y_t$ gives $t' = b^{y_t} t$. Alternatively, $t_i = h$ and $y_i = y_h$ gives $h' = b^{y_h} h$. From Eq.(74), we may write these as $t\xi^{y_t} = t'\xi'^{y_t}$ or

$h\xi^{y_h} = h'\xi'^{y_h}$. Therefore we identify $\xi \sim t^{-1/y_t}$ or $\xi \sim h^{-1/y_h}$, depending on whether $h = 0$ or $t = 0$. From Eq.(12), these give

$$\nu = \frac{1}{y_t} \quad \text{and} \quad \nu_c = \frac{1}{y_h}. \tag{78}$$

Now, y_t is related to the critical exponent α through Eq.(51). Combining this with Eq.(78) we obtain the hyperscaling relation (15). Eq.(53) then leads to Eq.(21) for the exponent ν_c.

In summary, the scaling relations (16), (17) (19) and (20) may be deduced from the Widom scaling hypothesis,[9] and Eqs.(15) and (21) can be derived from the Kadanoff block-spin construction[11] and ultimately from Wilson's RG.[31] Next we come to Fisher's scaling relation (18). There are at least two ways to do this in the literature, one more common than the other. We describe these in turn.

3.3. *Fisher's scaling relation*

With $h = 0$, fixing the remaining argument of the correlation function (11), we may write

$$\mathcal{G}_\infty(x,t,0) \sim \xi_\infty(t,0)^{-(d-2+\eta)} D\left(\frac{x}{\xi_\infty(t)}\right). \tag{79}$$

Following the original approach used by Fisher,[20] and writing the susceptibility as

$$\chi_\infty(t,0) = \int_0^{\xi_\infty(t,0)} d^dx \mathcal{G}_\infty(x,t), \tag{80}$$

one obtains

$$\chi_\infty(t,0) \sim \xi_\infty(t,0)^{2-\eta}. \tag{81}$$

The leading power-laws in Eqs.(10) and (12) then deliver the scaling relation (18).

An alternative argument presented in the literature[32] (see also Ref.[23,33]) attempts to relate the correlation function to the magnetization. If the spins decorrelate in the large-distance limit, one may expect that

$$\mathcal{G}_\infty(x,t) = \langle \vec{s}(0)\vec{s}(x)\rangle \to \langle \vec{s}(0)\rangle\langle \vec{s}(x)\rangle = m_\infty^2(t,0) \tag{82}$$

there. Using Eqs.(11) and (12) for the left hand side and matching with Eq.(9) on the right, one again obtains the standard scaling relation (18) from the leading exponents. Although this approach delivers the correct result in this instance, we will later see that the technique delivers a different result at the logarithmic level in general.

3.4. *The shift exponent*

The scaling relation for the shift exponent was given in Eq.(23) as the inverse of the correlation-length exponent ν. This is immediately derived from Eq.(48) which, with $b = L$ and in vanishing field, implies

$$c_L(t,0) = L^{-d} c(L^{\frac{1}{\nu}} t, 0), \tag{83}$$

having used Eq.(78). Setting the temperature derivative to zero (to maximise the specific heat) gives that the specific-heat peak scales as in Eq.(14) with $\lambda_{\rm shift}$ given by Eq.(23). However, it was also pointed out in Sec.2.2 that this relation does not always hold. One may ask whether there is a criterion for which to decide on the validity of Eq.(23). Such a criterion was recently derived.[26]

Although there is no FSS theory for the impact angle of Fisher zeros, for sufficiently large lattice size L one may expect that ϕ is approximated by the angle subtended by the lowest-lying zero on the real axis at the critical point,

$$\tan\phi \approx \frac{{\rm Im}[t_1(L)]}{{\rm Re}[t_1(L)]} \propto \frac{L^{-1/\nu}}{L^{-\lambda_{\rm shift}}} \sim C L^{\lambda_{\rm shift} - 1/\nu} + \dots . \tag{84}$$

Here we have used the FSS result that[7] ${\rm Im}[t_1(L)] \sim L^{-1/\nu}$. The angle ϕ is given by $L \to \infty$. If, as in most cases, $\lambda_{\rm shift} = 1/\nu$, then $\tan\phi$ is a finite value. If $\lambda_{\rm shift} < 1/\nu$ then $\tan\phi = 0$. But it is impossible for $\lambda_{\rm shift}$ to be less than $1/\nu$ – otherwise ν *becomes* $1/\lambda_{\rm shift}$. The only alternative to $\lambda_{\rm shift} = 1/\nu$, then, is $\lambda_{\rm shift} > 1/\nu$. In this case $\tan\phi$ diverges as $L \to \infty$, so that $\phi = \pi/2$ and the zeros impact onto the real axis vertically. Vertical impact implies a symmetry between the low- and high-temperature phases, which in turn implies that the specific heat amplitudes on either side of the critical point must coincide. In other words, only in circumstances where the specific-heat amplitude ratio (a universal quantity) is unity is it allowed to violate Eq.(23).

In the Ising model in $d = 2$ dimensions, vertical impact of the Fisher zeros and the coincidence of the specific-heat amplitudes is guaranteed by self-duality. In this case, Ferdinand and Fisher found $\lambda_{\rm shift} = 1/\nu = 1$ for the square lattice Ising model with periodic boundaries,[34] but other two-dimensional lattices with different topologies have different values of $\lambda_{\rm shift}$.[25]

4. Logarithmic Corrections

Logarithmic corrections are characteristic of marginal scenarios (see, e.g., Ref.[35] and references therein). The hyperscaling hypothesis $f_\infty(t,0) \sim \xi_\infty^{-d}(t,0)$ fails at and above the upper critical dimension d_c. While the leading scaling relation (15) holds at d_c itself, it fails for $d > d_c$, where mean-field behaviour prevails. This mean-field behaviour holds independent of d (provided $d > d_c$), so the critical exponents and scaling relations should also be d-independent there. At d_c itself, multiplicative logarithmic corrections to scaling appear.

The two-dimensional q–state Potts model has a first-order phase transition for $q > 4$ and a second-order one when $q \leq 4$. The borderline $q = 4$ case manifests multiplicative logarithmic corrections to scaling. The $q = 2$ version of the Potts model is the Ising model. In two dimensions this has a logarithmic divergence in the specific heat. Although no other thermodynamic quantity manifests such a logarithm in this model, this scenario also has to be accounted for in any theory of logarithmic corrections.

Staying in two dimensions, the Ising model with uncorrelated, quenched, random-site or bond disorder is another example where logarithmic corrections appear at a demarcation point. The Harris criterion[36] tells us that when the critical exponent α of a pure system is positive, random quenched disorder is relevant.[36] This means that critical exponents may change as disorder is added to the system. If α is negative in the pure system, the critical behaviour is not changed by such disorder. In the marginal $\alpha = 0$, no Harris prediction is possible, and there logarithmic corrections to the pure model may ensue.

Because of the ubiquity of these logarithms in critical phenomena, it is reasonable to seek scaling relations for their exponents in analogy to Eqs.(15)–(22) and (23) above. These are the scaling relations of Sec.2.3, which have only recently been developed.[2,29,30] Here, these theoretical developments are brought together and summarised, and their consequences are confronted with the literature. In many cases the values of logarithmic corrections derived in the literature using a multitude of disparate techniques are upheld. A few cases which conflict with the literature are highlighted as requiring further investigations. Finally, holes in the literature are filled, pointing the way for further research endeavours into the future.

The scaling theory presented in this section is entirely based on self-consistencies – independent of RG. The theory does not *predict* the existence of logarithmic corrections; rather, when they are known by other methods to be present, the theory restricts their form through scaling relations. For ab inito model-specific theories, the RG is more appropriate and the reader is referred to the literature.[31,37–39,41] Wegners analysis,[37] in particular, uncovered the role of marginal variables and nonlinear scaling fields. These were further developed by Huse and Fisher.[39] Excellent reviews on the origins of logarithmic corrections in the paradigmatic four-state Potts case are contained in Ref.[40].

4.1. *Static correction exponents*

We denote the jth Lee-Yang zero for a system of size L by

$$h_j(t, L) = r_j(t, L) \exp\left(i\phi(r_j(t, L))\right). \tag{85}$$

Here $h_j(t, L)$ is complex and $r_j(t, L)$ is real. The latter parameterises the position of the zero along the locus of all zeros which is given by the function $\phi(r)$. If the Lee-Yang theorem holds,[3] the angle $\phi(r)$ is $\pi/2$ and the zeros are on the imaginary field axis. In fact the validity of the Lee-Yang theorem is not required or assumed in what follows (and it does not hold for the Potts model, for example). Expressed in terms of these zeros, the finite-size partition function is

$$Z_L(t, h) \propto \prod_j (h - h_j(t, L)), \tag{86}$$

where the product is over all zeros and constant of proportionality (which is not displayed) contributes only to the regular part of the free energy. The logarithm in the free energy converts the product into a sum and

$$f_L(t, h) = \frac{1}{N} \sum_j \ln\left(h - h_j(t, L)\right), \tag{87}$$

up to an additive constant (not displayed). Defining

$$g_L(r, t) = \frac{1}{N} \sum_j \delta(r - r_j(t, L)), \tag{88}$$

we may write

$$f_L(t, h) = \int \ln\left(h - h(r, t)\right) g_L(r, t) dr, \tag{89}$$

where the integral is over the locus of zeros. Since these occur in complex conjugate pairs, and since the density of zeros vanishes up to $r = r_{\rm YL}(t)$, we express the free energy in the thermodynamic limit as

$$f_\infty(t, h) = 2{\rm Re} \int_{r_{\rm YL}(t)}^{R} \ln\left(h - h(r, t)\right) g_\infty(r, t) dr. \tag{90}$$

We have assumed that critical behaviour is dominated by the Lee-Yang zeros closest to the critical point and that the locus of these zeros can be approximated by $\phi(r, t) = \phi$, which is a constant. We have also inserted an integral cutoff R.

The susceptibility is the second derivative of the free energy with respect to the external field. It is convenient to substitute $r = x r_{\rm YL}(t)$, so that at $h = 0$ it is

$$\chi_\infty(t, 0) = -\frac{2\cos(2\phi)}{r_{\rm YL}(t)} \int_1^{\frac{R}{r_{\rm YL}(t)}} \frac{g_\infty(x r_{\rm YL}, t)}{x^2} dx. \tag{91}$$

Expanding Eq.(91) about $r_{\rm YL}(t)/R = 0$, which is reasonable near criticality, one finds

$$g_\infty(r,t) = \chi_\infty(t,0) r_{\rm YL}(t) \Phi\left(\frac{r}{r_{\rm YL}(t)}\right), \tag{92}$$

up to additive corrections in $r_{\rm YL}(t)/R$ and where Φ is an undetermined function. The ratio $r_{\rm YL}(t)/R$ is assumed small enough near criticality to drop additive corrections. Analogous deliberations for the magnetization in field give

$$m_\infty(t,h) = \chi_\infty(t,0) r_{\rm YL}(t) \Psi_\phi\left(\frac{h}{r_{\rm YL}(t)}\right), \tag{93}$$

where

$$\Psi_\phi\left(\frac{h}{r_{\rm YL}(t)}\right) = 2{\rm Re}\int_1^\infty \frac{\Phi(x)}{h/r_{\rm YL}(t) - xe^{i\phi}} dx. \tag{94}$$

Now, allowing $h \to 0$ in Eq.(93), and comparing to the assumed scaling form of Eq.(29), one recovers the leading scaling relation (22). One also arrives at the first scaling relation for logarithmic corrections, namely

$$\hat{\Delta} = \hat{\beta} - \hat{\gamma}. \tag{95}$$

Furthermore, fixing the argument of the function Ψ_ϕ in Eq.(93) leads to $t \sim h^{1/\Delta}|\ln h|^{-\hat{\Delta}/\Delta}$ using Eq.(35), so that Eq.(93) may be expressed as

$$m_\infty(t,h) \sim h^{1-\frac{\gamma}{\Delta}} |\ln h|^{\hat{\gamma}+\frac{\gamma\hat{\Delta}}{\Delta}} \Psi_\phi\left(\frac{h}{r_{\rm YL}(t)}\right). \tag{96}$$

Next taking the limit $t \to 0$ and comparing with the assumed form Eq.(30), recovers the leading edge behaviour (22). It also delivers the correction relation $\hat{\Delta} = \delta(\hat{\delta} - \hat{\gamma})/(\delta - 1)$. The former recovers the leading scaling relation (17), while the latter, together with Eq.(95), gives its logarithmic counterpart,

$$\hat{\beta}(\delta - 1) = \delta\hat{\delta} - \hat{\gamma}. \tag{97}$$

Note that the assumption of logarithmic corrections to the zeros in Eq.(35) necessarily leads to logarithms in the other observables. If one tries to omit them in other quantities (or vice versa), contradictions ensue. E.g., attempting to force $\hat{\gamma} = 0$ still necessitates a logarithmic correction to the magnetization through a non-zero value of $\hat{\Delta}$ in Eq.(96). Alternatively, attempting to force $\hat{\beta} = 0$ necessitates a non-vanishing $\hat{\gamma}$. By the same token, omission of logarithmic correction factors is the same as setting the relevant hatted exponents to zero and leads to contradictions. Therefore allowing for logarithmic corrections from the start reflects the philosophy of this exposition, which is based upon self-consistencies, rather than the Wegner approach[37,38] which predicts their existence.

We next introduce the cumulative distribution function of zeros:

$$G_\infty(r,t) = \int_{r_{\rm YL}(t)}^{r} g_\infty(s,t)ds \tag{98}$$

$$= \chi_\infty(t,0) r_{\rm YL}^2(t) I\left(\frac{r}{r_{\rm YL}(t)}\right), \tag{99}$$

where $I(y) = \int_1^y \Phi(z)dz$. Applying integration by parts to the free energy (90) then gives its singular part to be

$$f_\infty(t,h) = -2{\rm Re}\int_{r_{\rm YL}(t)}^{R} \frac{G_\infty(r,t)\exp{(i\phi)}dr}{h - r\exp{(i\phi)}}. \tag{100}$$

Again substituting $r = x r_{\rm YL}(t)$,

$$f_\infty(t,h) = \chi_\infty(t,0) r_{\rm YL}^2(t) \mathcal{F}_\phi\left(\frac{h}{r_{\rm YL}(t)}\right), \tag{101}$$

where

$$\mathcal{F}_\phi(y) = -2{\rm Re}\int_1^\infty \frac{I(x)dx}{y\exp{(-i\phi)} - x}, \tag{102}$$

and we have taken the limit $R/{\rm YL}(t) \to \infty$. The internal energy and specific heat are given by the first and second derivatives of the free energy with respect to t, respectively. These are

$$e_\infty(t,h) = \chi_\infty(t,0) r_{\rm YL}^2(t) t^{-1} \mathcal{F}_\phi\left(\frac{h}{r_{\rm YL}(t)}\right), \tag{103}$$

$$c_\infty(t,h) = \chi_\infty(t,0) r_{\rm YL}^2(t) t^{-2} \mathcal{F}_\phi\left(\frac{h}{r_{\rm YL}(t)}\right). \tag{104}$$

Inverting Eq.(35), we may write

$$t \sim r_{\rm YL}^{\frac{1}{\Delta}} |\ln r_{\rm YL}|^{-\frac{\hat{\Delta}}{\Delta}}, \tag{105}$$

which with Eq.(31) gives

$$\chi_\infty(t,0) \sim r_{\rm YL}^{-\frac{\gamma}{\Delta}} |\ln r_{\rm YL}|^{\frac{\gamma\hat{\Delta}}{\Delta}+\hat{\gamma}}. \tag{106}$$

Together, these give

$$e_\infty(t,h) = r_{\rm YL}^{2-\frac{\gamma}{\Delta}-\frac{1}{\Delta}} |\ln r_{\rm YL}|^{\frac{(\gamma+1)\hat{\Delta}}{\Delta}+\hat{\gamma}} \mathcal{F}_\phi\left(\frac{h}{r_{\rm YL}}\right), \tag{107}$$

$$c_\infty(t,h) = r_{\rm YL}^{2-\frac{\gamma}{\Delta}-\frac{2}{\Delta}} |\ln r_{\rm YL}|^{\frac{(\gamma+2)\hat{\Delta}}{\Delta}+\hat{\gamma}} \mathcal{F}_\phi\left(\frac{h}{r_{\rm YL}}\right). \tag{108}$$

Comparing Eq.(108) with the form (27), one obtains $\alpha = 2 + \gamma - 2\Delta$ and $\hat{\alpha} = \hat{\gamma} + 2\hat{\Delta}$. From Eqs.(17) and (22), the former is the standard scaling law (16). From

Eq.(95), the latter can be expressed as a third relation between the logarithmic-correction exponents, namely

$$\hat{\alpha} = 2\hat{\beta} - \hat{\gamma}. \tag{109}$$

Eqs.(107) and (108) may be rewritten as

$$e_\infty(t,h) = h^{2-\frac{\gamma}{\Delta}-\frac{1}{\Delta}} |\ln h|^{\frac{(\gamma+1)\hat{\Delta}}{\Delta}+\hat{\gamma}} \mathcal{F}'_\phi\left(\frac{h}{r_{\rm YL}}\right), \tag{110}$$

$$c_\infty(t,h) = h^{2-\frac{\gamma}{\Delta}-\frac{2}{\Delta}} |\ln h|^{\frac{(\gamma+2)\hat{\Delta}}{\Delta}+\hat{\gamma}} \mathcal{F}'_\phi\left(\frac{h}{r_{\rm YL}}\right). \tag{111}$$

Letting $t \to 0$ so that $r_{\rm YL}(t) \to 0$, one finds

$$e_\infty(h) = h^{2-\frac{\gamma}{\Delta}-\frac{1}{\Delta}} |\ln h|^{\frac{(\gamma+1)\hat{\Delta}}{\Delta}+\hat{\gamma}}, \tag{112}$$

$$c_\infty(h) = h^{2-\frac{\gamma}{\Delta}-\frac{2}{\Delta}} |\ln h|^{\frac{(\gamma+2)\hat{\Delta}}{\Delta}+\hat{\gamma}}. \tag{113}$$

From the leading behaviour one recovers Eqs.(19) and (20). From the logarithmic corrections, one obtains the counterpart scaling relations (43) and (44).

4.2. *Hyperscaling for logarithms*

The logarithmic analogue of the hyperscaling relation (15) has a rather different status than the other critical exponents, in that it necessitates consideration of finite-size effects (this aspect will be further discussed below). Consider, therefore, a system of finite volume $N = L^d$. The finite-size scaling (FSS) of Lee-Yang edge is given by

$$\frac{r_1(L)}{r_{\rm YL}(t)} = \mathcal{F}\left(\frac{\xi_L(0,0)}{\xi_\infty(t,0)}\right), \tag{114}$$

where $\xi_L(0)$ is the correlation length of the finite-size system at $t = 0$. To allow for logarithmic corrections to this quantity too, we assume the form (38). For finite systems, the cumulative density of zeros is the fractional total of zeros up to a given point,[42]

$$G_L(r_j(L)) = \frac{2j-1}{2L^d}. \tag{115}$$

Fixing the ratio $r/r_{\rm YL}(t)$ in (99), and using the scaling relations previously derived, and then taking the limit $t \to 0$, one arrives at an expression for the critical cumulative distribution function,

$$G_\infty(r,0) \sim r^{\frac{2-\alpha}{\Delta}} |\ln r|^{\hat{\alpha}-\frac{(2-\alpha)}{\Delta}\hat{\Delta}}. \tag{116}$$

At $t = 0$ and for sufficiently large L, the FSS expression (115) must converge to the infinite-volume expression (116). Equating them gives the FSS of the first Lee-Yang zero at criticality to be

$$r_1(L) \sim L^{-\frac{d\Delta}{2-\alpha}} (\ln L)^{\hat{\Delta}-\frac{\Delta\hat{\alpha}}{2-\alpha}} . \tag{117}$$

Inserting (33), (35), (38) and (117) into (114) recovers (15) and yields the logarithmic equivalent to the hyperscaling relation, namely

$$\hat{\alpha} = d\hat{q} - d\hat{\nu}, \tag{118}$$

The logarithmic scaling relations (97) and (109) but not (118) can be derived starting with a suitable modification to the phenomonological Widom ansatz,[43,44] namely

$$f(t,h) = b^{-d} f(b^{y_t} (\ln b)^{\hat{y}_t} t, b^{y_h} (\ln b)^{\hat{y}_h} h). \tag{119}$$

Following the approach of Sec.3.1, this recovers the static scaling equations (16), (17), (19) and (20) as well as their logarithmic counterparts (40), (41), (43) and (44) with

$$\hat{y}_t = \frac{\hat{\beta}}{\beta} - \frac{\delta\hat{\delta}}{\beta(1+\delta)} = \frac{2\hat{\beta}-\hat{\gamma}}{2\beta+\gamma} \tag{120}$$

and

$$\hat{y}_h = \frac{\delta\hat{\delta}}{1+\delta} = \frac{\gamma\hat{\beta}+\beta\hat{\gamma}}{2\beta+\gamma}. \tag{121}$$

The logarithmic counterparts of Eqs.(51) and (52) are then

$$\hat{\alpha} = d\frac{\hat{y}_t}{y_t}, \quad \hat{\beta} = \beta\hat{y}_t + \hat{y}_h, \quad \hat{\gamma} = -\gamma\hat{y}_t + 2\hat{y}_h \tag{122}$$

and

$$\hat{\delta} = d\frac{\hat{y}_h}{y_h}, \quad \hat{\epsilon} = \hat{y}_t + \epsilon\hat{y}_h, \quad \hat{\alpha}_c = 2\hat{y}_t - \alpha_c\hat{y}_h, \tag{123}$$

while

$$\hat{\Delta} = \Delta\hat{y}_t - \hat{y}_h. \tag{124}$$

4.3. *Logarithmic counterpart to Fisher's relation*

With multiplicative logarithmic corrections, the correlation function is given by Eq.(36). Fixing the argument of the function D, we may rewrite this as

$$\mathcal{G}_\infty(x,t,0) \sim \xi_\infty(t,0)^{-(d-2+\eta)}(\ln \xi_\infty(t,0))^{\hat{\eta}} D\left(\frac{x}{\xi_\infty(t,0)}\right). \tag{125}$$

Following Fisher[20] and writing the susceptibility in terms of the correlation function (Eq.(80)), one obtains

$$\chi_\infty(t,0) \sim \xi_\infty(t,0)^{2-\eta}(\ln \xi_\infty(t,0))^{\hat{\eta}}. \tag{126}$$

As in Sec.3.3, the leading power-laws in Eqs.(31) and (33) then deliver the scaling relation (18). Matching the logarithmic corrections exponents too yields

$$\hat{\eta} = \hat{\gamma} - \hat{\nu}(2-\eta). \tag{127}$$

This approach was used in Ref.[32] for the case of the $d = 2$ four-state Potts model. An alternative argument presented there (see Sec.3.3) assumes that the spins decorrelate in the large-distance limit,

$$\mathcal{G}_\infty(x,t) = \langle \vec{s}(0)\vec{s}(x)\rangle \to \langle \vec{s}(0)\rangle\langle \vec{s}(x)\rangle = m_\infty^2(t,0). \tag{128}$$

Using Eqs.(33) and (125) for the left hand side and matching with Eq.(29) on the right, one again obtains the standard scaling relation (18) from the leading exponents (see also Refs.23,33). However, it is interesting to note that matching the logarithmic exponents gives, instead of (127), the relation $\hat{\eta} = 2\hat{\beta} + \hat{\nu}(d-2+\eta)$. From the logarithmic scaling relation (40) this would mean

$$\hat{\eta} = d\hat{q} + \hat{\gamma} - \hat{\nu}(2-\eta), \tag{129}$$

which is, in general, *different* to Eq.(127).

When $\hat{q}$ vanishes, as is the case in the $d = 2$, four-state Potts model,[32] for example, this is actually identical to Eq.(127). To decide between Eqs.(127) and (129), we need a model with non-zero $\hat{q}$ value. Models at the upper critical dimension provide suitable determinators and one finds that the Eq.(129) fails in these cases. Eq.(127) holds in each case we have examined.

4.4. *Corrections to the logarithmic scaling relations*

Although it holds in most models which manifest multiplicative logarithmic corrections, there is an immediate and obvious problem with scaling relation (118) when it is confronted with the Ising model in two dimensions, which has been solved exactly. There, the specific heat has a logarithmic divergence so that $\alpha = 0$

and $\hat{\alpha} = 1$. There are, however, no logarithmic corrections in any of the other thermodynamic or correlation functions in this model, and, since $\hat{\nu} = \hat{q} = 0$, Eq.(118) fails immediately. The relation (118) also fails in the uncorrelated, quenched, random disordered version of the Ising model in two dimensions, where $\hat{q} = 0$,[45,46] $\hat{\alpha} = 0$, $\hat{\nu} = 1/2$ and[2,47–51]

$$C_\infty(t,0) \sim \ln|\ln t|. \tag{130}$$

This famous double logarithm in the specific heat of the diluted Ising model in two dimensions has been the source of great controversy throughout the years. This controversy has only recently been convincingly resolved, partly with the aid of logarithmic-correction theories,[52,53] which should be able to account for them. In this section, besides resolving the puzzle as to the value of $\hat{\alpha}$ in the pure and random Ising models in $d = 2$ we will see how this double logarithm emerges quite naturally from the general scheme.

Since the problem is associated with the even sector of the model (namely with the t-dependency of the specific heat), one may argue that a Lee-Yang analysis, which focuses on an odd (magnetic) scaling field, is not the best approach to fully extract the general relationship between the correction exponents appearing in Eq.(118). One may appeal to complex-temperature (Fisher) zeros for further insight, as they are appropriate to the even sector. We will see that the puzzle is resolved after consideration of two special properties of the pure and random two-dimensional Ising models, namely the vanishing of α and the angle at which their Fisher zeros impact onto the real axis.

An FSS theory for Fisher zeros is obtained[7] by writing the finite-size partition function in terms of the scaling ratio ξ_L/ξ_∞

$$Z_L(t,0) = Q\left(\frac{\xi_L(0)}{\xi_\infty(t,0)}\right). \tag{131}$$

This vanishes at a Fisher zero. Labeling the jth finite-size Fisher zero as $t_j(L)$, one therefore has

$$\frac{\xi_L(0)}{\xi_\infty(t_j(L))} = Q_j^{-1}(0), \tag{132}$$

where $Q_j^{-1}(0)$ is the jth root of the function Q. Using Eqs.(35) and (38), this equation can be solved to give the FSS form of the jth zero,

$$t_j(L) \sim L^{-\frac{1}{\nu}}(\ln L)^{\frac{\hat{\nu}-\hat{q}}{\nu}}. \tag{133}$$

So far, no assumptions other than the validity of FSS have been involved. Note that this is the same scaling form as Eq.(47), which is sensible, since the real part of the lowest zero is also a pseudocritical point.

The full expression for the scaling of the $j^{\rm th}$ Fisher zero is[7,42,54] a function of a fraction of the total number of zeros $\mathcal{N}$. Since this is proportional to the lattice volume $N = L^d$, Eq.(133) is more appropriately written as

$$t_j(L) \sim \left(\frac{j-1/2}{L^d}\right)^{\frac{1}{\nu d}} \left(\ln\left(\frac{j-1/2}{L^d}\right)\right)^{\frac{\hat{\nu}-\hat{q}}{\nu}} \exp\left(i\phi_j(r_j(L))\right), \qquad (134)$$

where $\phi_j(r_j(L))$ is the argument of the jth Fisher zero.

In all cases known from the literature, the Fisher zeros for isotropic models on homopolygonal lattices fall on curves in the complex plane. They impact onto the real axis along so-called singular lines.[4,55,56] We also assume this scenario in the present case, and we denote the impact angle onto the real axis in the thermodynamic limit by ϕ.

Similar to the Lee-Yang case, we may write the finite-size partition function in terms of Fisher zeros,

$$Z_L(t,0) \propto \prod_{j=1}^{\mathcal{N}} \left(t - t_j(L)\right)\left(t - t_j^*(L)\right), \qquad (135)$$

where we have been careful to identify $t_j(L)$ and $t_j^*(L)$ as complex conjugate pairs. Provided that the $\mathcal{M} \propto \mathcal{N}$ zeros which dominate scaling behavior in the vicinity of the critical point are described by the scaling form (134), and differentiating appropriately, one finds the FSS for the specific heat at $t = 0$ to be

$$C_L(0) \sim -L^{-d}{\rm Re}\sum_{j=1}^{\mathcal{M}} t_j^{-2}(L). \qquad (136)$$

We wish to compare this expression with that resulting from a direct application of the modified FSS hypothesis (114) to the specific heat, which yields

$$\frac{C_L(0)}{C_\infty(t,0)} = \mathcal{F}_C\left(\frac{\xi_L(0,0)}{\xi_\infty(t,0)}\right). \qquad (137)$$

Fixing the scaling ratio on the right hand side so that $t \sim L^{-1/\nu}(\ln L)^{(\hat{\nu}-\hat{q})/\nu}$, one obtains from (27) the FSS behaviour

$$C_L(0) \sim L^{\frac{\alpha}{\nu}}(\ln L)^{\hat{\alpha}-\alpha\frac{\hat{\nu}-\hat{q}}{\nu}}. \qquad (138)$$

We now match Eq.(136) to Eq.(138).

In the case where $\alpha \neq 0$, Eq. (136) gives

$$C_L(0) \sim L^{\frac{\alpha}{\nu}}\left(\ln L\right)^{-2\frac{\hat{\nu}-\hat{q}}{\nu}}, \qquad (139)$$

and comparison with Eq.(138) leads to the recovery of the previously derived logarithmic scaling relation (118).

If, however, $\alpha = 0$, the FSS expression (136) for the specific heat becomes

$$C_L(0) \sim \sum_{j=1}^{\mathcal{M}} \frac{\cos(2\phi_j(L))}{j-1/2} \left(\ln \left(\frac{j-1/2}{L^d} \right) \right)^{-2\frac{\hat{\nu}-\hat{q}}{\nu}} . \tag{140}$$

For large L and close enough to the transition point that $\phi_j(r_j(L)) \simeq \phi$, the cosine term becomes a non-vanishing constant if $\phi \neq \pi/4$. This is what happens in the pure Ising model in $d = 2$ dimensions, where $\phi = \pi/2$ is assured by duality on the square lattice[4] as well as for its symmetric random-bond counterpart.[6] Universality of ϕ ensures this happens for other lattice configurations too and continuity arguments lead one to also expect $\phi \neq \pi/4$ in the general random-bond and random-site two-dimensional Ising case.

In these cases, application of the Euler-Maclaurin formula gives that the leading FSS behavior when $\alpha = 0$ is

$$C_L(0) \sim \begin{cases} (\ln L)^{1-2\frac{\hat{\nu}-\hat{q}}{\nu}} & \text{if } 2(\hat{\nu}-\hat{q}) \neq \nu \\ \ln\ln L & \text{if } 2(\hat{\nu}-\hat{q}) = \nu \,. \end{cases} \tag{141}$$

In the thermodyamic limit, one may simply replace L and $C_L(0)$ by t and $C_\infty(t,0)$ in Eq.(141), respectively. Then, comparing Eq.(138) with Eq.(141) and using standard hyperscaling (15), one finds

$$\hat{\alpha} = 1 + d\hat{q} - d\hat{\nu}. \tag{142}$$

This expression replaces Eq.(118) when the model has both $\alpha = 0$ and $\phi \neq \pi/4$. Thus the logarithmic scaling relation (39) is established. Furthermore, we prefer to write Eq.(40) in terms of $\hat{q}$ and $\hat{\nu}$ rather than in terms of $\hat{\alpha}$ as in Eq.(109) because, unlike $\hat{\alpha}$, $\hat{q}$ and $\hat{\nu}$ the $\hat{q} - \hat{\nu}$ combination does not exhibit the subtleties discussed here.

We can perform similar considerations for the Lee-Yang zeros, replacing Eq.(131) by

$$Z_L(h) = Q\left(\frac{\xi_L(0)}{\xi_\infty(0,h)} \right). \tag{143}$$

We find

$$\hat{\delta} = d(\hat{q} - \hat{\nu}_c), \tag{144}$$

except when $\gamma = 0$ and the impact angle is not $\pi/4$, in which case an extra logarithm appears and Eq.(144) becomes

$$\hat{\delta} = 1 + d(\hat{q} - \hat{\nu}_c). \tag{145}$$

That the impact angle is $\pi/2$ is guaranteed by the Lee-Yang theorem, but we know of no cases with $\gamma = 0$, so we omit Eq.(144) from Eq.(45).

To summarize so far, we have explained where the standard scaling relations come from using the standard approaches of Widom scaling and the Kadanoff block-spin RG. These standard scaling laws for the leading critical exponents are well established in the literature. Analogous relations for the logarithmic corrections (39)–(46) have also been derived using a self-consistency approach. Next we confront the logarithmic scaling relations with results from the literature in a variety of models. Where predictions for, or measurements of, these logarithmic corrections already exist in the literature, we can test the new scaling relations. Indeed, we will show that they are upheld (barring at least 3 exceptional circumstances which require further research). Where there are gaps in the literature regarding the values of the logarithmic corrections, we can use the scaling relations to make predictions for them. These new predictions need to be independently tested in future research.

4.5. *The logarithmic shift exponent*

Finally we mention that the finite-size scaling of the pseudocritical point may be determined using the Widom static hypothesis (119) modified to include logarithms. In a similar manner to Sec.3.4, one finds that the specific heat peaks scales as Eq.(37) with

$$\hat{\lambda}_{\text{shift}} = -\hat{y}_t = \frac{\hat{\nu} - \hat{q}}{\nu}. \tag{146}$$

This is the final scaling relation for logarithmic corrections Eq.(47).

5. Fisher Renormalization for Logarithmic Corrections

Fisher renormalization concerns systems under some form of constraint. For such systems, the critical exponents take values which differ from their "ideal" counterparts. The systems we have been dealing with so far are "ideal", in the sense that they are not subject to constraints of this type. Typically, the theoretical power-law divergence of the specific heat in an ideal system, for example, is replaced by a finite cusp in a "real" experimental realization. Fisher[57] explained this as being due to the effect of hidden variables and established elegant relations between the ideal exponents and their constrained counterparts.

Phase transitions which exhibit Fisher renormalization include those in constrained magnetic and fluid systems (e.g., with fixed levels of impurities), the superfluid λ transition in ^{3}He-^{4}He mixtures in confined films,[58] the order-disorder transition in compressible ammonium chloride,[59,60] the critical behaviour at nematic-smectic-A transitions in liquid-crystal mixtures[61] and in emulsions.[62]

When multiplicative logarithmic corrections are present in the ideal system, these migrate to the Fisher-renormalized system in a non-trivial manner which has recently been determined.[63] Here, we summarize the Fisher renormalization for the logarithmic exponents. We later use this scheme to deduce the scaling behaviour of lattice animals at their upper critical dimension.

For a system under constraint, Fisher established that if the specific-heat exponent for the ideal system α is positive, it is altered in the constrained system. The magnetization, susceptibility and correlation-length critical exponents are also changed. If the subscript "X" represents the real or constrained system, its critical exponents are related to the ideal ones by the transformations

$$\alpha_X = \frac{-\alpha}{1-\alpha}, \quad \beta_X = \frac{\beta}{1-\alpha}, \quad \gamma_X = \frac{\gamma}{1-\alpha}, \quad \nu_X = \frac{\nu}{1-\alpha}. \tag{147}$$

The exponent δ and the anomalous dimension η are not renormalized:

$$\delta_X = \delta, \quad \text{and} \quad \eta_X = \eta. \tag{148}$$

Expressing Eqs.(53) in terms of the above exponents, and Fisher renomalizing, gives the RG eigenvalues for the constrained system as

$$y_{tX} = (1-\alpha)y_t, \quad \text{and} \quad y_{hX} = y_h. \tag{149}$$

One may then use the scaling relations (21), (22), (23) and (52) to determine the Fisher-renormalization formulae

$$\Delta_X = \frac{\Delta}{1-\alpha}, \quad \nu_{cX} = \nu_c, \quad \epsilon_X = \epsilon + \frac{\alpha}{\Delta}, \quad \alpha_{cX} = \alpha_c - 2\frac{\alpha}{\Delta}. \tag{150}$$

as well as

$$\lambda_{\text{shiftX}} = d - \lambda_{\text{shift}}. \tag{151}$$

These formulae have two attractive properties:

- If the ideal exponents obey the standard scaling relations then the Fisher renormalized exponents do likewise.
- Fisher renormalization is involutory in the sense that Fisher renormalization of the constrained exponents returns the ideal ones. This means that two successive applications of the transformation gives the identity.

Fisher renormalization applied to the most commonly encountered logarithmic correction exponents results in[63]

$$\hat{\alpha}_X = -\frac{\hat{\alpha}}{1-\alpha}, \tag{152}$$

$$\hat{\beta}_X = \hat{\beta} - \frac{\beta\hat{\alpha}}{1-\alpha}, \tag{153}$$

$$\hat{\gamma}_X = \hat{\gamma} + \frac{\gamma\hat{\alpha}}{1-\alpha}, \tag{154}$$

$$\hat{\nu}_X = \hat{\nu} + \frac{\nu\hat{\alpha}}{1-\alpha}. \tag{155}$$

As for the leading indices, no Fisher renormalization occurs for the logarithmic-correction exponents for the in-field magnetization or for the correlation function, since it is defined exactly at the critical point. Similarly $\hat{q}$ is unchanged.

$$\hat{\delta}_X = \hat{\delta}, \quad \hat{\eta}_X = \hat{\eta}, \quad \hat{q}_X = \hat{q}. \tag{156}$$

Each of these obey the scaling relations for logarithmic corrections, and Eqs.(120) and (121) give

$$\hat{y}_{tX} = (1-\alpha)\hat{y}_t - \hat{\alpha}, \tag{157}$$

$$\hat{y}_{hX} = \hat{y}_h, \tag{158}$$

while Eqs.(43)–(47) also lead to

$$\hat{\epsilon}_X = \hat{\epsilon} - \hat{\alpha} - \frac{\alpha\hat{\Delta}}{\Delta}, \tag{159}$$

$$\hat{\alpha}_{cX} = \hat{\alpha}_c - 2\left(\hat{\alpha} + \frac{\alpha\hat{\Delta}}{\Delta}\right), \tag{160}$$

$$\hat{\nu}_{cX} = \hat{\nu}_c, \tag{161}$$

$$\hat{\Delta}_X = \hat{\Delta} - \frac{\Delta\hat{\alpha}}{1-\alpha}, \tag{162}$$

$$\hat{\lambda}_{\text{shiftX}} = (1-\alpha)\hat{\lambda}_{\text{shift}} + \hat{\alpha}. \tag{163}$$

Fisher renormalization is also involutory at the logarithmic level: re-renormalizing the Fisher-renormalized logarithmic exponents returns their ideal counterpart. In other words, Fisher renormalization is its own inverse at the logarithmic as well as leading level.

Fisher renormalization at the logarithmic level has been tested in lattice animals and the Lee-Yang problem at their upper critical dimensions, which also led to new predictions for logarithmic corrections are made. These are discussed in Sec.6.6 below.

6. Logarithmic Correction Exponents for Various Models

In this section we confront the logarithmic scaling relations (39)–(46) with the literature for various models on a case-by-case basis.

6.1. $q = 4$ $d = 2$ *Potts model*

The leading critical exponents for the 4-state Potts model in $d = 2$ dimensions are[40,64]

$$\alpha = \frac{2}{3}, \quad \beta = \frac{1}{12}, \quad \gamma = \frac{7}{6}, \quad \delta = 15, \quad \nu = \frac{2}{3},$$
$$\eta = \frac{1}{4}, \quad \epsilon = \frac{4}{15}, \quad \alpha_c = \frac{8}{15}, \quad \nu_c = \frac{8}{15}, \quad \Delta = \frac{5}{4}. \tag{164}$$

The hitherto-known logarithmic-correction exponents are[32,40,65]

$$\hat{\alpha} = -1, \quad \hat{\beta} = -\frac{1}{8}, \quad \hat{\gamma} = \frac{3}{4}, \quad \hat{\delta} = -\frac{1}{15}, \quad \hat{\nu} = \frac{1}{2}, \quad \hat{\eta} = -\frac{1}{8}. \tag{165}$$

FSS of the thermodynamic functions[32,66] means that

$$\hat{q} = 0. \tag{166}$$

The correction relations (39)–(42) therefore hold, while (22) gives $\Delta = 5/4$ for the leading scaling of the Lee-Yang edge. We can also use the remaining static scaling laws (43)–(45) to predict

$$\hat{\epsilon} = -\frac{23}{30}, \quad \hat{\alpha}_c = -\frac{22}{15}, \quad \hat{\nu}_c = \frac{1}{30}, \quad \hat{\Delta} = -\frac{7}{8}. \tag{167}$$

6.2. $O(N)$ ϕ^4_d *theory*

The upper critical dimension for $O(N)$-symmetric ϕ^4_d theories – of crucial importance to the Higgs sector of the standard model – is $d_c = 4$. Here hyperscaling fails and the leading critical exponents take on their mean-field values which are

$$\alpha = 0, \quad \beta = \frac{1}{2}, \quad \gamma = 1, \quad \delta = 3, \quad \nu = \frac{1}{2},$$
$$\eta = 0, \quad \epsilon = \frac{2}{3}, \quad \alpha_c = 0, \quad \nu_c = \frac{1}{3}, \quad \Delta = \frac{3}{2}. \tag{168}$$

The RG predictions for the corrections are already known to be[44,67–69]

$$\hat{\alpha} = \frac{4-N}{N+8}, \quad \hat{\beta} = \frac{3}{N+8}, \quad \hat{\gamma} = \frac{N+2}{N+8}, \quad \hat{\delta} = \frac{1}{3}, \tag{169}$$

$$\hat{\nu} = \frac{N+2}{2(N+8)}, \quad \hat{\eta} = 0, \quad \hat{\Delta} = \frac{1-N}{N+8}, \quad \hat{q} = \frac{1}{4}. \tag{170}$$

The correction relations (39)–(42) and (46) therefore hold and from formulae (44)–(45) we predict

$$\hat{\epsilon} = \frac{10-N}{3(N+8)}, \quad \hat{\alpha}_c = \frac{4-N}{N+8}, \quad \hat{\nu}_c = \frac{1}{6}. \tag{171}$$

In the four-dimensional Ising case it is known[7,54,69] that the impact angle for Fisher zeros is $\phi = \pi/4$. Assuming the same for its $O(N)$ generalization, means that Eq.(142) does not follow from Eq.(140) and Eq.(118) remains valid there instead. The same is expected to be true for $O(N)$ theories with long-range interactions, which we next address.

6.3. *Long-range $O(N)$ ϕ^4_d theory*

The introduction of long-range interactions alters the universality class of $O(N)$ spin models. If the interactions decay as $x^{-(d+\sigma)}$, the upper critical dimension becomes $d_c = 2\sigma$. The critical exponents for the N-component long-range system are[70,71]

$$\alpha = 0, \quad \beta = \frac{1}{2}, \quad \gamma = 1, \quad \delta = 3, \quad \nu = \frac{1}{\sigma}, \tag{172}$$

$$\eta = 2-\sigma, \quad \epsilon = \frac{2}{3}, \quad \alpha_c = 0, \quad \nu_c = \frac{1}{3}, \quad \Delta = \frac{3}{2}, \tag{173}$$

which obey the leading scaling relations. The Privman-Fisher form for the free energy[71] gives the RG values for the logarithmic corrections to be

$$\hat{\alpha} = \frac{4-N}{N+8}, \quad \hat{\beta} = \frac{3}{N+8}, \quad \hat{\gamma} = \frac{N+2}{N+8},$$
$$\hat{\delta} = \frac{1}{3}, \quad \hat{\nu} = \frac{N+2}{\sigma(N+8)}, \quad \hat{\eta} = 0. \tag{174}$$

The correction relations (40) – (42) are obeyed, and the remaining relations (39) and (43)–(46) predict

$$\hat{\epsilon} = \frac{10-N}{3(N+8)}, \quad \hat{\alpha}_c = \frac{4-N}{N+8}, \quad \hat{\nu}_c = \frac{6-\sigma}{12\sigma}, \quad \hat{\Delta} = \frac{1-N}{N+8}, \quad \hat{q} = \frac{1}{2\sigma}. \tag{175}$$

The prediction $\hat{q} = 1/2\sigma$ recovers the known value[67] $\hat{q} = 1/4$ for $O(N)$ ϕ^4_4 theory in the $\sigma = 2$ case. It also leads to agreement with long-range Ising FSS in two dimensions when $\sigma = 1$.[72] The remaining predictions for long-range systems have yet to be verified.

6.4. *Spin glasses in 6 dimensions*

Spin glasses, percolation, the Lee-Yang edge, and lattice animals problems are each related to ϕ^3 field theory. The leading exponents are[73–75]

$$\alpha = -1, \quad \beta = 1, \quad \gamma = 1, \quad \delta = 2, \quad \nu = \frac{1}{2}, \quad \eta = 0, \tag{176}$$

and obey the standard scaling relations provided

$$\Delta = 2, \quad \epsilon = 1, \quad \alpha_c = -\frac{1}{2}, \quad \nu_c = \frac{1}{4}, \tag{177}$$

predictions which still need numerical verification. Ruiz-Lorenzo used RG methods to derive the critical scaling exponents of the correlation length, susceptibility and specific heat for these models at their upper critical dimensions as[75]

$$\hat{\alpha} = \frac{2(2b-3a)}{4b-a}, \hat{\gamma} = \frac{2a}{4b-a}, \hat{\nu} = \frac{5a}{6(4b-a)}, \tag{178}$$

where the values of (a, b) depend upon which problem one is considering.

In the m-component spin-glass case, the upper critical dimension is $d_c = 6$ and values of (a, b) are $(-4m, 1-3m)$. There, the leading exponents are given by Eq.(176) while Eq.(178) gives the logarithmic correction exponents as

$$\hat{\alpha} = -\frac{3m+1}{2m-1}, \quad \hat{\gamma} = \frac{2m}{2m-1}, \quad \hat{\nu} = \frac{5m}{6(2m-1)}. \tag{179}$$

These satisfy the scaling relations (39) – (46) provided that

$$\begin{gathered}\hat{\beta} = \frac{1+m}{2(1-2m)}, \quad \hat{\delta} = \frac{1-3m}{4(1-2m)}, \quad \hat{\eta} = \frac{m}{3(2m-1)}, \quad \hat{\epsilon} = \frac{1}{2}\frac{1+2m}{1-2m}, \\ \hat{\alpha}_c = \frac{1}{4}\frac{3+7m}{1-2m}, \quad \hat{\nu}_c = \frac{1}{24}\frac{5m-3}{2m-1}, \quad \hat{\Delta} = \frac{1+5m}{2(1-2m)}, \quad \hat{q} = \frac{1}{6}.\end{gathered} \tag{180}$$

Independent numerical investigations of these logarithmic corrections are required. In particular, Ruiz-Lorenzo's prediction for the finite-size correlation-length correction exponent is $\hat{q} = 1/3$.[75] This is the first instance where we encounter a clash with results in the literature regarding the exponent $\hat{q}$, the other two being in the percolation and the Lee-Yang/lattice-animal problems. These cases are discussed next.

6.5. *Percolation in 6 dimensions*

The percolation problem at its upper critical dimension of $d_c = 6$ has[73,74] the same leading, mean-field critical exponents as in Eqs.(176) and (177). With

$(a, b) = (-1, -2)$, in the percolation case the following correction exponents are known[74–76]

$$\hat{\alpha} = \frac{2}{7}, \quad \hat{\beta} = \frac{2}{7}, \quad \hat{\gamma} = \frac{2}{7}, \quad \hat{\delta} = \frac{2}{7}, \quad \hat{\nu} = \frac{5}{42}, \quad \hat{\eta} = \frac{1}{21}. \tag{181}$$

The scaling relations for logarithmic corrections now allow the prediction

$$\hat{\epsilon} = \frac{2}{7}, \quad \hat{\alpha}_c = \frac{2}{7}, \quad \hat{\nu}_c = \frac{5}{42}, \quad \hat{\Delta} = 0, \quad \hat{q} = \frac{1}{6}. \tag{182}$$

While other works[77] contain an implicit assumption that $\hat{q} = 0$, Ruiz-Lorenzo's prediction[75] for this quantity is $\hat{q} = 1/3$. Again, further investigations are needed.

6.6. *Lee-yang problem in 6 dimensions and lattice animals in 8 dimensions*

The lattice-animal problem[78,79] is closely linked to the Lee-Yang edge problem.[6] The former has upper critical dimensionality 8, while for the latter it is 6. The mean-field values of the critical exponents for the Lee-Yang edge problem are again given by Eqs.(176) and (177), while the values $(a, b) = (-1, -1)$ for Lee-Yang singularities[75] lead to

$$\hat{\alpha} = -\frac{2}{3}, \quad \hat{\gamma} = \frac{2}{3}, \quad \hat{\nu} = \frac{5}{18}. \tag{183}$$

The scaling relations for logarithmic corrections then predict

$$\begin{aligned} &\hat{\beta} = 0, \quad \hat{\delta} = \frac{1}{3}, \quad \hat{\eta} = \frac{1}{9}, \quad \hat{\epsilon} = 0, \\ &\hat{\alpha}_c = -\frac{1}{3}, \quad \hat{\nu}_c = \frac{1}{9}, \quad \hat{\Delta} = -\frac{2}{3}, \quad \hat{q} = \frac{1}{6}. \end{aligned} \tag{184}$$

Fisher-renormalizing both the mean-field leading critical exponents and the logarithmic corrections, one obtains (omiting the subscript X)

$$\begin{aligned} &\alpha = \frac{1}{2}, \quad \beta = \frac{1}{2}, \quad \gamma = \frac{1}{2}, \quad \delta = 2, \quad \nu = \frac{1}{4}, \\ &\eta = 0, \quad \Delta = 1, \quad \epsilon = \frac{1}{2}, \quad \alpha_c = -\frac{1}{2}, \quad \nu_c = \frac{1}{2}. \end{aligned} \tag{185}$$

and

$$\begin{aligned} &\hat{\alpha} = \frac{1}{3}, \quad \hat{\beta} = \frac{1}{3}, \quad \hat{\gamma} = \frac{1}{3}, \quad \hat{\delta} = \frac{1}{3}, \quad \hat{\nu} = \frac{1}{9}, \\ &\hat{\eta} = \frac{1}{9}, \quad \hat{\Delta} = 0, \quad \hat{\epsilon} = \frac{1}{3}, \quad \hat{\alpha}_c = \frac{1}{3}, \quad \hat{\nu}_c = \frac{1}{9}. \end{aligned} \tag{186}$$

together with

$$\hat{q} = \frac{1}{6}. \tag{187}$$

These deliver our theoretical predictions for the lattice-animal problem at its upper critical dimensionality $d = 8$.

A lattice animal is a cluster of connected sites on a regular lattice. Variants include clusters of connected bonds as well as weakly embedded and strongly embedded trees. It is believed that these models belong to the same universality class. The enumeration of lattice animals is a combinatorial problem of interest to mathematicians. In physics, they are closely linked to percolation and clustering in spin models. In chemistry they form a basis for models of branched polymers in good solvents. Lattice animals linked by translations are considered as belonging to the same equivalence class, and as such are considered to be essentially the same. One is interested in Z_N, the number of animals containing N sites, and the radius of gyration R_N, which is the average distance of occupied sites to the centre of mass of the animal. Allowing for logarithmic corrections, these behave as[80]

$$Z_N \sim \mu^N N^{-\theta} (\ln N)^{\hat{\theta}}, \tag{188}$$

$$R_N \sim N^{\nu} (\ln N)^{\hat{\nu}}. \tag{189}$$

The exponent θ may be identified with $3 - \alpha$, which, from the above Fisher-renormalized values is $5/2$. There is a famous scaling relation due to Parisi and Sourlas which predicts that θ and ν are related by[81]

$$\theta = (d - 2)\nu + 1. \tag{190}$$

This is essentially hyperscaling (15) with the d dimensionally reduced by 2 and has been numerically verified in dimensions $d = 2$ to $d = 9$.[82] Identifying $\hat{\theta} = \hat{\alpha}$ The scaling relation (39) may be written

$$\hat{\theta} = (d - 2)(\hat{q} - \hat{\nu}) = 6(\hat{q} - \hat{\nu}), \tag{191}$$

having reduced the dimensionality term from $d = d_c = 8$ to 6, appropriately. This is the logarithmic counterpart to the Parisi-Sourlas relation (190). The above values $\hat{\alpha} = 1/3$, $\hat{q} = 1/6$, $\hat{\nu} = 1/9$ satisfy Eq.(191) by construction. An alternative set of values in the literature[75] is $\hat{\alpha} = 1/3$, $\hat{q} = 1/3$, $\hat{\nu} = -1/72$. This set comes directly from an RG-based calculation and does not satisfy the new scaling relation. Moreover, a recent numerical study,[30] though not conclusive, indicates that the set of exponents developed here is more likely to be the correct one. Clearly more research is needed to explain the disparity of this set with the RG.

6.7. *Ising model in 2 dimensions*

The Ising model two dimensions has critical exponents

$$\alpha = 0, \quad \beta = \frac{1}{8}, \quad \gamma = \frac{7}{4}, \quad \delta = 15, \quad \nu = 1,$$
$$\hat{\eta} = \frac{1}{4}, \quad \Delta = 0, \quad \epsilon = 0, \quad \alpha_c = 0, \quad \nu_c = 0. \tag{192}$$

The specific heat has a logarithmic divergence with temperature in this model so that $\hat{\alpha} = 1$. However, this well-studied model is unusual in that there are no other logarithmic divergences and all the remaining hatted exponents vanish. The subtle nature of the emergence of this unusual behaviour was explained in Sec.4.4: a logarithm arises from summing over the Fisher-zero indices, a summation which is necessary because these zeros approach the real-temperature axis at an angle other than $\pi/4$. In fact the angle of approach is $\pi/4$. This is because the self-dual nature of the two-dimensional Ising model assures a symmetry between the high- and low- temperature sectors, which demands that the impact angle $\phi = \pi/2$.

6.8. *Quenched-disordered Ising model in 2 dimensions*

The uncorrelated, quenched, random removal of sites or the randomisation of bond strengths on a lattice is expected to immitate the presence of impurities in real physical systems. We usually appeal to the Harris criterion[36] to answer the question of how such randomisation affects the critical exponents.[36] If $\alpha > 0$ in the pure system, quenched disorder is deemed relevant and the critical exponents change as the disorder is increased. On the other hand, if $\alpha < 0$ in the pure model, then disorder of this type does not change the critical behaviour and the exponents are unaltered. In the case of $\alpha = 0$, which as we have seen describes the Ising model in two dimensions, the Harris criterion does not provide a clear answer.

For this reason, the bond- and site-diluted Ising models in two dimensions have been controversial over the years. The notion that some of the leading critical exponents change as the lattice structure is randomised, but where combinations which appear in terms of the correlation length such as β/ν and γ/ν are unchanged, became known as the *weak universality hypothesis*.

Gradually this gave way to the *strong universality hypothesis*,[52] which is now mostly believed to hold, although agreement is not universal.[53] This predicts that the two-dimensional diluted models have the same leading critical exponents as in the pure case, but that there are multiplicative logarithmic corrections[47,83]

$$\hat{\alpha} = 0, \quad \hat{\beta} = -\frac{1}{16}, \quad \hat{\gamma} = \frac{7}{8}, \quad \hat{\delta} = 0, \quad \hat{\nu} = \frac{1}{2}, \quad \hat{\eta} = 0. \tag{193}$$

With $\hat{q} = 0$,[45] these correction exponents obey the scaling relations for logarithmic corrections (39)–(42). The remaining scaling relations (43)–(46) predict

$$\hat{\Delta} = -\frac{15}{16}, \quad \hat{\epsilon} = -\frac{1}{2}, \quad \hat{\alpha}_c = -1, \quad \hat{\nu}_c = 0. \tag{194}$$

Moreover Eq.(141) means that there is a double logarithm in the specific heat[2,47–51]

$$c_\infty(t, 0) \sim \ln|\ln t|. \tag{195}$$

The power of the scaling relations for logarithmic corrections is well illustrated in this model as Eq.(39) connects the hitherto most elusive and controversial quantity $\hat{\alpha}$ directly to other exponents, which are more clearly established. For a review of the two-dimensional disordered Ising model, see Ref.[53].

6.9. *Ashkin-Teller model in 2 dimensions*

The new relation (127) also holds in the N-colour Ashkin-Teller model which consists of N coupled Ising models. the leading exponents are the same as Eq.(192) and[47,84]

$$\hat{\alpha} = -\frac{N}{N-2}, \; \hat{\beta} = -\frac{n-1}{8(n-2)}, \; \hat{\gamma} = \frac{7(N-1)}{4(N-2)}, \; \hat{\delta} = 0, \; \hat{\nu} = \frac{N-1}{N-2}, \; \hat{\eta} = 0. \tag{196}$$

If $\hat{q} = 0$, these values also support the scaling relations (39)–(42). Eqs.(43)–(46) further lead to the predictions

$$\hat{\Delta} = -\frac{15}{8}\frac{n-1}{n-2}, \quad \hat{\epsilon} = -\frac{n-1}{n-2}, \quad \hat{\alpha}_c = -2\frac{n-1}{n-2}, \quad \hat{\nu}_c = 0. \tag{197}$$

6.10. *Spin models on networks*

Here we consider another set of phase transitions which exhibit multiplicative logarithmic corrections to scaling, namely spin systems on scale free networks. There are both academic and practical motivations for studying critical phenomena on complex networks.[85] Phenomena such as opinion formation in social networks[86] are expected to be modelled by such systems, but one is also interested in realistic physics on complex geometries, such as for integrated nanoparticle systems.[87] Some complex networks are characterised by so-called scale-free behavior. The degree of a node in the network is the number of links emanating from it and is denoted by k. With $P(k)$ symbolising the degree probability distribution (the likelihood that an arbitrary chosen node has a certain degree value), this is power-law in scale-free systems,

$$P(k) \sim k^{-\lambda}. \tag{198}$$

For the critical phenomena previously described, dimensionality and length scales play crucial roles. Near the critical points itself, order-parameter fluctuations tend to be strongly correlated, the correlation length diverges and the pair-correlation function changes from an exponential to a power-law. The FSS hypothesis (114) depends upon the ratio of length scales and the dimensionality enters the scaling relations through hyperscaling (15) and (39).

However this notion of metrics is lost when we move from a lattice substrate to a complex network. For Euclidean lattices the coordination number is dimension dependent (it is $2d$ for a d-dimensional hypercube). For networks the coordination numbers become the degrees associated with nodes and λ plays the role of d. The node degree distribution function exponent λ in Eq.(198) controls the non-homogeneity manifest in a network due to its internal structure. This is a principal difference between the origin of logarithmic corrections on regular lattices and on networks. It turns out that if λ exceeds a critical value $\lambda_c = 5$ the phase transition has the usual mean-field critical exponents. As λ decreases the node-degree distribution becomes increasingly fat-tailed and the relative amount of high-degree nodes (so-called hubs) increases. This leads to non-trivial critical behavior. As a result, systems with intermediate degree distributions where $\lambda_s < \lambda < \lambda_c$ have critical exponents which are generally λ-dependent. Here $\lambda_s = 3$. Decreasing λ still further to $\lambda < \lambda_s$, the system becomes ordered at any finite temperature. Here only an infinite temperature field is capable of destroying the order.

Just as in the regular-lattice case, where logarithmic corrections arise at $d = d_c$, so too in the network case can they emerge at a marginal value $\lambda = \lambda_c$ for a number of classical spin models. In Ref.[29], the field and temperature dependencies of critical thermodynamic functions were investigated for a system with two coupled order parameters on a scale-free network. Models of this type are frequently used to describe systems with two different types of ordering. Physical examples of such systems include ferromagnetic and antiferromagnetic, ferroelectric and ferromagnetic, structural and magnetic ordering. A sociophysics application[86] may be opinion formation where there is a coupling between the preferences for a candidate and a party in an election, for example.

A comprehensive description the system with two coupled scalar order parameters on a scale-free network was given in Ref.[29], where Eq.(44) was derived. The leading exponents (there are no exponents associated with correlation functions or the correlation length) when $\lambda = 5$ are[29]

$$\alpha = 0, \quad \beta = 1/2, \quad \gamma = 1, \quad \delta = 3, \quad \epsilon = \frac{2}{3}, \quad \alpha_c = 0, \quad \Delta = \frac{3}{2}. \tag{199}$$

In addition, the logarithmic-correction exponents are[29]

$$\hat{\alpha} = -1, \quad \hat{\beta} = -\frac{1}{2}, \quad \hat{\gamma} = 0, \quad \hat{\delta} = -\frac{1}{3}, \quad \hat{\epsilon} = -\frac{2}{3}, \quad \hat{\alpha_c} = -1, \quad \hat{\Delta} = -\frac{1}{2}. \tag{200}$$

This completes our list of systems exhibiting multiplicative logarithmic corrections to scaling. The list is not exhaustive and the reader is invited to test the validity of the logarithmic scaling relations in other models which exhibit these phenomena. For example, three-dimensional anisotropic dipolar ferromagnets, in which spatial and spin degrees of freedom are linked, are experimentally accessible syetems which exhibit logarithmic corrections[89] Systems with tricriticality also involve important logarithmic factors in three dimensions.[90]

The four-dimensional diluted Ising model offers an example of a system with logarithmic corrections to leading scaling behaviour which is not power law. The XY model in two dimensions is an analagous example for a system with an infinite-order phase transition.

7. Conclusions

Over the past few years, a set of relations which link the exponents of logarithmic corrections to scaling at higher-order phase transitions has been developed. These are the logarithmic analogues of the famous scaling relations between the leading critical exponents which were developed in the 1960's and which are now pivotal in modern statistical mechanics as well as in related areas such as lattice quantum field theory. In this review, the logarithmic scaling relations are presented and tested in a number of models. In cases where there are gaps in the literature these scaling relations allow us to make predictions.

With hindsight it is perhaps surprising that these logarithmic scaling relations have not been developed earlier. As mentioned in the text, while the relations between static exponents may be derived from a suitably modified Widom hypothesis, this is not the case for the remaining exponents. One case which may have hindered earlier attempts to develop logarithmic scaling relations is that of the Ising model in two dimensions. There, only the specific-heat dependency on the reduced temperature has a logarithm. The fact that there appears, at first sight, to be no other logarithms to relate the specific-heat logarithm to may be a reason why scaling relations for logarithmic corrections have not been developed earlier.

Another stymying factor for an earlier development of the logarithmic scaling relations is the rather enigmatic, new, logarithmic-correction critical exponent $\hat{q}$ which characterises the FSS of the correlation length. It was previously thought[28] that the finite-size correlation length could not exceed the actual length of the system. However, Eq.(39) shows not just that $\hat{q}$ may exceed zero, but that it is *universal*. This is clear because the exponents $\hat{\alpha}$ and $\hat{\nu}$, with which $\hat{q}$ is related, are universal. This means that (a) sensible definitions of the finite-size correlation length and (b) sensible boundary conditions must respect the universal value of $\hat{q}$. Any definition of the correlation length or any implementation of boundary conditions not respecting the universality of $\hat{q}$ can only be, in some sense, artificial.

At the upper critical dimension, where $\hat{q}$ is non-trivial, the leading critical exponents take on their mean-field values. Therefore they cannot be used to distinguish the universality class (e.g., as we have seen, the leading critical exponents for the $O(n)$ model in four dimensions are all the same). Instead, the exponents of the multiplicative logarithmic corrections may be used, so that these have a similar status there to the leading exponents for $d < d_c$.

The logarithmic-correction exponents given and derived above are listed in Tables 1 and 2. In the Table 2, bold-face symbols indicate the current gaps in the literature, which are filled by the theory presented herein and which now require independent verification.

Acknowledgements

The author is indebted to Bertrand Berche, Christian von Ferber, Reinhard Folk, Damien Foster, Antonio Gordillo-Guerrero, Yurij Holovatch, Hsiao-Ping Hsu, Wolfhard Janke, Des Johnston, Christian Lang, Juan Ruiz-Lorenzo, Vasyl Palchykov and Jean-Charles Walter with whom he has collaborated on the area of logarithmic corrections. He is especially grateful to Yurij Holovatch for careful reading of the manuscript. This work was supported by the 7th FP, IRSES project No. 269139 "Dynamics and cooperative phenomena in complex physical and biological environments" and IRSES project No. 295302 "Statistical physics in diverse realizations".

Appendix A. Homogeneous Functions

A function of a single variable $f(x)$ is said to be *homogeneous* if multiplicative rescaling of x by an amount λ results in a multiplicative rescaling of $f(x)$ by a factor $g(\lambda)$, where g is a function of λ only:

$$f(\lambda x) = g(\lambda) f(x). \tag{A.1}$$

Thus power-laws are homogeneous functions, but logarithmic functions are not. Indeed, if (A.1) holds, it turns out that both $f(x)$ and $g(x)$ have to be power laws.

Table 1. Leading critical exponents for the models discussed herein.

	α	β	γ	δ	ϵ	α_c	ν	ν_c	Δ	η
Pure Ising model (2D)	0	$\frac{1}{8}$	$\frac{7}{4}$	15	0	0	1	0	0	$\frac{1}{4}$
Random-bond/site Ising model (2D)	0	$\frac{1}{8}$	$\frac{7}{4}$	15	0	0	1	0	0	$\frac{1}{4}$
$O(n)$ ϕ^4 (4D)	0	$\frac{1}{2}$	1	3	$\frac{2}{3}$	0	$\frac{1}{2}$	$\frac{1}{3}$	$\frac{3}{2}$	0
4−State Potts model (2D)	$\frac{2}{3}$	$\frac{1}{12}$	$\frac{7}{6}$	$\frac{1}{15}$	$\frac{4}{15}$	$\frac{8}{15}$	$\frac{2}{3}$	$\frac{8}{15}$	$\frac{5}{4}$	$\frac{1}{4}$
Long-range models (2σD)	0	$\frac{1}{2}$	1	3	$\frac{2}{3}$	0	$\frac{1}{\sigma}$	$\frac{1}{3}$	$\frac{3}{2}$	$2-\sigma$
$m-$ Component Spin Glasses (6D)	-1	1	1	2	1	$-\frac{1}{2}$	$\frac{1}{2}$	$\frac{1}{4}$	2	0
Percolation (6D):	-1	1	1	2	1	$-\frac{1}{2}$	$\frac{1}{2}$	$\frac{1}{4}$	2	0
YL Edge (6D)	-1	1	1	2	1	$-\frac{1}{2}$	$\frac{1}{2}$	$\frac{1}{4}$	2	0
Lattice Animals (8D)	$\frac{1}{2}$	$\frac{1}{2}$	$\frac{1}{2}$	2	$\frac{1}{2}$	$-\frac{1}{2}$	$\frac{1}{4}$	$\frac{1}{2}$	1	0
n-colour Ashkin-Teller model (2D):	0	$\frac{1}{8}$	$\frac{7}{4}$	15	0	0	1	0	0	$\frac{1}{4}$
Networks ($\lambda = 5$)	0	$\frac{1}{2}$	1	3	$\frac{2}{3}$	0			$\frac{3}{2}$	

To see this, consider a second rescaling – this time by a factor μ. Rescaling first by λ and then by μ leads to

$$f(\mu\lambda x) = g(\mu)f(\lambda x) = g(\mu)g(\lambda)f(x), \tag{A.2}$$

Table 2. Logarithmic-correction critical exponents for a variety of models. Bold face indicates gaps in the literature filled by the theory presented herein - i.e., bold-face values are predictions coming from the scaling relations for logarithmic corrections and remain to be verified independently.

	$\hat{\alpha}$	$\hat{\beta}$	$\hat{\gamma}$	$\hat{\delta}$	$\hat{\epsilon}$	$\hat{\alpha}_c$	$\hat{\nu}$	$\hat{\nu}_c$	$\hat{\Delta}$	$\hat{q}$	$\hat{\eta}$
Pure Ising model (2D)	1	0	0	0	0	0	0	0	0	0	0
Random-bond/site Ising model (2D)	0	$-\frac{1}{16}$	$\frac{7}{8}$	0	$-\mathbf{\frac{1}{2}}$	-1	$\frac{1}{2}$	$\mathbf{0}$	$-\mathbf{\frac{15}{16}}$	$\mathbf{0}$	0
$O(n)$ ϕ^4 (4D)	$\frac{4-n}{n+8}$	$\frac{3}{n+8}$	$\frac{n+2}{n+8}$	$\frac{1}{3}$	$\mathbf{\frac{10-N}{3(N+8)}}$	$\mathbf{\frac{4-N}{2N+8}}$	$\frac{n+2}{2(n+8)}$	$\mathbf{\frac{1}{6}}$	$\frac{1-n}{n+8}$	$\frac{1}{4}$	0
4−State Potts model (2D)	-1	$-\frac{1}{8}$	$\frac{3}{4}$	$-\frac{1}{15}$	$-\mathbf{\frac{23}{30}}$	$-\mathbf{\frac{22}{15}}$	$\frac{1}{2}$	$\mathbf{\frac{1}{30}}$	$-\mathbf{\frac{7}{8}}$	0	$-\frac{1}{8}$
Long-range models (2σD)	$\frac{4-n}{n+8}$	$\frac{3}{n+8}$	$\frac{n+2}{n+8}$	$\frac{1}{3}$	$\mathbf{\frac{10-N}{3(N+8)}}$	$\mathbf{\frac{4-N}{N+8}}$	$\frac{n+2}{\sigma(n+8)}$	$\mathbf{\frac{6-\sigma}{12\sigma}}$	$\mathbf{\frac{1-n}{n+8}}$	$\mathbf{\frac{1}{2\sigma}}$	0
$m-$ Component Spin Glasses (6D)	$-\frac{3m+1}{2m-1}$	$\mathbf{\frac{1+m}{2(1-2m)}}$	$\frac{2m}{2m-1}$	$\mathbf{\frac{1-3m}{4(1-2m)}}$	$\mathbf{\frac{1}{2}\frac{1+2m}{1-2m}}$	$\mathbf{\frac{1}{4}\frac{7m+3}{1-2m}}$	$\frac{5m}{6(2m-1)}$	$\mathbf{\frac{1}{24}\frac{5m-3}{2m-1}}$	$\mathbf{\frac{1+5m}{2(1-2m)}}$	$\mathbf{\frac{1}{6}}$	$\mathbf{\frac{m}{3(2m-1)}}$
Percolation (6D):	$\frac{2}{7}$	$\frac{2}{7}$	$\frac{2}{7}$	$\frac{2}{7}$	$\mathbf{\frac{2}{7}}$	$\mathbf{\frac{2}{7}}$	$\frac{5}{42}$	$\mathbf{\frac{5}{42}}$	$\mathbf{0}$	$\mathbf{\frac{1}{6}}$	$\frac{1}{21}$
YL Edge (6D)	$-\frac{2}{3}$	$\mathbf{0}$	$\frac{2}{3}$	$\mathbf{\frac{1}{3}}$	$\mathbf{0}$	$-\mathbf{\frac{1}{3}}$	$\frac{5}{18}$	$\mathbf{\frac{1}{9}}$	$-\mathbf{\frac{2}{3}}$	$\mathbf{\frac{1}{6}}$	$\mathbf{\frac{1}{9}}$
Lattice Animals (8D)	$\frac{1}{3}$	$\mathbf{\frac{1}{3}}$	$\frac{1}{3}$	$\mathbf{\frac{1}{3}}$	$\mathbf{\frac{1}{3}}$	$\mathbf{\frac{1}{3}}$	$\frac{1}{9}$	$\mathbf{\frac{1}{9}}$	$\mathbf{0}$	$\mathbf{\frac{1}{6}}$	$\mathbf{\frac{1}{9}}$
n-colour Ashkin-Teller model (2D):	$\frac{-n}{n-2}$	$\frac{-(n-1)}{8(n-2)}$	$\frac{7(n-1)}{4(n-2)}$	0	$\mathbf{\frac{1-n}{n-2}}$	$\mathbf{2\frac{1-n}{n-2}}$	$\frac{n-1}{n-2}$	$\mathbf{0}$	$\mathbf{\frac{-15(n-1)}{8(n-2)}}$	$\mathbf{0}$	0
Networks	-1	$-\frac{1}{2}$	0	$-\frac{1}{3}$	$-\mathbf{\frac{2}{3}}$	-1			$-\frac{1}{2}$		

while rescaling by the combined factor $\mu\lambda$ yields

$$f(\mu\lambda x) = g(\mu\lambda)f(x). \tag{A.3}$$

Equating the right-hand sides of these two equations gives

$$g(\mu\lambda) = g(\mu)g(\lambda). \tag{A.4}$$

Differentiating this with respect to μ gives $\lambda g'(\mu\lambda) = g'(\mu)g(\lambda)$, which, after chosing $\mu = 1$ gives

$$\lambda g'(\lambda) = pg(\lambda), \tag{A.5}$$

in which $p = g'(1)$. Solving for g then gives $g(\lambda) \propto \lambda^p$, so that $g(\lambda)$ is power -law. Inserting this into Eq.(A.1) and chosing $\lambda = x^{-1}$ then gives $f(1) = x^{-p}f(x)$, or

$$f(x) \propto x^p, \tag{A.6}$$

so that $f(x)$ is also power-law.

A function of two variables $f(x, y)$ is called homogeneous if

$$f(\lambda x, \lambda y) = g(\lambda)f(x, y). \tag{A.7}$$

Following similar considerations one can easily show that $g(\lambda)$ is again power-law. A more generalised homogeneous function obeys

$$f(\lambda^r x, \lambda^s y) = \lambda^p f(x, y). \tag{A.8}$$

Renaming $\lambda^p \to \lambda$ one arrives at the following property for a generalised homogeneous function:

$$f(\lambda^a x, \lambda^b y) = \lambda f(x, y). \tag{A.9}$$

References

1. B. Berche, M. Henkel and R. Kenna, *J. Phys. Studies* **13** 3201, (2009).
2. R. Kenna, D. A. Johnston, and W. Janke, *Phys. Rev. Lett.* **96**, 115701, (2006); *ibid.* **97**, 155702, (2006).
3. C.N. Yang and T.D. Lee, *Phys. Rev.* **87**, 404, (1952); *ibid.* 410, (1952).
4. M.E. Fisher, in *Lecture in Theoretical Physics VIIC*, edited by W.E. Brittin (University of Colorado Press, Boulder, 1965), p.1.
5. W. van Saarlos and D. Kurtze, *J. Phys. A* **17**, 1301, (1984); J. Stephenson and R. Cuzens, *Physica A* **129**, 201, (1984).
6. M.E. Fisher, *Phys. Rev. Lett.* **40**, 1610, (1978).
7. C. Itzykson, R.B. Pearson, and J.B. Zuber, *Nucl. Phys. B* **220**, 415, (1983).
8. V. Privman, P.C. Hohenberg, and A. Aharony, in *Phase Transitions and Critical Phenomena* **14**, Edited by C. Domb, J.L. Lebowitz, Ch.1, (Academic, NY, 1991) p.1.

9. B. Widom, *J. Chem. Phys.* **43**, 3892, (1965); *ibid.* 3898, (1965).
10. R.B. Griffiths, *Phys. Rev.* **158**, 176, (1967).
11. L.P. Kadanoff, *Physics* **2**, 263, (1966).
12. B.D. Josephson, *Proc. Phys. Soc.* **92**, 269, (1967); *ibid* 276, (1967).
13. J.J. Binney, N.J. Dowrick, A.J. Fisher and M.E.J. Newman, *The Theory of Critical Phenomena.* (Oxford University Press, 1992).
14. J.W. Essam and M.E. Fisher, *J. Chem. Phys.* **38**, 802, (1963).
15. G.S. Rushbrooke, *J. Chem. Phys.* **39** 842, (1963).
16. B. Widom, *J. Chem. Phys.* **41**, 1633, (1964).
17. R.B. Griffiths, *Phys. Rev. Lett.* **14**, 623, (1965).
18. R. Abe, Prog. *Theor. Phys.* **38**, 72, (1967).
19. M. Suzuki, Prog. *Theor. Phys.* **38**, 289, (1967); *ibid.* 1225, (1967).
20. M.E. Fisher, *J. Math. Phys.* **5**, 944, (1964).
21. M.J. Buckingham and J.D. Gunton, *Phys. Rev.* **178**, 848, (1969).
22. C. Domb and D.L. Hunter, *Proc. Phys. Soc.* **86**, 1147, (1965) ; C. Domb, Proc. 1966 Low Temperature Calorimetry Conf., ed. by O.V. Lounasmaa, Ann. Acad. Sci. Fennicae A **VI**, 167, (1966).
23. R. Abe, *Prog. Theor. Phys.* **38**, 568, (1967).
24. A.Z. Patashinsky and V.L. Pokrovsky, *Sov. Phys. JETP* **23**, 292, (1966); *ibid.* **50**, 439, (1966).
25. W. Janke and R. Kenna, *Phys. Rev. B* **65**, 064110, (2002).
26. A. Gordillo-Guerrero, R. Kenna, J.J. Ruiz-Lorenzo, *JSTAT* P09019, (2011).
27. M. Le Bellac, *Quantum and Statistical Field Theory*. (Oxford Science Publications, 1991.)
28. P.G. Watson, in *Phase Transitions and Critical Phenomena* (eds Domb C. & Green M.S.) Vol. 2 101-159 (Academic Press, London, 1972); J. Rudnick, G. Gaspari, and V. Privman, *Phys. Rev. B* **32**, 7594, (1985).
29. V. Palchykov, C. von Ferber, R. Folk, Yu. Holovatch and R. Kenna, *Phys. Rev. E* **82**, 011145, (2010).
30. C. von Ferber, D. Foster, H.-P. Hsu and R. Kenna, arXiv 1107.1187, to appear in *Eur. Phys. J. B.* (2011).
31. K.G. Wilson, *Phys. Rev. B* **4**, 3174, (1971); *ibid.* 3184, (1971).
32. J. Salas and A.D. Sokal, *J. Stat. Phys.* **88**, 567, (1997).
33. M. Suzuki, *Prog. Theor. Phys.* **39**, 349, (1968).
34. A. E. Ferdinand and M. E. Fisher, *Phys. Rev.* **185**, 832, (1969).
35. F.J. Wegner, in *Phase Transitions and Critical Phenomena*, VI, ed. by C. Domb and M.S. Green (Academic Press, London, 1976), p.8.
36. A.B. Harris, *J. Phys. C* **7**, 1661, (1974).
37. F.J. Wegner, *Phys. Rev. B* **4**, 4529, (1972).
38. F. J. Wegner, in *Phase Transitions and Critical Phenomena*, Vol. VI, edited by C. Domb and M. S. Green (Academic Press, London, 1976), p. 8.
39. D.A. Huse and M.E. Fisher, *J. Phys. C* **15**, L585, (1982); A. Aharony and M.E. Fisher, *Phys. Rev. B* **27**, 4394, (1983).
40. L.N. Shchur, B. Berche and P. Butera, *EPL* **81** 30008, (2008); *Nucl. Phys. B* **811** 491, (2009); B. Berche, P. Butera, W. Janke and L.N. Shchur, *Comp. Phys. Comput.* **180** 493, (2009); B. Berche, L.N. Shchur and P. Butera, in Computer Simulation Studies

in Condensed Matter Physics XX, eds. D.P. Landau, S.P. Lewis and H.B. Schüttler (Springer, 2010). A. Aharony, Phys. Rev. B 22, 400, (1980).
41. M.E. Fisher, *Rev. Mod. Phys.* **70**, 653, (1998).
42. W. Janke and R. Kenna, *J. Stat. Phys.* **102**, 1211, (2001).
43. N. Aktekin, *J. Stat. Phys.* **104**, 1397, (2001).
44. R. Kenna, *Nucl. Phys. B* **691**, 292, (2004).
45. F.D.A. Aarão Reis, S.L.A. de Queiroz and R.R. dos Santos, *Phys. Rev. B* **54**, R9616, (1996); *ibid.* **56**, 6013, (1997).
46. P. Lajkó and F. Iglói, Phys. Rev. E **61**, 147, (2000).
47. B.N. Shalaev, *Sov. Phys. Solid State* **26**, 1811, (1984); *Phys. Rep.* **237**, 129, (1994); R. Shankar, *Phys. Rev. Lett.* **58**, 2466, (1987); **61**, 2390, (1988); A.W.W. Ludwig, *ibid.* **61**, 2388, (1988); *Nucl. Phys. B* **330**, 639, (1990); G. Jug and B.N. Shalaev, *Phys. Rev. B* **54**, 3442, (1996).
48. Vik. S. Dotsenko and Vl. S. Dotsenko, *JETP Lett.* **33**, 37, (1981); *Adv. Phys.* **32**, 129, (1983).
49. V.B. Andreichenko, Vl. S. Dotsenko, W. Selke, and J.-S. Wang, *Nucl. Phys. B* **344**, 531, (1990); J.-S. Wang, W. Selke, Vl. S. Dotsenko, and V.B. Andreichenko, *Europhys. Lett.* **11**, 301, (1990); *Physica A* **164**, 221, (1990); W. Selke, L.N. Shchur, and O.A. Vasilyev, *Physica A* **259**, 388, (1998).
50. G. Jug, in *Multicritical Phenomena*, edited by R. Pynn and A. Skyeltorp, Proceedings of the 1983 Geilo School [NATO ASI held April 10-21 1983 in Geilo, Norway] (NATO ASI Series B: Physics, Vol. 106, Plenum Press, New York 1984), 329.
51. G. Jug, *Phys. Rev. Lett.* **53**, 9, (1984).
52. R. Kenna and J.J. Ruiz-Lorenzo, *Phys. Rev. E* **78**, 031134, (2008).
53. A. Gordillo-Guerrero, R. Kenna and J.J. Ruiz-Lorenzo, *AIP Conf. Proc.* **1198**, 42, (2009).
54. A. Caliri and D.C. Mattis, *Phys. Lett. A* **106**, 74, (1984); M.L. Glasser, V. Privman, and L.S. Schulman, *J. Stat. Phys.* **45**, 451, (1986); *Phys. Rev. B* **35**, 1841, (1987).
55. R. Abe, *Prog. Theor. Phys.* **37**, 1070, (1967).
56. V. Matveev and R. Shrock, *J. Phys. A* **28**, 5235, (1995).
57. M.E. Fisher, *Phys. Rev.* **176**, 257, (1968).
58. I.D. Lawrie and S. Sarbach, in *Phase Transitions and Critical Phenomena*, edited by C. Domb and J.L. Lebowitz (Academic Press, London, 1984), Vol.9, p.1; Y. Deng and H.W.J. Blöte, *Phys. Rev. E* **70**, 046111, (2004); M.O. Kimball and F.M. Gasparini, *Phys. Rev. Lett.* **95**, 165701, (2006).
59. Y. Imry, O. Entin-Wohlman and D.J. Bergman, *J. Phys. C: Solid State Phys.* **6**, 2846, (1973).
60. C.W. Garland and B.B. Weiner, *Phys. Rev. B* **3**, 1634, (1971); A. Aharony, *Phys. Rev. B* **8**, 4314, (1973).
61. *The Physics of Liquid Crystals*, P.-G. de Gennes and J. Prost, (Oxford University Press, 1995).
62. A.M. Bellocq in *Handbook of Microemulsion Science and Technology*, edited by P. Kumar and K.L. Mittal (CRC Press, 1999) p.139 .
63. R. Kenna, H.-P Hsu and C. von Ferber, *J. Stat. Mech. (JSTAT): Theor. Exp.* L10002, (2008).
64. M.P.M. den Nijs, *J. Phys. A* **12**, 1857, (1979); B. Nienhuis, E.K. Riedel and M. Schick,

J. Phys. A **13**, L189, (1980).
65. M. Nauenberg and D.J. Scalapino, *Phys. Rev. Lett.* **44**, 837, (1980); J.L. Cardy, M. Nauenberg and D.J. Scalapino, *Phys. Rev. B* **22**, 2560, (1980).
66. J.L. Black and V.J. Emery, *Phys. Rev. B* **23**, 429, (1981).
67. E. Brézin, *J. Physique* **43**, 15, (1982).
68. E. Brézin, J.C. Le Guillou and J. Zinn-Justin, in *Phase Transitions and Critical Phenomena*, VI, ed. by D. Domb and M.S. Green (Academic Press, London, 1976), p. 127; M. Lüscher and P. Weisz, Nucl. Phys. B **318**, 705, (1989).
69. R. Kenna and C.B. Lang, *Phys. Lett. B* **264**, 396, (1991); *Phys. Rev. E* **49**, 5012, (1994); *Nucl. Phys. B* **393**, 461, (1993); **411**, 340, (1994).
70. M.E. Fisher, S.-K. Ma and B.G. Nickel, *Phys. Rev. Lett.* **29**, 917, (1972).
71. E. Luijten and H.W.J. Blöte, *Phys. Rev. B* **56**, 8945, (1997).
72. D. Grüneberg and A. Hucht, *Phys. Rev. E* **69**, 036104, (2004).
73. S.F. Edwards and P.W. Anderson, *J. Phys. F* **5**, 965, (1975); A.B. Harris, T.C. Lubensky and J.-H. Chen, *Phys. Rev. Lett.* **36**, 415, (1976).
74. A.B. Harris, J.C. Lubensky, W.K. Holcomb, and C. Dasgupta, *Phys. Rev. Lett.* **35**, 327, (1975); I.W. Essam, D.S. Gaunt, and A.J. Guttmann, *J. Phys. A* **11**, 1983, (1978).
75. J.J. Ruiz-Lorenzo, *J. Phys. A* **31**, 8773, (1998).
76. O. Stenull and H.K. Janssen, *Phys. Rev. E* **68**, 036129, (2003).
77. S. Fortunato, A. Aharony, A. Coniglio and D. Stauffer, *Phys. Rev. E* **70**, 056116, (2004).
78. T.C. Lubensky and J. Isaacson, *Phys. Rev. Lett* **41**, 829 (1978); *ibid.* **42**, 410, (1979) (erratum).
79. D. Stauffer and A. Aharony, *An Introduction to Percolation Theory* (Taylor & Francis, London, 1994).
80. T.C. Lubensky and J. Isaacson, *Phys. Rev. A* **20**, 2130, (1979).
81. G. Parisi and N. Sourlas, *Phys. Rev. Lett.* **46**, 871, (1981).
82. H.-P Hsu, W. Nadler and P. Grassberger, *J. Phys. A* **38**, 775, (2005).
83. G. Jug, *Phys. Rev. B* **27**, 609, (1983); *ibid* 4518, (1983).
84. G.S. Grest and M. Widom, *Phys. Rev. B* **24**, 6508, (1981).
85. R. Albert and A.-L. Barabasi, *Rev. Mod. Phys.* **74**, 47, (2002); S. N. Dorogovtsev and J. F. F. Mendes, *Adv. Phys.* **51**, 1079, (2002); M. E. J. Newman, *SIAM Review* **45**, 167, (2003); Yu. Holovatch, C. von Ferber, A. Olemskoi, T. Holovatch, O. Mryglod, I. Olemskoi, and V. Palchykov, *J. Phys. Stud.* **10**, 247, (2006) (in Ukrainian); S. N. Dorogovtsev and A. V. Goltsev, *Rev. Mod. Phys.* **80**, 1275, (2008).
86. S. Galam, Physica A **274**, 132 (1999); *Int. J. Mod. Phys. C* **19**, 409, (2008); K. Sznajd-Weron and J. Sznajd, *ibid.* **11**, 1157, (2000); K. Sznajd-Weron, *Acta Phys. Pol. B* **36**, 2537, (2005); D. Stauffer and S. Solomon, *Eur. Phys. J. B* **57**, 473, (2007); K. Kułakowski and M. Nawojczyk, *e-print* arXiv:0805.3886.
87. B. Tadić, K. Malarz, and K. Kułakowski, *Phys. Rev. Lett.* **94**, 137204, (2005).
88. M. Leone, A. Vázquez, A. Vespignani, and R. Zecchina, *Eur. Phys. J. B* **28**, 191, (2002); S. N. Dorogovtsev, A. V. Goltsev, and J. F. F. Mendes, *Phys. Rev. E* **66**, 016104, (2002); F. Igloi and L. Turban, *Phys. Rev. E* **66**, 036140, (2002).
89. K. Ried, Y. Millev, M. Fähnle and H. Kronmüller, *Phys. Rev. B* **51** 15229, (1995).
90. M.J. Stephen, *Phys. Rev. B* **12** 1015, (1975).

Chapter 2

Phase Behaviour and Criticality in Primitive Models of Ionic Fluids

Oksana Patsahan* and Ihor Mryglod†

Institute for Condensed Matter Physic, 1 Svientsitskii Str., 79011 Lviv, Ukraine

For the last two decades, both phase diagrams and the critical behaviour of ionic fluids with dominant Coulomb interactions have been intensively studied using experimental and theoretical methods. In this chapter we present the overview of the progress made in this field. The main attention is focused on the theoretical results obtained for the primitive models of ionic fluids. We discuss the results of the mean field theories, in particular the Debye-Hückel theory and the mean spherical approximation as well as their generalizations related to the explicit inclusion of the ion association. We also review the recent results of statistical field theory that uses functional methods and allows one to go beyond the mean field approximation.

Contents

*oksana@icmp.lviv.ua
†mryglod@icmp.lviv.ua

1. Introduction

It is well-known that electrostatic forces determine the major properties of various systems: physical as well as chemical or biological. In particular, the Coulomb interactions are of great importance when dealing with ionic fluids i.e., fluids consisting of dissociated cations and ions. Electrolyte solutions, molten salts, colloidal systems and room temperature ionic liquids (ILs) are examples of systems in which the dominant interparticle forces are the Coulomb interactions. Besides being of fundamental interest, ionic fluids deserve great attention from practical point of view. For example, ILs are becoming increasingly popular subjects of investigation as "neoteric" solvents for catalysis, separations, fuel cells and many other applications. Reviews of the history of ILs development and their expanding applications are available in Refs. [1,2]

The fundamental aspect in theoretical studies of ionic fluids consists in the long-range nature of the Coulomb interactions. In most cases the electrostatic interaction is a dominant interaction and due to its long-range character can substantially affect the critical properties and the phase behaviour of ionic systems.

It is commonly accepted that both types of critical points (gas-liquid and liquid-liquid/fluid-fluid) of fluid mixtures with dominant short-range interactions belong to the universality class of three-dimensional Ising model. In the vicinity of a critical point the order parameter obeys the universal power law of the form

$$\varphi = \pm B_0 \tau^{\beta}(1 + B_1 \tau^{\Delta} + \ldots), \tag{1}$$

where φ is the difference between density and critical density or between concentration and critical concentration along the coexistence curve, $\tau = |T - T_c|/T$, T is the temperature, T_c is the critical temperature, β and Δ are universal critical exponents, and B_0, B_1 are system-dependent amplitudes. The leading term of this expansion represents the asymptotic critical limit, while the next terms represent corrections to the asymptotic behaviour. Expansion (1) is called the Wegner series.[3] For the Ising universality class $\beta = 0.325(5) \pm 0.002$, $\Delta = 0.52 \pm 0.02$.[4]

In contrast to nonionic fluids, the situation in the ionic systems is more challenging owing to interplay between strong but screened long-range interactions and diverging critical density fluctuations. It is worth noting that in accordance with the theory of phase transitions the mean-field behaviour with $\beta = 0.5$ is expected for the systems with long-range interactions.[5] Thus, the critical properties and the phase behaviour of ionic fluids are of special interest.

In these notes we review the current state of the research of the phase and critical behaviour of the systems dominated by Coulomb interactions. In particular, the main attention is focused on the theoretical studies of the gas-liquid equilibrium of primitive models of ionic fluids.

2. Historical Background

Over the last two decades, both phase diagrams and the critical behaviour of ionic solutions have been intensively studied using experimental techniques, numerical simulations and theoretical methods. These studies were stimulated by controversial experimental results, demonstrating the three types of the critical behaviour in electrolyte solutions:[6–8] (i) Ising-like, (ii) classical (or mean-field) behaviour as well as (iii) crossover between the two. In accordance with these peculiarities, ionic solutions were conventionally divided into two classes:[9]

- The first class includes *solvophobic systems* with Ising-like critical behaviour in which Coulomb forces are not supposed to play a major role. In this case the solvents are generally characterized by a high dielectric constant. The most of weakly dissociated aqueous mixtures showing partial miscibility near their consolute point belong to this type of systems. The common feature of these mixtures is that the solute consists of fairly large organic molecules. Examples of such systems are given by the aqueous solutes of organic compounds C_3H_7COOH, $(C_2H_5)_3N$, $(C_2H_{11})_4NBr$.
- The second class includes *Coulombic systems* in which the phase separation is primarily driven by Coulomb interactions and the solvents are characterized by a low dielectric constant. These are mixtures of complex, often organic salts, with non-aqueous solvents in the vicinity of the consolute critical point. Triethyl n-hexyl ammonium triethyl n-hexyl borate ($N^+_{2226}B^-_{2226}$) in diphenyl ether provides an example of such systems.

In fact, Ref.[9] initiated active investigations of the nature of the phase and critical behaviour of ionic fluids. Among experimental works regarding this issue one should point out the works by Narayanan and Pitzer,[10–12] Japas and Levelt Sengers,[13] Anisimov and co-workers[4,14] as well as the investigations performed by Schröer, Weingärtner, Wiegand et al.[15–20]

What do we know about criticality in ionic fluids today? We can summarize that most experiments suggest that the gas-liquid and liquid-liquid critical points in ionic fluids belong to the Ising universality class. Experiments that supported the expectation of mean-field critical behaviour could not be reproduced in later works.[16] Strong evidence for the Ising universal class has been also found by recent simulations[21–23] and theoretical studies.[24,25] However, there are still unresolved problems. In particular, more recent accurate experiments indicate a crossover to the mean-field behaviour with crossover temperatures much closer to the critical temperature than it is observed in nonionic fluids.[19,20,26]

The simplest theoretical model that demonstrates the phase separation driven exclusively by Coulomb forces is a restricted primitive model (RPM). In this model, the ionic fluid is described as an equimolar mixture of positively and negatively charged hard spheres of equal diameter immersed in a structureless dielectric continuum with dielectric constant ϵ. The pair potential is

$$u_{\alpha\beta}(r) = \begin{cases} \infty, & r < \sigma \\ \dfrac{q_\alpha q_\beta}{\epsilon r}, & r \geqslant \sigma \end{cases}, \tag{2}$$

where σ is the hard sphere diameter, $q_\alpha = \pm q$ is the charge of ion α, and r is the interionic distance. For this system we define the reduced temperature by $T^* = k_B T \epsilon\sigma/q^2$ and the reduced ion number density by $\rho^* = \rho\sigma^3$. It is remarkable that the critical parameters of the RPM are in general agreement with those of the liquid-liquid phase transitions found in many solutions of low-melting salts in organic solvents.[20,27,28]

Because of its relative simplicity, the RPM has been intensively studied by both simulations and theoretical methods. In 1968 Stillinger and Lovett[29] predicted the existence of a gas-liquid-like critical point in the RPM giving only a schematic phase diagram. Vorontsov-Vel'yaminov et al.[30] were the first to propose the location of the gas-liquid critical point for this model using constant pressure Monte Carlo (MC) simulations. They found the following values for the critical parameters: $T_c^* \approx 0.095$, $\rho_c^* \approx 0.17$. In 1976 Stell, Wu and Larsen[31] undertook a systematic theoretical investigation of thermodynamics of the RPM using three different techniques. All the approximations confirmed the presence of a simple gas-liquid-like critical point at low temperature and low density. Active investigations of the gas-liquid critical point of the RPM were renewed in the 90ies of the last century.

The RPM is the simplest model of ionic fluids with dominant Coulomb interactions. However, the investigation of more complex models is very important in understanding the nature of phase and critical behaviour of real ionic fluids which demonstrate both the size and charge asymmetry as well as other complexities such as short-range attractions. The study of the phase behaviour of two-component size- and charge-asymmetric primitive models (PMs) has been started more recently. The PM pair potential for two ions α and β at distance r apart is

$$u_{\alpha\beta}(r) = \begin{cases} \infty, & r < \sigma_{\alpha\beta} \\ \dfrac{q_\alpha q_\beta}{\epsilon r}, & r \geqslant \sigma_{\alpha\beta} \end{cases}, \tag{3}$$

where an ion of species α has a diameter σ_α, and charge q_α, $\sigma_{\alpha\beta} = \frac{1}{2}(\sigma_\alpha + \sigma_\beta)$ and ϵ is the dielectric constant.

Considerable efforts have been made in order to develop the theories for the description of phase behaviour and criticality in ionic systems. Summarizing one should stress that a theoretical description of the gas-liquid critical point of the PMs was mainly accomplished at a mean-field level. In particular, several theoretical methods have been proposed in which the ion association is explicitly taken into account. The main of them are the generalized Debye-Hückel (GDH) theory developed by Fisher and Levin[32–34] and the approach that is based on the mean spherical approximation (MSA) proposed by Stell and his co-workers.[35,36] More recently, new approaches that use the functional integration methods have been developed. Among them one should mention the scalar field KSSHE (Kac-Siegert-Stratonovich-Hubbard-Edwards) theory developed by Caillol,[37,38] mesoscopic field theory proposed by Ciach and Stell[39,40] and the theory which uses the collective variables (CVs) method.[41–43]

To summarise, theoretical investigations of the phase and critical behaviour of ionic fluids have been mainly aimed at solving such problems as: (i) the location of the gas-liquid critical point of the RPM; (ii) the full phase diagram of the RPM; (iii) the effects of size and charge asymmetry on the gas-liquid phase diagram of the PM; (iv) the behaviour of the charge-charge correlation function near the gas-liquid critical point; (v) the universality class of the gas-liquid critical point and the peculiarities of the crossover behaviour in an ionic fluid.

3. Location of the Gas-Liquid Critical Point of the RPM

3.1. *Computer simulations*

Active simulation studies focused on locating the gas-liquid critical point of the RPM started in the early 90ies of the last century.[44–49] They confirmed the existence of a gas-liquid phase transition at low temperatures and low densities. As it appeared, the coexistence curve of the RPM in a ρ versus T plot is of a strong asymmetric shape. Over the years the figures for critical parameters have changed appreciably. A methodological advance in calculations of critical points of fluids was related to the mixed-field finite-size scaling (MFFSS) approach proposed by Bruce and Wilding.[50] Using different modifications of the Monte Carlo method the MFFSS analysis has been used by most simulation studies of the gas-liquid critical point of the RPM. The location of the gas-liquid critical point was obtained by Caillol et al.[21,51,52] in hyperspherical boundary conditions and Orcoulas and Panagiotopoulos[53] in standard cubic boundary conditions. Yan and de Pablo[54] used this approach in combination with hyper-parallel tempering Monte Carlo. A summary of recent results for critical parameters of the RPM is presented in Table 1. As is seen, there is generally a good agreement between simulations performed by different teams.

Table 1. The gas-liquid critical parameters of the RPM.

$T_c^* = k_B T_c \epsilon\sigma/q^2$	$\rho_c^* = \rho_c\sigma^3$	References
0.04917 ± 0.00002	0.080 ± 0.005	Ref.[21]
0.0490 ± 0.0003	0.070 ± 0.005	Ref.[53]
0.0492 ± 0.0003	0.062 ± 0.005	Ref.[54]
0.0489 ± 0.0003	0.076 ± 0.003	Ref.[55]
0.04933 ± 0.00005	0.075 ± 0.001	Ref.[56]

3.2. *Generalized Debye-Hückel theory*

The theory of ionic systems has a long history going back to the 19th century.[57] A major advance in the comprehension of strong electrolytes was the theory proposed by Debye and Hückel[58] (DH) in 1923. It is based on the linearization of the Poisson-Boltzmann (PB) equation. The theory appeared to be successful for the description of the behaviour of ionic systems at low density (the DH limiting law (DHLL)) being independent on of the size and structure of the ions. The full DH theory[59] that makes allowance for the finite sizes of ions modeled by hard spheres of diameter σ is less recognized. A generalization of this early theory was proposed by Fisher and Levin in Refs.[32,33] We briefly dwell on the main ideas and results of these works.

First, the authors considered the expression for the density of the Helmholtz free energy, $\bar{f} = -F/Vk_BT$, in the approximation of the DH full theory supplementing it by an ideal-gas free energy:

$$\bar{f}(T,\rho) = \bar{f}^{id}(T,\rho) + \bar{f}^{DH}(T,\rho), \tag{4}$$

where

$$\bar{f}^{id}(T,\rho) = \rho\left(1 - \ln[\rho\Lambda^3(T)]\right) \tag{5}$$

is the ideal-gas free energy,

$$\overline{f}^{DH}(T,\rho) = [\ln(1+x) - x + \frac{1}{2}x^2]/4\pi\sigma^3$$
$$\simeq \frac{\kappa_D^3}{12\pi}(1 - \frac{3}{4}x + \frac{3}{5}x^2 + \ldots) \quad (6)$$

is the DH approximation of an electroneutral system of charged hard spheres. Here

$$\kappa_D^2 = 4\pi \sum_\alpha \rho_\alpha q_\alpha^2/\epsilon k_B T \quad (7)$$

represents inverse squared Debye length, $x = \kappa_D\sigma$, $\rho = \rho_+ + \rho_-$, ρ_α is the ionic density of the αth species, $\Lambda(T)$ is the de Broglie thermal wavelength.

Based on Eq. (4), one can easily find the values for the critical parameters: $T_c^* = \frac{1}{16} \simeq 0.0625$, $\rho_c^* \simeq 0.00497$. An extension of the model by taking into account the hard core in the approximation of the second virial coefficient ($B_2^{HC} = \frac{2}{3}\pi\sigma^3$) slightly changes the parameters of the critical point giving:[33] $T_c^* \simeq 0.06133$, $\rho_c^* \simeq 0.004551$. However, the value of the critical density still remains much lower than that obtained in simulations.

A further development of the theory goes beyond the original DH theory. Following the ideas of Bjerrum[60] the theory incorporates ion pairing and takes into account the interactions of free ions and ion pairs by macroscopic electrostatics as the DH theory does for ion-ion interaction. The Debye-Hückel-Bjerrum (DHBj) theory[60] keeps the critical temperature unchanged comparing with the standard DH theory but substantially improves the value of the total critical density ($\rho_c^* = 0.045238$). However, the coexistence curve obtained within the framework of this theory assumes an unphysical "banana" shape. It is worth noting that an inclusion of hard-core terms changes the results for the worse. On the other hand, an inclusion of ion-dipole (DI) interactions, the DHBj-DI theory, leads to the correct shape of the coexistence curve but to the worse result for the critical density:[33] $T_c^* \simeq 0.0574$, $\rho_c^* \simeq 0.0277$ (without inclusion of hard-core terms) and $T_c^* \simeq 0.0554$, $\rho_c^* \simeq 0.0259$ (with inclusion of hard-core terms in the free-volume approximation corresponding to body-centered cubic close packing).

3.3. *Associating mean-spherical approximation (AMSA)*

Another group of the mean-field theories is based on the mean-spherical approximation (MSA).[61] The MSA relies on a simple and attractive approximation for the direct correlation functions combined with the Ornstein-Zernike equation. In the case of the RPM it allows one to get an analytical solution.[62] In contrast to the DH theory, the MSA provides a self-consistent description of the hard sphere part of the interaction potential: when the charges vanish the theory reduces to the Percus-Yevick (PY) equation for hard spheres.[61] The ionic contribution to the density of the reduced Helmholtz free energy, derived via the equation for internal energy, is as follows:[62]

$$\overline{f}^{MSA}(T,\rho) = [2 + 6x + 3x^2 - 2(1+2x)^{\frac{3}{2}}/12\pi\sigma^3 \simeq \frac{\kappa_D^3}{12\pi}[1 - \frac{3}{4}x + \frac{3}{4}x^2 - \ldots]. \quad (8)$$

Here we use the same notations as in Eqs (5)-(6). Adding the expression for the ideal gas free energy to equation (8) ($\overline{f} = \overline{f}^{id} + \overline{f}^{MSA}$) one obtains the following values of the critical point parameters: $T_c^* \simeq 0.0858$, $\rho_c^* \simeq 0.0398$. Equation (8) supplemented by the hard-core free energy term in the free-volume approximation (as for bcc close packing) yields $T_c^* \simeq 0.08212$, $\rho_c^* \simeq 0.0221$. The critical parameters obtained within the framework of the so-called full MSA theory that takes into account the hard-core in the PY approximation are as follows: $T_c^* = 0.079$, $\rho_c^* = 0.014$.[35]

The MSA can also be improved by inclusion of the effects connected with association. Adding the Bjerrum pairs to the MSA and taking into account the effects of hard-core in the free-volume approximation yield a reasonable shape of the coexistence curve and the following values of the critical parameters: $T_c^* \simeq 0.08237$, $\rho_c^* \simeq 0.03307$.[35]

In the 1980s, Ebeling and Grigo[63] proposed an elegant extension of the Bjerrum theory of ionic association. They combined ion pairing with the MSA through the mass action law and the second ionic virial coefficient. As a result, the critical temperature and critical density are as follows: $T_c^* \simeq 0.08368$, $\rho_c^* \simeq 0.018$.

Stell and coworkers[64,65] proposed the pairing MSA on the basis of a simple interpolation scheme (SIS). In the SIS the cavity correlation function for an associating fluid remains nearly the same as that of the dissociated fluid. This approximation is equivalent to the Wertheim first order thermodynamic perturbation theory that has been widely used to study thermodynamic properties of associating fluids and chain-like macromolecules.

Several theoretical methods were proposed in which the ion pairing was explicitly taken into account.[66–68] In particular, two approximations were proposed to model the phase behaviour of the RPM electrolyte.[67] First, the effect of ion association was included using the binding mean-spherical approximation (BIMSA) based on the Wertheim-Ornstein-Zernike integral equation formalism. The second method was the combination of the BIMSA with a simple interpolation scheme (SIS/BIMSA). Four different association constants were used to calculate the degree of dissociation, the critical point and the gas-liquid coexistence curve. As was shown, an increase in the association constant led to a lower critical temperature and a higher critical density, and better agreement with computer simulations. Compared with the BIMSA, in general, the SIS/BIMSA gave a better critical point and coexistence curve.

Thus, ion association being taken into account results in some improvement of quantitative estimates for the critical parameters of the RPM, in particular, the critical density. In the RPM, the clusters (dipols) appear to a strong Coulomb attraction between ions of opposite sign when they are close to one another. The DH theory and the MSA theory for strong electrolytes are essentially linearized PB theories. They yield the correct Coulomb screening of long-range interactions but cannot capture the nonlinear effects of ion association.[35] It is worth noting that the both groups of theories are of a mean-field type and therefore, they cannot correctly describe the critical behaviour, in particular, they are not capable of producing correct critical exponents.

4. The KSSHE Theory

More recent developments of the theory of phase and critical behaviour of ionic fluids are related to the functional integration methods. Among them a field theoretical approach (the KSSHE theory) developed by Caillol deserves a special consideration.[37,38] The approach is based on the Hubbard-Stratonovich transformation of the Boltzmanns factor connected with the Coulomb potential. Thanks to this transformation, the grand-canonical partition function of the PM is equal to the grand-canonical partition function of a fluid of bare hard spheres in the presence of an imaginary random Gaussian field $\mathrm{i}\varphi$ with a covariance given by the Coulomb potential.[37]

In Ref.[38] the theory is formulated for the PM in which the number m of species as well as the charges q_α are arbitrary but all the ions have the same diameter σ (the so-called *special PM* or SPM). In the case of equal valences ($|q_\alpha| = |q_\beta| = q$) a two-component SPM reduces to the RPM.

The exact functional representation of the grand-canonical partition function of the SPM derived in Ref.[38] has the form:

$$\Xi[\nu_\alpha] = \mathcal{N}_{v_C}^{-1} \int \mathcal{D}\varphi \exp\left(-\frac{1}{2}\langle\varphi|v_C^{-1}|\varphi\rangle + \ln \Xi_{\mathrm{HS}}[\{\bar{\nu}_\alpha + \mathrm{i}q_\alpha\phi\}]\right), \quad (9)$$

where

$$\mathcal{N}_{v_C} \equiv \int \mathcal{D}\varphi\ \exp\left(-\frac{1}{2}\langle\varphi|v_C^{-1}|\varphi\rangle\right)$$

and the Dirac's notations are used

$$\langle\varphi|A|\varphi\rangle = \int \mathrm{d}\mathbf{r}_1 \int \mathrm{d}\mathbf{r}_2\, \varphi(\mathbf{r}_1)A(\mathbf{r}_1,\mathbf{r}_2)\varphi(\mathbf{r}_2)$$
$$\equiv \varphi(1)A(1,2)\varphi(2).$$

The following notations are also introduced in Eq. (9): $\Xi_{\mathrm{HS}}[\{\bar{\nu}_\alpha + \mathrm{i}q_\alpha\phi\}]$ is the grand partition function of the bare hard spheres in the presence of local chemical potential $\bar{\nu}_\alpha + \mathrm{i}q_\alpha\phi$; $\bar{\nu}_\alpha = \nu_\alpha + \nu_\alpha^s$, where $\nu_\alpha = \beta\mu_\alpha$ is the dimensionless chemical potential of the αth species and ν_α^s describes the self-energy of the αth species being a well-defined positive and finite quantity for nonpoint-like ions; $\varphi(\mathbf{r})$ denotes the real scalar field conjugate to the charge density; the field variable $\phi(\mathbf{r})$ is defined to be the space convolution

$$\phi(\mathbf{r}) = \beta^{1/2}\int \mathrm{d}\,\mathbf{r}\tau(|\mathbf{r}-\mathbf{r}'|)\varphi(\mathbf{r}'),$$

$\tau(r)$ specifies a spherically symmetric distribution of the charge q_α of each ion smeared out inside its volume, $\tau_\alpha(r) = 0$ for $r \geq \sigma_\alpha/2$; v_C^{-1} is the inverse of the positive operator $v_C(r) = 1/r$.

For the SPM, a loop expansion was developed and the Helmholtz free energy up to the second loop order was calculated. In this case the free energy can be presented as follows[38]

$$\beta f(\{\rho_\alpha\}) = \beta f^{(0)}(\{\rho_\alpha\}) + \beta f^{(1)}(\{\rho_\alpha\}) + \beta f^{(2)}(\{\rho_\alpha\}) + \ldots, \tag{10}$$

where $f^{(0)}$ is the zero-loop free energy:

$$\beta f^{(0)}(\{\rho_\alpha\}) = \beta f_{\mathrm{HS}}(\{\rho_\alpha\}) - \frac{\beta}{2}\rho_\alpha q_\alpha^2 v_C(0), \tag{11}$$

$f^{(1)}$ is the one-loop free energy:

$$\beta f^{(1)}(\{\rho_\alpha\}) = \beta f^{(0)}(\{\rho_\alpha\}) + \frac{1}{2}\sum_{\mathbf{k}} \ln\left(1 + \beta[\rho_\alpha q_\alpha^2]\widetilde{v}_C(k)\right), \tag{12}$$

and $f^{(2)}$ is free energy at the second-loop order:

$$\beta f^{(2)}(\{\rho_\alpha\}) = \beta f^{(1)}(\{\rho_\alpha\}) - \frac{\beta^2}{4}[\rho_\alpha q_\alpha^2]^2 \int \mathrm{d}\mathbf{r} \times h_{\mathrm{HS},\rho}(r)\Delta^2(r) + \frac{\beta^3}{12}[\rho_\alpha q_\alpha^3]^2 \int \mathrm{d}\mathbf{r}\Delta^3(r). \tag{13}$$

Here the following notations are introduced: $\widetilde{v}_C(k) = 4\pi\widetilde{\tau}^2(k)/k^2$, $h_{\mathrm{HS},\rho}(r)$ denotes the usual pair correlation function of a fluid of a one-component hard sphere system at the total number density $\rho = \sum_\alpha \rho_\alpha$ and $\Delta(r)$ is the propagator with Fourier transform being of the form

$$\widetilde{\Delta}(k) = \frac{\widetilde{v}_C(k)}{1 + \beta\rho_\alpha q_\alpha^2 \widetilde{v}_C(k)}. \tag{14}$$

In Eqs. (11)-(14) summation over repeated indices is meant.

The regularization of the potential $v_C(r)$ inside the hard core (the physically inaccessible region) is arbitrary to some extent. In Ref.[38] different schemes of a regularization of the Coulomb potential inside the hard core were considered. In particular,

a) the optimized random phase approximation (ORPA) that coincides with MSA:[62]

$$v_C(r) = \begin{cases} \dfrac{B}{\sigma}\left(2 - \dfrac{Br}{\sigma}\right), & r < \sigma \\ \dfrac{1}{r}, & r \geqslant \sigma \end{cases}, \tag{15}$$

where

$$B = \frac{x^2 + x - x(1+2x)^{1/2}}{x^2}, \qquad x = \kappa\sigma; \tag{16}$$

b) the optimized mean field (OMF) scheme proposed by Caillol[37]

$$v_C(r) = \begin{cases} \dfrac{2\sigma - r}{\sigma^2}, & r < \sigma \\ \dfrac{1}{r}, & r \geqslant \sigma \end{cases}; \tag{17}$$

c) the Weeks-Chandler-Andersen (WCA) regularization scheme[69]

$$v_C(r) = \begin{cases} \dfrac{1}{\sigma}, & r < \sigma \\ \dfrac{1}{r}, & r \geqslant \sigma \end{cases}. \tag{18}$$

In Ref.[38] the coexistence curves and the corresponding critical parameters of both the RPM and the binary SPM were determined numerically for the one-loop and two-loop expressions of the free energy for various regularization schemes discussed above. The results obtained for the critical temperature $T_c^* = k_B T_c \epsilon\sigma/q^2$ and the critical density $\rho_c^* = \rho_c\sigma^3$ of the RPM are summarized in Table 2. As was shown, the critical parameters of the RPM strongly depend upon the regularization scheme adopted for the Coulomb potential inside the hard sphere core. The inclusion of the two-loop correction yields a slight decrease of the critical temperature in all cases (except for the MSA case) as well as an increase of the critical density.

As regards the charge asymmetry, the gas-liquid critical parameters $T_c^* = k_B T_c \epsilon\sigma/zq^2$ and $\rho_c^* = \rho_c\sigma^3$ of the binary SPM obtained at the one-loop order are independent of the charge asymmetry factor z where $z = q_+/|q_-|$. At the two-loop order, the critical temperature is found to increase slightly with z in contradiction with the MC simulation data.[70–72] This trend holds for all the regularization schemes considered. However, the critical density, despite very low value, is found to increase in qualitative agreement with the MC results.

Table 2. Critical temperature T_c^* and critical density ρ_c^* of the RPM for various regularization schemes of the Coulomb potential:[38] $T_{c,1}^*$ and $\rho_{c,1}^*$ (one-loop), $T_{c,2}^*$ and $\rho_{c,2}^*$ (two-loop).

Regularization	$T_{c,1}^*$	$\rho_{c,1}^*$	$T_{c,2}^*$	$\rho_{c,2}^*$
HS+DHLL	0.3271	0.12267	-	-
OMF	0.1150	0.01834	0.1133	0.0316
WCA	0.08446	0.00880	0.08428	0.0137
MSA	0.07858	0.01449	0.07993	0.01722

Summarizing, in Ref.[38] the gas-liquid critical parameters of the SPM where studied by means of various mean-field theories. These theories were obtained in the framework of the KSSHE field-theoretical representation of charged hard sphere systems by means of a loop expansion of the free energy of a homogeneous system. Some of these mean-field theories are equivalent to the well known approximations of the theory of liquids such as the hard-sphere term plus Debye-Hückel limiting law (HS+DHLL), or MSA (ORPA) theory. The other theories were considered in this work for the first time. The results obtained for critical parameters are in all cases in poor agreement with the MC simulations.

5. The Method of Collective Variables. Links to Other Theories

In this section we present the statistical-field theory that uses a method of collective variables (CVs). Recent results obtained within the framework of this theory for two-component PMs will be presented in the next Sections.

The ideas of the collective variable (CV) method were formulated initially in the 1950s.[73,74] Later on the method was successfully developed for the study of a number of problems of statistical physics, in particular, for the second order phase transition.[75,76] Recently the statistical field theory that uses the method of CVs has been proposed for the study of the phase and critical behaviour of the ionic systems (see Ref.[41] and the references herein). The approach allowed one to derive the exact functional representation of the grand partition function and formulate, on this basis, the perturbation theory.

The ideas of the CV method are based on: (i) the concept of collective coordinates being appropriate for the physics of the system considered; (ii) the integral identity allowing one to derive an exact functional representation for the configurational Boltzmann factor and (iii) the concept of the reference system, one of the basic ideas of the modern liquid state theory.[61] The idea consists in splitting the interaction potential into two parts

$$u_{\alpha\beta}(r) = v^0_{\alpha\beta}(r) + v_{\alpha\beta}(r),$$

where $v^0_{\alpha\beta}(r)$ is a potential of a short-range repulsion which describes the mutual impenetrability of the particles, while $v_{\alpha\beta}(r)$, on the contrary, mainly describes the behaviour at moderate and large distances. The equilibrium properties of the system interacting via the potential $v^0_{\alpha\beta}(r)$ are assumed to be known. Therefore, this system can be regarded as the "reference" system. Within the framework of the CV method the interaction connected with potential $v_{\alpha\beta}(r)$ is described in the phase space of CVs. The fluid of hard spheres is the most frequently used as the reference system in the liquid state theory since its thermodynamic and structural properties are well known.[61] However, alternative approaches are also proposed.[77] Here we present the main grounds of the theory based on the method of CVs.

Let us start with a general case of a classical two-component system consisting of N particles among which there are N_1 particles of species 1 and N_2 particles of species 2. The pair interaction potential is assumed to be of the following form:

$$u_{\alpha\beta}(r) = v^0_{\alpha\beta}(r) + v^C_{\alpha\beta}(r) + v^S_{\alpha\beta}(r), \tag{19}$$

where $v^0_{\alpha\beta}(r)$ is the interaction potential between the two additive hard spheres of diameters σ_α and σ_β. We call the two-component hard sphere system a reference system. Thermodynamic and structural properties of the reference system are assumed to be known. $v^C_{\alpha\beta}(r)$ is the Coulomb potential: $v^C_{\alpha\beta}(r) = q_\alpha q_\beta v^C(r)$, where $v^C(r) = 1/(\epsilon r)$, ϵ is the dielectric constant and hereafter we put $\epsilon = 1$. The solution is made of both positive and negative ions so that the electroneutrality condition is satisfied, i.e. $\sum_{\alpha=1}^{2} q_\alpha c_\alpha = 0$, where c_α is the concentration of the species α, $c_\alpha = N_\alpha/N$. The ions of the species $\alpha = 1$ are characterized by their hard sphere diameter σ_1 and their electrostatic charge $+zq$ and those of species $\alpha = 2$, characterized by diameter σ_2, bear opposite charge $-q$ (q is elementary charge and z is the parameter of charge asymmetry). $v^S_{\alpha\beta}(r)$ is the potential of the short-range interaction: $v^S_{\alpha\beta}(r) = v^R_{\alpha\beta}(r) + v^A_{\alpha\beta}(r)$, where $v^R_{\alpha\beta}(r)$ is used to mimic the soft core asymmetric repulsive interaction, $v^R_{\alpha\beta}(r)$ is assumed to have a Fourier transform. $v^A_{\alpha\beta}(r)$ describes a van der Waals-like attraction.

The fluid is considered at equilibrium in the grand canonical ensemble. The grand partition function of the model (19) can be written as follows:

$$\Xi[\nu_\alpha] = \sum_{N_+\geq 0}\sum_{N_-\geq 0}\prod_{\alpha=+,-}\frac{\exp(\nu_\alpha N_\alpha)}{N_\alpha!}\int(\mathrm{d}\Gamma) \times\exp\left[-\frac{\beta}{2}\sum_{\alpha\beta}\sum_{ij}u_{\alpha\beta}(r_{ij})\right], \tag{20}$$

where the following notations are used: ν_α is the dimensionless chemical potential, $\nu_\alpha = \beta\mu_\alpha - 3\ln\Lambda_\alpha$ and μ_α is the chemical potential of the αth species; β is the reciprocal temperature; $\Lambda_\alpha^{-1} = (2\pi m_\alpha\beta^{-1}/h^2)^{1/2}$ is the inverse de Broglie thermal wavelength; $(\mathrm{d}\Gamma)$ is the element of configurational space of the particles.

Let us introduce the Fourier transforms of the microscopic number density of the species α

$$\hat{\rho}_{\mathbf{k},\alpha} = \sum_i \exp(-\mathrm{i}\mathbf{k}\mathbf{r}_i^\alpha).$$

In this case the part of the Boltzmann factor in Eq. (20) which does not include the reference system interaction can be presented as follows:

$$-\frac{\beta}{2}\sum_{\alpha,\beta}\sum_{i,j}(v_{\alpha\beta}^C(r_{ij}) + v_{\alpha\beta}^S(r_{ij})) = -\frac{1}{2}\sum_{\alpha,\beta}\sum_{\mathbf{k}}\widetilde{V}_{\alpha\beta}(k)(\hat{\rho}_{\mathbf{k},\alpha}\hat{\rho}_{-\mathbf{k},\beta} - N_\alpha\delta_{\alpha\beta}), \tag{21}$$

with

$$\widetilde{V}_{\alpha\beta}(k) = \widetilde{V}_{\alpha\beta}^C(k) + \widetilde{V}_{\alpha\beta}^S(k),$$

where $\widetilde{V}_{\alpha\beta}^{C/S}(k) = \frac{\beta}{V}\tilde{v}_{\alpha\beta}^{C/S}(k)$, $\tilde{v}_{\alpha\beta}^{C/S}(k)$ is the Fourier transform of the corresponding interaction potential.

In order to introduce CVs we use the identity

$$\exp\left[-\frac{1}{2}\sum_{\alpha,\beta}\sum_{\mathbf{k}}\widetilde{V}_{\alpha\beta}(k)\hat{\rho}_{\mathbf{k},\alpha}\hat{\rho}_{-\mathbf{k},\beta}\right] = \int(\mathrm{d}\rho)\prod_\alpha\delta_{\mathcal{F}}[\rho_{\mathbf{k},\alpha} - \hat{\rho}_{\mathbf{k},\alpha}] \times\exp\left[-\frac{1}{2}\sum_{\alpha,\beta}\sum_{\mathbf{k}}\widetilde{V}_{\alpha\beta}(k)\rho_{\mathbf{k},\alpha}\rho_{-\mathbf{k},\beta}\right], \tag{22}$$

where $\delta_{\mathcal{F}}[\rho_{\mathbf{k},\alpha} - \hat{\rho}_{\mathbf{k},\alpha}]$ denotes the functional "delta function"[78]

$$\delta_{\mathcal{F}}[\rho_{\mathbf{k},\alpha} - \hat{\rho}_{\mathbf{k},\alpha}] \equiv \int(\mathrm{d}\omega)\exp\left[\mathrm{i}\sum_{\mathbf{k}}\omega_{\mathbf{k}}(\rho_{\mathbf{k},\alpha} - \hat{\rho}_{\mathbf{k},\alpha})\right],$$

$\rho_{\mathbf{k},\alpha} = \rho^c_{\mathbf{k},\alpha} - \mathrm{i}\rho^s_{\mathbf{k},\alpha}$ is the CV which describes the value of the $\mathbf{k}$-th fluctuation mode of the number density of the αth species, the indices c and s denote real and imaginary parts of $\rho_{\mathbf{k},\alpha}$; $\omega_{\mathbf{k},\alpha}$ is conjugate to the CV $\rho_{\mathbf{k},\alpha}$ and each of $\rho^{c/s}_{\mathbf{k},\alpha}$ $(\omega^{c/s}_{\mathbf{k},\alpha})$ takes all the real values from $-\infty$ to $+\infty$; $(\mathrm{d}\rho)$ and $(\mathrm{d}\omega)$ denote volume elements of the CV phase space

$$(\mathrm{d}\rho) = \prod_{\alpha} \mathrm{d}\rho_{0,\alpha} {\prod_{\mathbf{k}\neq 0}}' \mathrm{d}\rho^c_{\mathbf{k},\alpha}\mathrm{d}\rho^s_{\mathbf{k},\alpha}, \quad (\mathrm{d}\omega) = \prod_{\alpha} \mathrm{d}\omega_{0,\alpha} {\prod_{\mathbf{k}\neq 0}}' \mathrm{d}\omega^c_{\mathbf{k},\alpha}\mathrm{d}\omega^s_{\mathbf{k},\alpha}$$

and the product over $\mathbf{k}$ is performed in the upper semi-space ($\rho_{-\mathbf{k},\alpha} = \rho^*_{\mathbf{k},\alpha}$, $\omega_{-\mathbf{k},\alpha} = \omega^*_{\mathbf{k},\alpha}$).

Using Eq. (22), we obtain the exact functional representation of the grand canonical partition function (20)

$$\Xi[\nu_\alpha] = \int (\mathrm{d}\rho)(\mathrm{d}\omega) \exp\left(-\mathcal{H}[\nu_\alpha, \rho_\alpha, \omega_\alpha]\right), \tag{23}$$

where the action $\mathcal{H}$ reads as

$$\begin{aligned} \mathcal{H}[\nu_\alpha, \rho_\alpha, \omega_\alpha] = & \frac{1}{2}\sum_{\alpha,\beta}\sum_{\mathbf{k}} \widetilde{V}_{\alpha\beta}(k)\rho_{\mathbf{k},\alpha}\rho_{-\mathbf{k},\beta} - \mathrm{i}\sum_{\alpha}\sum_{\mathbf{k}} \omega_{\mathbf{k},\alpha}\rho_{\mathbf{k},\alpha} \\ & - \ln \Xi_{\mathrm{HS}}[\bar{\nu}_\alpha - \mathrm{i}\omega_\alpha]. \end{aligned} \tag{24}$$

Here $\Xi_{\mathrm{HS}}[\bar{\nu}_\alpha; -\mathrm{i}\omega_\alpha]$ is the grand canonical partition function of a two-component system of the bare hard spheres with the renormalized chemical potential $\bar{\nu}_\alpha$ in the presence of the local field $\psi_\alpha(r_i)$

$$\begin{aligned} \Xi_{\mathrm{HS}}[\ldots] = & \sum_{N_1\geq 0}\sum_{N_2\geq 0}\prod_{\alpha} \frac{\exp(\bar{\nu}_\alpha N_\alpha)}{N_\alpha!} \int (\mathrm{d}\Gamma) \\ & \times \exp\left[-\frac{\beta}{2}\sum_{\alpha,\beta}\sum_{i,j} v^0_{\alpha\beta}(r_{ij}) + \sum_{\alpha}\sum_{i}^{N_\alpha} \psi_\alpha(r_i)\right], \end{aligned} \tag{25}$$

where

$$\bar{\nu}_\alpha = \nu_\alpha + \frac{1}{2}\sum_{\mathbf{k}} \widetilde{V}_{\alpha\alpha}(k), \tag{26}$$

$$\psi_\alpha(r_i) = -\mathrm{i}\omega_\alpha(r_i). \tag{27}$$

Mean-field approximation. The mean-field approximation of functional (23) is defined by

$$\Xi_{\mathrm{MF}}[\nu_\alpha] = \exp(-\mathcal{H}[\nu_\alpha, \bar{\rho}_\alpha, \bar{\omega}_\alpha]), \tag{28}$$

where $\bar{\rho}_\alpha$ and $\bar{\omega}_\alpha$ are the solutions of the saddle point equations:

$$\left.\frac{\delta\,\mathcal{H}\left[\nu_\alpha,\rho_\alpha,\omega_\alpha\right]}{\delta\rho_{\mathbf{k},\alpha}}\right|_{(\bar{\rho}_\alpha,\bar{\omega}_\alpha)} = \left.\frac{\delta\,\mathcal{H}\left[\nu_\alpha,\rho_\alpha,\omega_\alpha\right]}{\delta\omega_{\mathbf{k},\alpha}}\right|_{(\bar{\rho}_\alpha,\bar{\omega}_\alpha)} = 0\,.$$

The replacement of the CV action by its expression (24) in the above equations yields a set of two coupled implicit equations for $\bar{\rho}_\alpha$ and $\bar{\omega}_\alpha$

$$\bar{\rho}_\alpha = \langle N_\alpha[\bar{\nu}_\alpha - \mathrm{i}\bar{\omega}_\alpha]\rangle_{\mathrm{HS}}, \quad \mathrm{i}\bar{\omega}_\alpha = \sum_\beta \bar{\rho}_\beta \widetilde{V}_{\alpha\beta}(0). \tag{29}$$

Substituting (29) in Eq. (28) one obtains

$$\Xi_{\mathrm{MF}} = \exp\left[\frac{1}{2}\sum_{\alpha,\beta}\bar{\rho}_\alpha\bar{\rho}_\beta\widetilde{V}_{\alpha\beta}(0)\right]\Xi_{\mathrm{HS}}[\bar{\nu}_\alpha - \mathrm{i}\bar{\omega}_\alpha].$$

Taking fluctuations into account. In order to take fluctuations into account we present CVs $\rho_{\mathbf{k},\alpha}$ and $\omega_{\mathbf{k},\alpha}$ in the form:

$$\rho_{\mathbf{k},\alpha} = \bar{\rho}_\alpha\delta_{\mathbf{k}} + \delta\rho_{\mathbf{k},\alpha}, \quad \omega_{\mathbf{k},\alpha} = \bar{\omega}_\alpha\delta_{\mathbf{k}} + \delta\omega_{\mathbf{k},\alpha},$$

where $\delta_{\mathbf{k}}$ is the Kronecker symbol and the quantities with a bar are given by Eq. (29).

The function $\ln\Xi_{\mathrm{HS}}[\bar{\nu}_\alpha; -\mathrm{i}\omega_\alpha]$ in Eq. (24) can be presented in the form of the cumulant expansion

$$\ln\Xi_{\mathrm{HS}}[\ldots] = \sum_{n\geq 0}\frac{(-\mathrm{i})^n}{n!}\sum_{\alpha_1,\ldots,\alpha_n}\sum_{\mathbf{k}_1,\ldots,\mathbf{k}_n}\mathfrak{M}_{\alpha_1\ldots\alpha_n}[\bar{\nu}_\alpha - \mathrm{i}\bar{\omega}_\alpha](k_1,\ldots,k_n)$$
$$\times\delta\omega_{\mathbf{k}_1,\alpha_1}\ldots\delta\omega_{\mathbf{k}_n,\alpha_n}\delta_{\mathbf{k}_1+\ldots+\mathbf{k}_n}, \tag{30}$$

where $\mathfrak{M}_{\alpha_1\ldots\alpha_n}[\bar{\nu}_\alpha - \mathrm{i}\bar{\omega}_\alpha](k_1,\ldots,k_n)$ is the nth cumulant which is defined by

$$\mathfrak{M}_{\alpha_1\ldots\alpha_n}[\ldots](k_1,\ldots,k_n) = \left.\frac{\partial^n\ln\Xi_{HS}[\ldots]}{\partial\delta\omega_{\mathbf{k}_1,\alpha_1}\ldots\partial\delta\omega_{\mathbf{k}_n,\alpha_n}}\right|_{\delta\omega_{\mathbf{k}_i,\alpha_i}=0}. \tag{31}$$

Here the nth cumulant coincides with the nth connected correlation function of the reference system.[61] After the substitution of the cumulant expression in Eqs (23)-(24) one can develop the perturbation theory.

5.1. *A two-component charge-asymmetric PM*

Let us consider a two-component SPM, consisting of charged hard spheres of the same diameter

$$\sigma_+ = \sigma_- = \sigma \tag{32}$$

which differ by their respective charges ($z \neq 1$). The interaction potential of the SPM can be presented by Eq. (19) under conditions (32) and $v^S(r) = 0$.

Introducing CVs $\rho_{\mathbf{k},N}$ and $\rho_{\mathbf{k},Q}$

$$\rho_{\mathbf{k},N} = \rho_{\mathbf{k},+} + \rho_{\mathbf{k},-}, \qquad \rho_{\mathbf{k},Q} = q_+\rho_{\mathbf{k},+} + q_-\rho_{\mathbf{k},-}, \tag{33}$$

which describe fluctuations of the total number density and charge density, respectively we can rewrite the functional of the grand partition function (23)-(24) in the form:

$$\Xi[\nu_\alpha] = \int (\mathrm{d}\rho_N)(\mathrm{d}\rho_Q)(\mathrm{d}\omega_N)(\mathrm{d}\omega_Q)\exp\left(-\mathcal{H}[\nu_\alpha,\rho_N,\rho_Q,\omega_N,\omega_Q]\right), \tag{34}$$

where

$$\mathcal{H}[\nu_\alpha,\rho_N,\rho_Q,\omega_N,\omega_Q] = \frac{\beta}{2V}\sum_{\mathbf{k}}\widetilde{v}^C(k)\rho_{\mathbf{k},Q}\rho_{-\mathbf{k},Q} - \mathrm{i}\sum_{\mathbf{k}}(\omega_{\mathbf{k},N}\rho_{\mathbf{k},N} \\ +\omega_{\mathbf{k},Q}\rho_{\mathbf{k},Q}) - \ln\Xi_{\mathrm{HS}}[\bar{\nu}_\alpha; -\mathrm{i}\omega_N, -\mathrm{i}q_\alpha\omega_Q]. \tag{35}$$

In Eq. (35) $\widetilde{v}^C(k)$ is the Fourier transform of the Coulomb potential $v^C(r) = 1/(\epsilon r)$ and CVs $\omega_{\mathbf{k},Q}$ and $\omega_{\mathbf{k},N}$ are linear combinations of the initial CVs $\omega_{\mathbf{k},+}$ and $\omega_{\mathbf{k},-}$.[41]

As is seen from Eq. (35), the Hamiltonian of the charge asymmetric PM, as for RPM, does not include a direct pair interaction of number density fluctuations. Integration over $\rho_{\mathbf{k},N}$ and $\omega_{\mathbf{k},N}$ in (34) is trivial in this case and leads to the KSSHE action given by Eqs. (9). We recall here that the z dependence of the critical temperature calculated using the two-loop free energy contradicts the results of the MC simulations. Below we will present an alternate way of taking into account the fluctuation effects near the gas-liquid critical point.

The cumulant expansion (30) in terms of variables $\omega_{\mathbf{k},Q}$ and $\omega_{\mathbf{k},N}$ can be rewritten as

$$\ln\Xi_{\mathrm{HS}}[\ldots] = \sum_{n\geq 0}\frac{(-\mathrm{i})^n}{n!}\sum_{i_n\geq 0}\sum_{\mathbf{k}_1,\ldots,\mathbf{k}_n}\mathfrak{M}_n^{(i_n)}(k_1,\ldots,k_n)\delta\omega_{\mathbf{k}_1,Q}\ldots\delta\omega_{\mathbf{k}_{i_n},Q} \\ \times\delta\omega_{\mathbf{k}_{i_{n+1}},N}\ldots\delta\omega_{\mathbf{k}_n,N}\delta_{\mathbf{k}_1+\ldots+\mathbf{k}_n}. \tag{36}$$

The calculation of the correlation functions of a reference system is a separate task. Here we consider the hard sphere system of equal diameters with the local chemical potential $\nu_\alpha^*(i) = \bar{\nu}_\alpha - \mathrm{i}q_\alpha\gamma$ as a reference system. In this case, the following recurrence formulas for the cumulants are obtained[41]

$$\mathfrak{M}_n^{(0)}(k_1,\ldots) = \widetilde{G}_n(k_1,\ldots), \tag{37}$$

$$\mathfrak{M}_n^{(1)}(k_1,\ldots) = 0, \tag{38}$$

$$\mathfrak{M}_n^{(2)}(k_1,\ldots) = q_\alpha^2 c_\alpha \widetilde{G}_{n-1}(k_1,\ldots), \tag{39}$$

$$\mathfrak{M}_n^{(3)}(k_1,\ldots) = q_\alpha^3 c_\alpha \widetilde{G}_{n-2}(k_1,\ldots), \tag{40}$$

$$\mathfrak{M}_n^{(4)}(k_1,\ldots) = 3\left[q_\alpha^2 c_\alpha\right]^2 \widetilde{G}_{n-2}(k_1,\ldots) + \left(q_\alpha^4 c_\alpha - 3\left[q_\alpha^2 c_\alpha\right]^2\right)\widetilde{G}_{n-3}(k_1,\ldots) \tag{41}$$

In Eqs. (37)-(41) $\widetilde{G}_n(k_1,\ldots)$ is the Fourier transform of the n-particle connected correlation function of a one-component hard sphere system and summation over repeated indices is meant. As is seen, the charge asymmetry does not manifest itself to the third order cumulant.

In the charge symmetric case ($z = 1$) corresponding to the RPM the above expressions reduce to the simple form:

$$\mathfrak{M}_n^{(0)} = \widetilde{G}_n, \qquad \mathfrak{M}_n^{(1)} \equiv 0, \qquad \mathfrak{M}_n^{(2)} = q^2\widetilde{G}_{n-1}$$
$$\mathfrak{M}_n^{(3)} \equiv 0, \qquad \mathfrak{M}_n^{(4)} = q^4(3\widetilde{G}_{n-2} - 2\widetilde{G}_{n-3}). \tag{42}$$

In the Gaussian approximation, which corresponds to taking into account in Eq. (36) only the terms with $n \leq 2$, the integration in Eqs. (34)-(35) is trivial and leads to the following expression for the logarithm of the grand partition function

$$\ln \Xi_G(\nu_\alpha) = \ln \Xi_{\mathrm{HS}}(\bar{\nu}_\alpha) - \frac{1}{2}\sum_{\mathbf{k}} \ln\left[1 + \tilde{\Phi}^C(k)\mathfrak{M}_2^{(2)}(\bar{\nu}_\alpha)\right]. \tag{43}$$

After the Legendre transform of $\ln \Xi_G(\nu_\alpha)$ one obtains the well-known expression for the free energy in RPA (one-loop free energy given by Eq. (12))

$$\beta f_{\mathrm{RPA}} = \frac{\beta \mathcal{F}_{\mathrm{RPA}}}{V} = \beta f_{\mathrm{MF}} + \frac{1}{2V}\sum_{\mathbf{k}} \ln(1 + \kappa_D^2 \widetilde{v}^C(k)), \tag{44}$$

where the mean-field free energy βf_{MF} has the form

$$\beta f_{\mathrm{MF}} = \beta f_{\mathrm{HS}}(\rho_\alpha) - \frac{\beta}{2V}\sum_\alpha q_\alpha^2 \rho_\alpha \sum_{\mathbf{k}} \widetilde{v}^C(k),$$

κ_D^2 is the squared Debye number. It is worth noting that a use of momentum cutoff $|\mathbf{k}_\Lambda| = 2\pi/a$ in Eq. (44) leads to the same expression for the βf_{DH} as in Ref.[79] For point charge particles, Eq. (44) yields the free energy in the DHLL approximation $\beta f_{\mathrm{DHLL}} = \beta f_{id} - \kappa_D^3/12\pi$. Using the optimized regularization of the Coulomb potential inside the hard core (see Eq. (15)) we arrive at the free energy in ORPA (or MSA).

As is seen, βf_{RPA} does not explicitly depend on the charge asymmetry factor z. The same is true for MSA and the DH theories. As was indicated before, the z-dependent free energy can be found only in the higher-order approximations.

6. Gas-Liquid Separation in Asymmetric PMs: The Method of CVs

The effects of size and charge-asymmetry on the gas-liquid phase diagram of PMs have been studied using both the theoretical methods and computer simulations. The key findings from simulation studies of asymmetric models are as follows: the suitably normalized critical temperatures decrease with size and charge asymmetry while the critical densities increase with charge asymmetry but decrease with size asymmetry.[23,55,70–72,80–83] Comparison of simulated critical parameters and theoretical predictions for asymmetric models has revealed that several established theories, such as the MSA and the original DH theory are not capable of predicting the trends observed in simulations.[84,85] Moreover, both the original DH theory and the MSA predict no dependence on charge asymmetry in the equisize case. The exception are the theories that include the association effects explicitly.[86–88] The trends found from the GDH theory for the both critical parameters of an equisize (z:1) charge-asymmetric PM as a function of charge asymmetry qualitatively agree with simulation data for $z = 1 \div 3$.[87] As for the RPM, this theory is sensitive to the approximations used for the hard core. As regards the size asymmetry, the extensions of the DH theory for monovalent size-asymmetric PMs that describe the charge-unbalanced "border zones" surrounding each ion lead to the trends of the both critical parameters that qualitatively agree with simulation predictions.[85] However, this is true only for modest size asymmetries. More recently, the study of the effects of size or/and charge asymmetry on the gas-liquid phase separation has been accomplished within the framework of various field-theoretical approaches.[38,40,79,89] However, these approaches allow one to obtain the correct predictions only for some of the effects of asymmetry. On the other hand, the progress in the understanding of the issue was achieved within the framework of the CV based theory. In Ref.[42] the method is proposed for the study of the gas-liquid phase diagram of the binary SPM that allows one to calculate the trends of the both critical parameters that qualitatively agree with Monte Carlo simulation results. The method is based on determining the chemical potential conjugate to the order parameter and allows one to take into account the effects of higher-order correlations. The procedure described in Ref.[42] has been recently applied to the study of the gas-liquid separation in the size- and charge-asymmetric PM.[43]

6.1. *Charge-asymmetric PMs*

Let us consider the SPM. For this model Eqs. (32)-(41) were obtained in the previous section. We start with the grand partition function found in the Gaussian approximation (see Eq. (43)) and pass from the initial chemical potentials ν_+ and ν_- to their linear combinations

$$\nu_N = \frac{1}{1+z}\left(\nu_+ + z\nu_-\right), \quad \nu_Q = \frac{1}{q(1+z)}(\nu_+ - \nu_-). \tag{45}$$

New chemical potentials ν_N and ν_Q are conjugate to the total number density and charge density, respectively. Since near the gas-liquid critical point the fluctuations of the number density play a crucial role, ν_N is of special interest in this study.

Taking into account that $\ln \Xi_{\text{HS}}$ and $\mathfrak{M}_n^{(i_n)}$ are functions of the full chemical potentials we present ν_N and ν_Q as

$$\nu_N = \nu_N^0 + \varepsilon\Delta\nu_N, \qquad \nu_Q = \nu_Q^0 + \varepsilon\Delta\nu_Q,$$

with ν_N^0 and ν_Q^0 being the mean-field values of ν_N and ν_Q, respectively and $\Delta\nu_N$ and $\Delta\nu_Q$ being the solutions of the equations

$$\frac{\partial \ln \Xi_G(\nu_N, \nu_Q)}{\partial \Delta\nu_N} = \varepsilon\langle N\rangle_{HS}, \tag{46}$$

$$\frac{\partial \ln \Xi_G(\nu_N, \nu_Q)}{\partial \Delta\nu_Q} = 0. \tag{47}$$

Expanding r.h.s. of Eq. (43) in powers of $\Delta\nu_N$ and $\Delta\nu_Q$ one obtains

$$\ln \Xi_G(\nu_N, \nu_Q) = \sum_{n\geq 0}\sum_{i_n\geq 0}^{n} C_n^{i_n} \frac{\mathcal{M}_n^{(i_n)}(\nu_N^0, \nu_Q^0)}{n!} \Delta\nu_Q^{i_n} \Delta\nu_N^{n-i_n}, \tag{48}$$

where

$$\mathcal{M}_n^{(i_n)}(\nu_N^0, \nu_Q^0) = \left.\frac{\partial^n \ln \Xi_G(\nu_N, \nu_Q)}{\partial \Delta\nu_Q^{i_n} \partial \Delta\nu_N^{n-i_n}}\right|_{\Delta\nu_N=0, \Delta\nu_Q=0}.$$

The explicit expressions for the coefficients $\mathcal{M}_n^{(i_n)}$ are obtained elsewhere.[42]

Equations (46)-(47) are solved self-consistently for the relevant chemical potential $\Delta\nu_N$ by means of successive approximations taking into account Eq. (48) and keeping terms of a certain order in formal parameter ε. The procedure is as follows. First, $\Delta\nu_Q$ is calculated from Eq. (47) in the approximation which corresponds to a certain order of ε e.g., order s. Then, this $\Delta\nu_Q$ is substituted into Eq. (46) in order to find $\Delta\nu_N$ in the approximation corresponding to order $s+1$ of ε. In Eq. (46) only the linear terms with respect to $\Delta\nu_N$ are taken into account while terms with *all powers of* $\Delta\nu_Q$ are kept within a given approximation with respect to ε.

As is readily seen, the first nontrivial solution for $\Delta\nu_N$ is obtained in the approximation of the first order of ε. This is the result of substitution in Eq. (46) of the solution $\Delta\nu_Q = 0$. As a result, we have

$$\Delta\nu_N = -\frac{\mathfrak{M}_3^{(2)}}{2\mathfrak{M}_2^{(0)}} \sum_{\mathbf{k}} \tilde{g}(k), \tag{49}$$

where

$$\tilde{g}(k) = -\frac{\widetilde{V}^C(k)}{1 + \widetilde{V}^C(k)\mathfrak{M}_2^{(2)}}.$$

Taking into account Eqs. (37) and (39), the above expression for $\Delta\nu_N$ can be rewritten as

$$\Delta\nu_N = \frac{1}{2N} \sum_{\mathbf{k}} \frac{\kappa_D^2 \widetilde{v}^C(k)}{1 + \kappa_D^2 \widetilde{v}^C(k)} \tag{50}$$

which corresponds to RPA. As is seen, $\Delta\nu_N$ given by Eq. (50) does not depend explicitly on charge asymmetry factor z. In order to obtain the z-dependent expression for the chemical potential related to the number density fluctuations one should consider the next approximation in ε for $\Delta\nu_N$. To this end, we substitute in Eq. (46) the solution $\Delta\nu_Q$ as follows

$$\Delta\nu_Q = -\frac{\mathfrak{M}_3^{(3)}}{2\mathfrak{M}_2^{(2)}} \sum_{\mathbf{k}} \tilde{g}(k), \tag{51}$$

which is found from (47) in the first approximation of ε. It should be noted that $\Delta\nu_Q$ in (51) depends on the charge asymmetry factor through the cumulant $\mathfrak{M}_3^{(3)}$ (see Eq. (40)). As a result one gets from (46)

$$\Delta\nu_N = -\frac{1}{\mathfrak{M}_2^{(0)}} \left[\frac{1}{2} \sum_{\mathbf{k}} \tilde{g}(k)\mathfrak{M}_3^{(2)} + \frac{1}{2} \sum_{\mathbf{k}} \tilde{g}(k)\mathfrak{M}_4^{(3)} \Delta\nu_Q \right.$$
$$\left. + \frac{1}{2}\mathfrak{M}_3^{(2)} \Delta\nu_Q^2 + \frac{1}{3!}\mathfrak{M}_4^{(3)} \Delta\nu_Q^3 \right]. \tag{52}$$

The correlation effects of the order higher than the second order enter Eq. (52) through the cumulants $\mathfrak{M}_n^{(i_n)}$ for $n \geq 3$ and $i_n \neq 0$. The appearance of these cumulants reflects the fact that the terms proportional to $\omega\gamma^2$, γ^3 and $\omega\gamma^3$ are taken into account in the cumulant expansion (30) ($n \leq 4$). Summarizing, we obtained the expression for the chemical potential conjugate to the order parameter of the gas-liquid phase transition in which the effects of indirect correlations between the number density fluctuations are taken into consideration via a charge subsystem. Based on Eq. (52) supplemented by the Maxwell construction the coexistence curves and the corresponding critical parameters for different values of z can be calculated.

Using Eqs. (37)-(39) one can get the following explicit expression for (52)

$$\Delta\nu_N = \frac{i_1}{\pi}\left[1 + \frac{i_1(1-z)^2}{2z\pi}\left(1 - \frac{i_1(1-z)^2}{3z\pi}\right)\right], \tag{53}$$

where for the WCA regularization of the Coulomb potential inside the hard core (see Eq. (18)) is used, so that

$$i_1 = \frac{1}{T^*}\int_0^{\infty} \frac{x^2 \sin x dx}{x^3 + \kappa^{*2} \sin x} \tag{54}$$

with $\kappa^* = \kappa_D \sigma$ being the reduced Debye number. Let us recall that $T^* = k_B T\sigma/zq^2$ and $\rho^* = \rho\sigma^3$ stand for normalized temperature and density. Finally, the full chemical potential ν_N conjugate to the total number density has the form

$$\nu_N - 3\ln\Lambda/\sigma = \ln\rho^* + \frac{\eta(8 - 9\eta + 3\eta^3)}{(1-\eta)^3} + \frac{z}{1+z}\ln z - \ln(1+z)$$
$$-\frac{1}{2T^*} + \Delta\nu_N, \tag{55}$$

where $\Delta\nu_N$ is given in (53)-(54) and $\eta = \pi\rho^*/6$. In Eq. (55) the Carnahan-Starling approximation for the hard sphere system is used. Fig. 1 shows the coexistence curves for $z = 2 \div 4$ calculated based on the isotherms of chemical potential (53)-(55) supplemented with the Maxwell construction. A comparison of the coexistence curve forms for $z = 2$ and $z = 3$ (results for the coexistence curve for $z = 4$ are not available by now) with the GDH theory indicates their similarity.[87] The results obtained for the critical parameters are shown in Fig. 2(a) and Fig. 2(b). In Fig. 2(a) the critical temperature depending on z is displayed for z ranging from 2 to 4. As is seen, a qualitative agreement with the simulation data shown by the open circles is obtained. In Fig. 2(b) the dependence of the critical density on z is shown. Similar to the computer simulation findings our results indicate a sharp increase of the critical density with the increase of z. The critical parameters of the RPM obtained within the framework of this method but in the higher order approximation[90] are shown by the solid squares ($T_c^* = 0.0503$, $\rho_c^* = 0.042$).

We may conclude that the method outlined in this section allows one to obtain, without additional assumptions (such as the presence of the dipoles or the higher order clusters), the dependence of both the critical temperature and critical density on the charge asymmetry parameter z which qualitatively agrees with the results

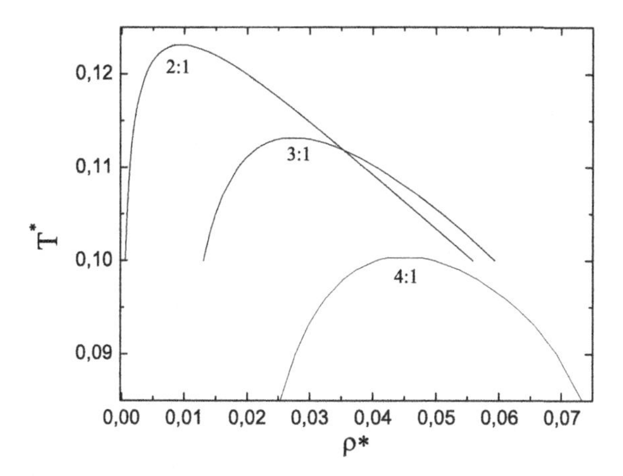

Fig. 1. Coexistence curves of the $(z{:}1)$ charge-asymmetric ionic model.

of the computer simulations. Moreover, the application of this method to the case of the RPM ($z = 1$) leads to the best theoretical estimations for the critical parameters.

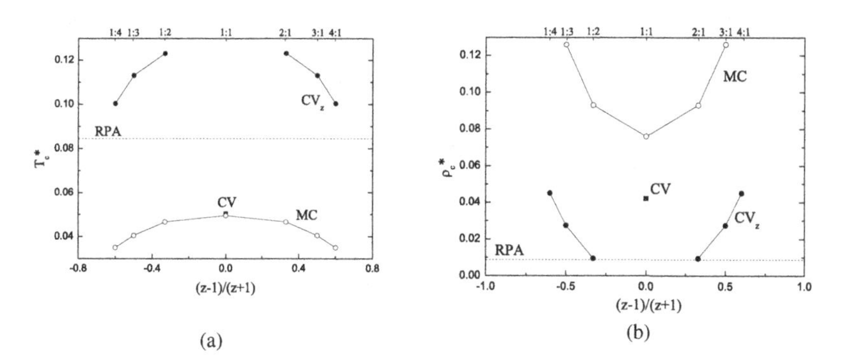

Fig. 2. Dependence of the critical parameters of the SPM on charge asymmetry. Solid symbols correspond to the results of the CVs based theory: CV ($z = 1$) and CV$_z$ ($z \neq 1$); open circles are the results of simulations: $z = 1$;[55] $z = 2, 3$;[23] $z = 4$.[70] (a) Dependence of the critical temperature. (b) Dependence of the critical density.

6.2. *Size and charge asymmetric PMs*

Now we turn back to a two-component PM with size and charge asymmetry. The interaction in the model is described by Eq. (3). Besides the parameter of charge asymmetry $z = q_+/|q_-|$, we introduce the parameter $\lambda = \sigma_+/\sigma_-$, characterizing the size asymmetry (σ_+ and σ_- are the hard sphere diameters of cations and anions, respectively). Let us consider the WCA regularization scheme for $v^C_{\alpha\beta}(r)$:[69]

$$v^C_{\alpha\beta}(r) = \begin{cases} q_\alpha q_\beta/\sigma_{\alpha\beta}, & r < \sigma_{\alpha\beta} \\ q_\alpha q_\beta/r, & r \geq \sigma_{\alpha\beta}. \end{cases}$$

In this case the Fourier transform of the Coulomb potential $\beta v^C_{\alpha\beta}(r)$ has the form:

$$\beta\tilde{v}^C_{++}(k) = \frac{4\pi z\sigma^3_\pm}{T^*(1+\delta)}\frac{\sin(x(1+\delta))}{x^3}, \tag{56}$$

$$\beta\tilde{v}^C_{--}(k) = \frac{4\pi\sigma^3_\pm}{T^* z(1-\delta)}\frac{\sin(x(1-\delta))}{x^3}, \tag{57}$$

$$\beta\tilde{v}^C_{+-}(k) = -\frac{4\pi\sigma^3_\pm}{T^*}\frac{\sin(x)}{x^3}, \tag{58}$$

where

$$T^* = \frac{k_B T\sigma_\pm}{q^2 z}, \tag{59}$$

is the dimensionless temperature,

$$\delta = \frac{\lambda - 1}{\lambda + 1} \tag{60}$$

and $x = k\sigma_\pm$, $\sigma_\pm = (\sigma_+ + \sigma_-)/2$.

Gaussian approximation. First, we consider the Gaussian approximation of $\Xi[\nu_\alpha]$ setting $\mathfrak{M}_{\alpha_1\ldots\alpha_n} \equiv 0$ for $n \geq 3$. After integration in (23)-(24) over $\delta\omega_{\mathbf{k},\alpha}$ we obtain

$$\Xi_{\mathrm{G}}[\nu_\alpha] = \Xi_{\mathrm{MF}}[\bar{\nu}_\alpha - \mathrm{i}\bar{\omega}_\alpha]\,\Xi' \int (\mathrm{d}\delta\rho) \exp\Big\{-\frac{1}{2}\sum_{\alpha,\beta}\sum_{\mathbf{k}} \tilde{\mathcal{C}}_{\alpha\beta}(k)$$
$$\times \delta\rho_{\mathbf{k},\alpha}\delta\rho_{-\mathbf{k},\beta}\Big\}, \tag{61}$$

where $\tilde{\mathcal{C}}_{\alpha\beta}(k)$ is the Fourier transform of the two-particle direct correlation function in the Gaussian approximation

$$\tilde{\mathcal{C}}_{\alpha\beta}(k) = \frac{\beta}{V}\tilde{v}^C_{\alpha\beta}(k) + \frac{1}{\sqrt{N_\alpha N_\beta}}\tilde{\mathcal{C}}^0_{\alpha\beta}(k). \tag{62}$$

$\tilde{\mathcal{C}}^0_{\alpha\beta}(k)$ is the Fourier transform of the direct correlation function of a two-component hard-sphere system. It is connected with $\mathfrak{M}_{\alpha\beta}(k)$ by the relation

$$\tilde{\mathcal{C}}^0_2(k)\mathfrak{M}_2(k) = \underline{1}, \tag{63}$$

where $\tilde{\mathcal{C}}_2^0(k)$ denotes the matrix of elements $\tilde{\mathcal{C}}^0_{\alpha\beta}(k)/\sqrt{N_\alpha N_\beta}$ and $\mathfrak{M}_2$ is the matrix of elements $\mathfrak{M}_{\alpha\beta}(k)$, $\underline{1}$ is the unit matrix. It should be noted that $\tilde{\mathcal{C}}_{\alpha\beta}(k)$ is connected to the ordinary direct correlation function $\tilde{c}_{\alpha\beta}(k)$ by[61]

$$\tilde{\mathcal{C}}_{\alpha\beta}(k) = \frac{\delta_{\alpha\beta}}{\rho_\alpha} - \tilde{c}_{\alpha\beta}(k),$$

where $\rho_\alpha = \langle N_\alpha \rangle / V$.

In order to determine the CV connected with the order parameter we diagonalize the square form in (61). This leads us to the equation

$$\Xi_G[\nu_\alpha] = \Xi_{HS}[\bar{\nu}_\alpha]\,\Xi' \int (\mathrm{d}\xi) \exp\Big\{ -\frac{1}{2} \sum_{\alpha=1,2} \sum_{\mathbf{k}} \tilde{\varepsilon}_\alpha(k) \xi_{\mathbf{k},\alpha} \xi_{-\mathbf{k},\alpha} \Big\}. \quad (64)$$

Eigenvectors $\xi_{\mathbf{k},1}$ and $\xi_{\mathbf{k},2}$ in the long-wavelength limit have the form:[43]

$$\begin{aligned} \xi_{0,1} &= \frac{1}{\sqrt{1+z^2}} \left(\frac{1+z^2}{1+z} \rho_{0,N} + \frac{1-z}{1+z} \rho_{0,Q} \right), \\ \xi_{0,2} &= \frac{1}{\sqrt{1+z^2}} \rho_{0,Q}, \end{aligned} \quad (65)$$

where CVs $\rho_{0,N} = \delta\rho_{0,+} + \delta\rho_{0,-}$ and $\rho_{0,Q} = z\delta\rho_{0,+} - \delta\rho_{0,-}$ describe long-wavelength fluctuations of the total number density and charge density, respectively (see Eqs. (33)). As is seen, CV $\xi_{0,2}$ describes fluctuations of the charge density. In the general case $z \neq 1$, $\xi_{0,1}$ is a linear combination of CVs $\rho_{0,N}$ and $\rho_{0,Q}$ with the z-dependent coefficients. At $z = 1$, CV $\xi_{0,1}$ describes solely fluctuations of the total number density. Thus, we suggest that CV $\xi_{0,1}$ is connected with the order parameter of the gas-liquid critical point.

Eigenvalues $\tilde{\varepsilon}_1(k)$ and $\tilde{\varepsilon}_2(k)$ in the long-wavelength limit are found to be[43]

$$\begin{aligned} \tilde{\varepsilon}_1(k=0) = \frac{1+z}{1+z^2} \Big(& -\frac{4\pi\rho^* z\delta^2}{3T^*(1+z)} + \tilde{c}_{++}^{HS}(0) + 2\sqrt{z}\tilde{c}_{+-}^{HS}(0) \\ & + z\tilde{c}_{--}^{HS}(0) \Big), \end{aligned} \quad (66)$$

$$\tilde{\varepsilon}_2(k=0) = \infty, \quad (67)$$

where T^* and δ are given by Eqs. (59)-(60) and

$$\rho^* = \rho\sigma_\pm^3 \quad (68)$$

is a reduced total number density. Equation (67) leads to $\tilde{G}_{QQ}(k=0) = 0$, where $\tilde{G}_{QQ}(k=0)$ is the Fourier transform of the charge-charge connected correlation function. It reflects the fact that the zeroth-moment Stillinger-Lovett (SL) rule is satisfied.[29,61]

Equation $\tilde{\varepsilon}_1(\delta \neq 0; k = 0) = 0$ leads to the gas-liquid spinodal curve in the Gaussian approximation

$$T_s^* = \frac{4\pi\rho^*\delta^2}{3(1+z)} \left(\tilde{c}_{++}^{\mathrm{HS}}(0)/z + 2\tilde{c}_{+-}^{\mathrm{HS}}(0)/\sqrt{z} + \tilde{c}_{--}^{\mathrm{HS}}(0)\right)^{-1}. \tag{69}$$

Equations (65)-(69) are analogous to those obtained by Ciach et al.[40] but for another type of the regularization of the Coulomb potential inside the hard core. The trends of the critical parameters calculated from the maximum point of spinodal (69) are also consistent with the corresponding trends found previously:[40] at the fixed z, the critical temperature T_c^* is a convex down function of δ while the critical density $\rho^*(\delta)$ is a convex up in δ; both T_c^* and ρ_c^* increase at a given $\delta > 0$ and decrease at a given $\delta < 0$ when z increases. Therefore, only some of the trends are correctly predicted within the framework of this approximation.

Finally, after integration in r.h.s. of Eq. (64) over $\xi_{\mathbf{k},\alpha}$ one arrives at the logarithm of the grand partition function in the Gaussian approximation

$$\ln \Xi_G[\nu_\alpha] = \ln \Xi_{\mathrm{HS}}[\bar{\nu}_\alpha] - \frac{1}{2}\sum_{\mathbf{k}} \ln \det\left[\underline{1} + V_C\mathfrak{M}_2\right] \tag{70}$$

where V_C and $\mathfrak{M}_2$ are matrices of elements $\beta\widetilde{v}_{\alpha\beta}^C(k)$ and $\mathfrak{M}_{\alpha\beta}(\bar{\nu}_\alpha; k)$, respectively.

It should be noted here that the issue of particular relevance concerns the behaviour of the charge-charge correlation function $G_{QQ}(k)$ at the gas-liquid critical point where the density-density correlation length diverges. The SL sum rules state that when $\mathbf{k} \to 0$, the Fourier transform of the charge-charge correlation function behaves as[29,61]

$$\widetilde{G}_{QQ}(k) \simeq 0 + \xi_D^2 k^2 - \sum_{p\geq 2}(-1)^p \xi_{Z,p}^{2p} k^{2p}$$

where the first vanishing term results from electroneutrality (zeroth-moment SL condition) while the second term yields the second-moment SL rule with $\xi_{Z,1} = \xi_D$ (ξ_D is the Debye length). As was shown above, the zeroth-moment SL rule for the charge- and size-asymmetric model at the critical point is satisfied at the level of RPA. The study of the validity of the second-moment SL rule for symmetric and asymmetric models is of considerable importance in understanding the nature of criticality in ionic fluids. We will address this issue in the last section.

Beyond the Gaussian approximation. In order to properly describe the effects of size and charge asymmetry on the critical parameters one should take into account the terms of the higher-order than the second order in the functional Hamiltonian (23)-(24). To this end, we follow the method applied to the SPM in the previous section.

We introduce

$$\nu_1 = \frac{\nu_+ + z\nu_-}{\sqrt{1+z^2}}, \qquad \nu_1 = \frac{z\nu_+ - \nu_-}{\sqrt{1+z^2}},$$

where ν_1 and ν_2 are the chemical potentials conjugate to CVs $\xi_{0,1}$ and $\xi_{0,2}$, respectively. We start with the expression (70) replacing the cumulants $\mathfrak{M}_{\alpha\beta}(\bar{\nu}_\alpha;k)$ by their values in the long-wavelength limit putting $\mathfrak{M}_{\alpha\beta}(\bar{\nu}_\alpha;k) = \mathfrak{M}_{\alpha\beta}(\bar{\nu}_\alpha;k=0) = \mathfrak{M}_{\alpha\beta}(\bar{\nu}_\alpha)$. Then, we self-consistently solve the equations

$$\frac{\partial \ln \Xi_G(\nu_1,\nu_2)}{\partial \Delta\nu_1} = \langle N_+\rangle_{HS} + z\langle N_-\rangle_{HS},$$
$$\frac{\partial \ln \Xi_G(\nu_1,\nu_2)}{\partial \Delta\nu_2} = 0$$

for the relevant chemical potential $\Delta\nu_1$ by means of successive approximations. The procedure of searching for a solution is described in the previous section.

The expression for the relevant chemical potential ν_1 found in the first nontrivial approximation corresponding to $\nu_2 = \nu_2^0$ with ν_2^0 being the mean-field value of ν_2 has the form[43]

$$\nu_1 = \nu_1^0 + \frac{\beta\sqrt{1+z^2}}{2\left[\mathfrak{M}_{++} + 2z\mathfrak{M}_{+-} + z^2\mathfrak{M}_{--}\right]}\frac{1}{V}\sum_{\mathbf{k}}\frac{1}{\det\left[\underline{1} + V_C\mathfrak{M}_2\right]}$$
$$\times\left(\beta\widetilde{v}_{++}^C(k)\mathcal{S}_1 + \beta\widetilde{v}_{--}^C(k)\mathcal{S}_2 + 2\beta\widetilde{v}_{+-}^C(k)\mathcal{S}_3\right), \qquad (71)$$

where

$$\nu_1^0 = \frac{1}{\sqrt{1+z^2}}\left(\nu_+^{\mathrm{HS}} + z\nu_-^{\mathrm{HS}} - \frac{\beta}{2V}\sum_{\mathbf{k}}\left[\widetilde{v}_{++}^{\mathrm{C}}(k) + z\widetilde{v}_{--}^{\mathrm{C}}(k)\right]\right) \qquad (72)$$

and ν_α^{HS} with $\alpha = +,-$ is the hard-sphere chemical potential of the αth species.

Apart from the second order cumulants, Eq. (71) includes the third order cumulants or equivalently the third order connected correlation functions of the hard sphere system:

$$\mathcal{S}_1 = \mathfrak{M}_{+++} + z\mathfrak{M}_{++-}, \quad \mathcal{S}_2 = \mathfrak{M}_{+--} + z\mathfrak{M}_{---}, \quad \mathcal{S}_3 = \mathfrak{M}_{++-} + z\mathfrak{M}_{+--}. \qquad (73)$$

In the case of a two-component hard sphere system, the analytical expressions for $\mathfrak{M}_{\alpha_1\alpha_2}$ and ν_α^{HS} can be obtained in the PY approximation using the Lebowitz' solution.[91,92] In order to derive the expressions for $\mathfrak{M}_{\alpha_1\alpha_2\alpha_3}$ one can use the recurrent relation.[43]

Below formulas (71)-(73) are used for the study of the gas-liquid phase equilibria of size- and charge-asymmetric PMs.

6.3. *Monovalent PMs with size asymmetry*

First we consider a monovalent PM with size asymmetry corresponding to $z = 1$ and $\lambda \neq 1$. Because of symmetry with respect to the exchange of $+$ and $-$ ions, only the range $\lambda < 1$ (or $\lambda > 1$) needs be considered in this case.

The data for coexistence curves were calculated using Eqs. (71)-(73) supplemented by the Maxwell construction.[43,93] Estimates of the critical temperature and the critical density were given by their values for which the maxima and minima of the van der Waals loops coalesce. Fig. 3 demonstrates the calculated coexis-

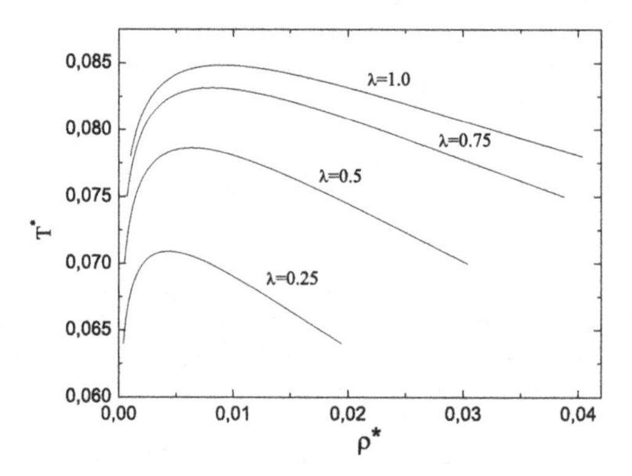

Fig. 3. Coexistence curves of the monovalent PM at different values of λ.

tence curves of the monovalent size-asymmetric PM at $\lambda = 1.0, 0.75, 0.5, 0.25$. As is seen, the region of coexistence narrows when the size asymmetry increases. This effect becomes increasingly pronounced as λ becomes smaller which is in qualitative agreement with the simulation findings.[80,81]

Fig. 4 and Fig. 5 demonstrate the effects of size asymmetry on the critical parameters of the monovalent PM. In Fig. 4 the critical temperature T_c^* depending on λ is shown by solid circles for λ ranging from 0.1 to 1. As is seen, a qualitative agreement with the simulation data shown by the open circles is obtained. The results obtained in the MSA are shown by the open squares. In Fig 5 the dependence of the critical density ρ_c^* on λ is shown. Similar to the computer simulation findings, the results obtained within the CV based theory indicate a decrease of the critical density with the increase of λ but the figures obtained in this approximation turn out to be rather small.

It should be noted that the RPM limit turns out to be a special case. When $z = \lambda = 1$, expression (71) reduces to the form that corresponds to RPA (see Eq. (50)).

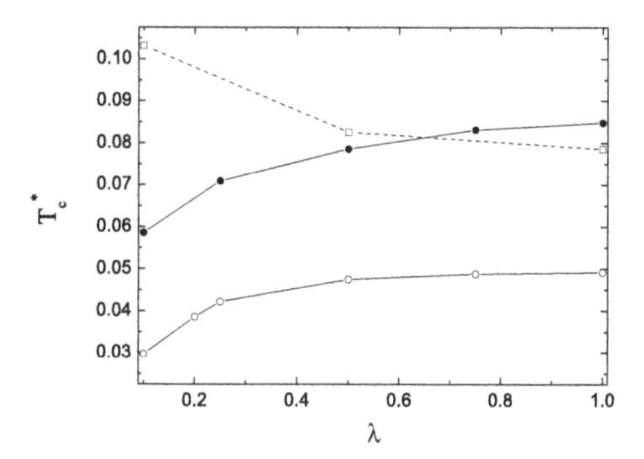

Fig. 4. Critical temperature T_c^* of the monovalent PM as a function of size asymmetry. Open circles correspond to the results of simulations;[71] open squares are MSA results via the energy route[84] and solid circles correspond to the results of the CV based theory.

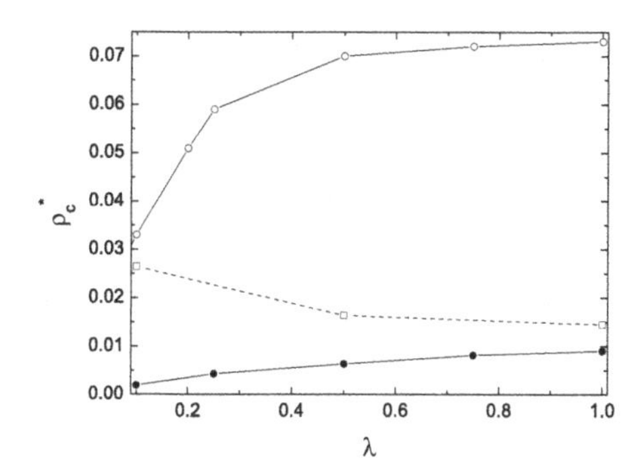

Fig. 5. Critical density ρ_c^* of the monovalent PM as a function of size asymmetry. The meaning of the symbols is the same as in Fig. 4

6.4. *Size- and charge-asymmetric PMs*

Now Eqs (71)-(73) are used for the study of the size- and charge-asymmetric PMs with $z = 2$ and $\lambda \neq 1$ (2:1 PM). As before, in order to calculate the coexistence curves and the corresponding critical parameters the Maxwell construction is applied. The phase coexistence envelopes for 2:1 systems with size asymmetries are presented in Fig. 6(a) for $\lambda < 1$ and in Fig. 6(b) for $\lambda > 1$. Similar to the monovalent case, the coexistence regions become narrower with the increase of size asymmetry. The corresponding critical parameters decrease when size asymmetry

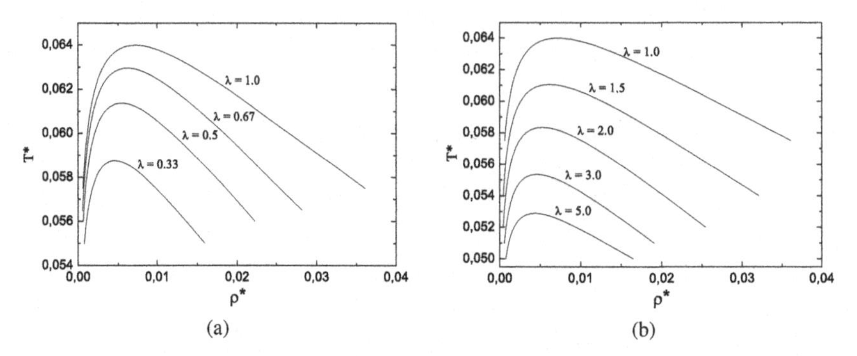

Fig. 6. Coexistence curves of the 2:1 PM at different values of λ. (a) $\lambda \leq 1$. (b) $\lambda \geq 1$.

increases which qualitatively agrees with the simulation results.[82] The dependence of critical parameters on the size asymmetry is shown graphically in Fig. 7 and Fig. 8, respectively, along with the results of simulations.[82] In general, the trends of T_c^* and ρ_c^* with δ are consistent with the simulation findings: 2:1 systems exhibit a maximum in both the critical temperature and the critical density when plotted as a function of size asymmetry. Similar to simulations, the theoretical results (especially the critical temperature) reveal a pronounced sensitivity to δ. However, both the critical temperature and the critical density found from Eqs. (71)-(73) demonstrate maxima at $\delta = 0$ ($\lambda = 1$) while the corresponding maxima obtained by simulations are shifted towards nonzero values of δ ($\delta > 0$). Interestingly, $\rho_c^*(\delta)$ demonstrates the general shape similar to that obtained for the dumbbell system.[82] As before, the theoretical estimates of the critical density obtained in the approximation (71)-(73) are more than an order of magnitude lower than those found in the simulations.[82]

In conclusion, the trends of the critical temperature and the critical density with size asymmetry obtained within the framework of the same approximation qualitatively agree with the Monte Carlo simulation findings: both T_c^* and ρ_c^* decrease with increasing size asymmetry at the fixed z. However, the problem of quantitative description is still open. The results obtained allow one to expect that the agreement between the theory and computer simulations can be significantly improved by taking into account the higher order correlations within the framework of the scheme presented above.

7. Critical Behaviour of the "Coulombic Systems": Theoretical Studies

The hypothesis of critical-point universality is based on the long-range nature of fluctuations of the order parameter associated with the phase transition. In the

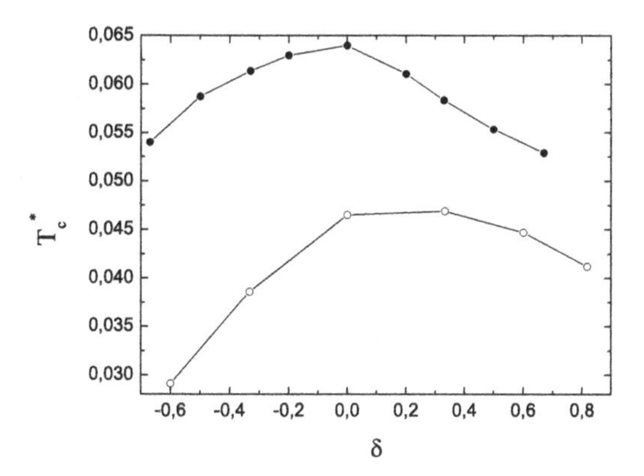

Fig. 7. Critical temperature of the (2:1) PM as a function of size asymmetry. Solid circles correspond to the results of the CV based theory; open circles are the results of simulations.[82]

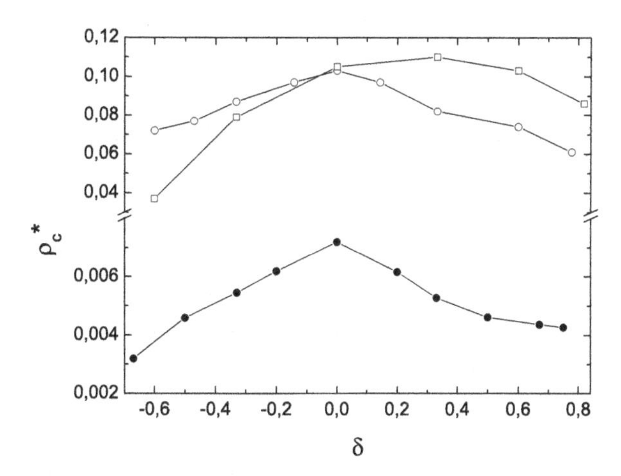

Fig. 8. Critical density of the (2:1) PM as a function of size asymmetry. Solid circles are the results of the CV based theory. Open symbols correspond to the results of simulations:[82] circles are spherical electrolytes; squares are dumbbell electrolytes.

vicinity of the critical point the correlation length of critical fluctuations becomes so large that microscopic details of short-range interactions become unimportant. Universality claims that all systems having the same spatial dimensionality, symmetry and range of interactions between the particles belong to the same universality class. If one presents the interaction potential in the power-law form $u(r) \sim -r^{-d-\sigma}$ (d is the dimensionality), the Ising-like critical exponents are expected for $\sigma > 2-\eta$ where η is the correlation function exponent ($\eta \approx 0.03$ for $d = 3$).[94–97] For simple fluids and their mixtures we have $d = \sigma = 3$, the critical behaviour of such systems belongs to the Ising model class of universality. For more slowly decaying potentials one must be prepared to find non-Ising critical exponents.

Since the bare Coulomb potential is long ranged, non-Ising critical behaviour may be expected in ionic fluids with dominant electrostatic interactions (*Coulombic systems*). However, inside the electrolyte solution the Coulomb interaction is screened by the surrounding particles and the effective interaction potential is short-ranged. Thus, the critical behaviour of *Coulombic systems* is not obvious *a priori*.

The nature of criticality in ionic fluids has been an outstanding experimental and theoretical issue for more than two decades. In this section we present a brief review of the research of critical behaviour of Coulomb-dominated systems that has been undertaken by the methods of analytical theory. The main attention is focused on the theoretical approaches that can provide an explanation for the experimental observations.

7.1. *The effective Hamiltonian of the RPM in the vicinity of the gas-liquid critical point*

According to the basic idea of universality, different systems which are described by the effective Hamiltonians with the same symmetry demonstrate the same critical behaviour. Thus, the knowledge of the effective Hamiltonian of the system is important in determining its critical behaviour. Several attempts to construct such a Hamiltonian for the RPM have been made so far. The major part of the studies was based on the mean-field theories and dealt with calculations of the Ginzburg temperature.[98–102]

On the other hand, the issue of criticality in the RPM was addressed using the functional integration methods,[25,103] In Ref. ,[103] the effective Ginzburg-Landau-Wilson (GLW) Hamiltonian was obtained in terms of a scalar field conjugate to the charge density density. Unfortunately, a non-perturbative renormalization group analysis of such Hamiltonian did not allow one to make an unambiguous statement on the nature of the critical behaviour of the RPM. As was shown within the framework of the CV based theory, the total number density is the strong fluctuating quantity in the vicinity of the gas-liquid critical point of the RPM (see Eqs.

(65)). In Ref.,[25] a microscopic based effective Hamiltonian of the RPM given in terms of the CVs $\rho_{\mathbf{k},N}$ related to the order parameter associated with the gas-liquid separation was derived. We outline here the main points of this work.

Let us start with Eqs. (34)-(36). In the case of RPM, the recurrence relations for the cumulants are given by Eqs (42). Since $\mathfrak{M}_2^{(0)}(0)$ and $\mathfrak{M}_2^{(2)}$ are positive and smooth functions in the region under consideration, we can integrate over CVs $\omega_{\mathbf{k},N}$ and $\omega_{\mathbf{k},Q}$ using the Gaussian density measures as basic ones. As a result, we get the action (Hamiltonian) $\mathcal{H}$ in terms of CVs $\rho_{\mathbf{k},N}$ and $\rho_{\mathbf{k},Q}$

$$
\begin{aligned}
\mathcal{H}[\nu_\alpha,\rho_N,\rho_Q] &= a_1^{(0)}\rho_{0,N} - \frac{1}{2!}\sum_{\mathbf{k}}\left(a_2^{(0)}\rho_{\mathbf{k},N}\rho_{-\mathbf{k},N\cdot} + a_2^{(2)}\rho_{\mathbf{k},Q}\rho_{-\mathbf{k},Q}\right) \\
&\quad +\frac{1}{3!}\sum_{\mathbf{k}_1,\mathbf{k}_2,\mathbf{k}_3}\left(a_3^{(0)}\rho_{\mathbf{k_1},N}\rho_{\mathbf{k_2},N}\rho_{\mathbf{k_3},N}\right. \\
&\quad \left.+3a_3^{(2)}\rho_{\mathbf{k_1},N}\rho_{\mathbf{k_2},Q}\rho_{\mathbf{k_3},Q}\right)\delta_{\mathbf{k}_1+\mathbf{k}_2+\mathbf{k}_3} \\
&\quad +\frac{1}{4!}\sum_{\mathbf{k}_1,\ldots,\mathbf{k}_4}\left(a_4^{(0)}\rho_{\mathbf{k_1},N}\rho_{\mathbf{k_2},N}\rho_{\mathbf{k_3},N}\rho_{\mathbf{k_4},N}\right. \\
&\quad +6a_4^{(2)}\rho_{\mathbf{k_1},N}\rho_{\mathbf{k_2},N}\rho_{\mathbf{k_3},Q}\rho_{\mathbf{k_4},Q} \\
&\quad \left.+a_4^{(4)}\rho_{\mathbf{k_1},Q}\rho_{\mathbf{k_2},Q}\rho_{\mathbf{k_3},Q}\rho_{\mathbf{k_4},Q}\right)\delta_{\mathbf{k}_1+\ldots+\mathbf{k}_4},
\end{aligned}
\tag{74}
$$

where super index i_n indicates the number of variables $\rho_{\mathbf{k},Q}$ at $a_n^{(i_n)}$. The following notations are introduced here:

$$
a_1^{(0)} = \Delta\nu_N, \qquad a_2^{(0)} = \frac{1}{\mathfrak{M}_2^{(0)}}, \qquad a_2^{(2)} = \tilde{V}^C(k) + \frac{1}{\mathfrak{M}_2^{(2)}}, \tag{75}
$$

$$
a_3^{(0)} = \frac{\mathfrak{M}_3^{(0)}}{(\mathfrak{M}_2^{(0)})^3}, \qquad a_3^{(2)} = \frac{\mathfrak{M}_3^{(2)}}{\mathfrak{M}_2^{(0)}(\mathfrak{M}_2^{(2)})^2}, \tag{76}
$$

$$
\begin{aligned}
a_4^{(0)} &= \frac{1}{(\mathfrak{M}_2^{(0)})^4}\left(\mathfrak{M}_4^{(0)} - 3\frac{(\mathfrak{M}_3^{(0)})^2}{\mathfrak{M}_2^{(0)}}\right), \\
a_4^{(2)} &= \frac{1}{(\mathfrak{M}_2^{(0)})^2(\mathfrak{M}_2^{(2)})^2}\left(\mathfrak{M}_4^{(2)} - \frac{\mathfrak{M}_3^{(0)}\mathfrak{M}_3^{(2)}}{\mathfrak{M}_2^{(0)}} - 2\frac{(\mathfrak{M}_3^{(2)})^2}{\mathfrak{M}_2^{(2)}}\right), \\
a_4^{(4)} &= \frac{1}{(\mathfrak{M}_2^{(2)})^4}\left(\mathfrak{M}_4^{(4)} - 3\frac{(\mathfrak{M}_3^{(2)})^2}{\mathfrak{M}_2^{(0)}}\right),
\end{aligned}
\tag{77}
$$

and $\Delta\nu_N = \bar{\nu}_N - \nu_N^0$ with ν_N^0 being the mean-field value of $\bar{\nu}_N$.

Since CVs $\rho_{\mathbf{k},Q}$ are not related with the order parameter we can integrate over $\rho_{\mathbf{k},Q}$ with the Gaussian density measure (under the condition $a_2^{(2)} > 0$). This leads to

$$\begin{aligned}\mathcal{H}^{eff}[\nu_\alpha, \rho_N] = a_1\rho_{0,N} - \frac{1}{2!}\sum_{\mathbf{k}} a_2\rho_{\mathbf{k},N}\rho_{-\mathbf{k},N}\cdot \\ +\frac{1}{3!}\sum_{\mathbf{k}_1,\mathbf{k}_2,\mathbf{k}_3} a_3\rho_{\mathbf{k_1},N}\rho_{\mathbf{k_2},N}\rho_{\mathbf{k_3},N}\delta_{\mathbf{k}_1+\mathbf{k}_2+\mathbf{k}_3} \\ +\frac{1}{4!}\sum_{\mathbf{k}_1,\ldots,\mathbf{k}_4} a_4\rho_{\mathbf{k_1},N}\rho_{\mathbf{k_2},N}\rho_{\mathbf{k_3},N}\rho_{\mathbf{k_4},N}\delta_{\mathbf{k}_1+\ldots+\mathbf{k}_4},\end{aligned} \tag{78}$$

where

$$a_n = a_n^{(0)} + \Delta a_n \tag{79}$$

and Δa_n are correction terms obtained as a result of integration over CVs $\rho_{\mathbf{k},Q}$. Restricting consideration here to the terms with a one sum over the wave vector $\mathbf{q}$ we get for Δa_n

$$\begin{aligned}\Delta a_1 &= \frac{1}{2\langle N\rangle}\sum_{\mathbf{q}}\widetilde{G}_{QQ}(q), \\ \Delta a_2 &= \frac{1}{\langle N\rangle^2}\sum_{\mathbf{q}}\widetilde{G}_{QQ}(q) + \frac{1}{2\langle N\rangle^2}\sum_{\mathbf{q}}\widetilde{G}_{QQ}(q)\widetilde{G}_{QQ}(\mid \mathbf{q}+\mathbf{k}\mid) + \ldots, \\ \Delta a_3 &= -\frac{3}{\langle N\rangle^3}\sum_{\mathbf{q}}\widetilde{G}_{QQ}(q)\widetilde{G}_{QQ}(\mid \mathbf{q}+\mathbf{k_1}\mid) + \ldots, \\ \Delta a_4 &= \frac{6}{\langle N\rangle^Q}\sum_{\mathbf{q}}\widetilde{G}_{QQ}(q)\widetilde{G}_{QQ}(\mid \mathbf{k_1}+\mathbf{k_2}-\mathbf{q}\mid) + \ldots,\end{aligned} \tag{80}$$

where $\widetilde{G}_{QQ}(q)$ is the Fourier transform of a charge-charge connected correlation function of the RPM determined in the Gaussian approximation. It is worth noting that in the vicinity of the gas-liquid critical point $\widetilde{G}_{QQ}(q)$ remains a smooth function and behaves as

$$\widetilde{G}_{QQ}(q) = \frac{q^2}{q^2+\kappa_D^2}.$$

Now let us look more closely at the coefficient a_2. Expanding this coefficient at small k one can readily see that the linear term vanishes. Thus, a_2 is given by

$$a_2 = a_{2,0} + \frac{1}{2}k^2 a_{2,2} + \ldots, \tag{81}$$

where

$$a_{2,0} = a_2^{(0)} + \frac{1}{\langle N \rangle^2}\sum_{\mathbf{q}} \widetilde{G}_{QQ}(q) + \frac{1}{2\langle N \rangle^2}\sum_{\mathbf{q}} \widetilde{G}_{QQ}(q)\widetilde{G}_{QQ}(\mid \mathbf{q}+\mathbf{k} \mid),$$

$$a_{2,2} = \frac{1}{2\langle N \rangle^2}\sum_{\mathbf{q}} \widetilde{G}_{QQ}(q)\frac{\partial^2 \widetilde{G}_{QQ}(q)}{\partial q^2}. \tag{82}$$

It is worth noting that the dependence on $\mathbf{k}$ results from the second order of the perturbation theory in $\widetilde{G}_{QQ}(q)$. Inserting expansion (81) into Eq. (78) one arrives at the effective Hamiltonian of the RPM in the neighborhood of the gas-liquid critical point. The coefficients of the Hamiltonian are explicitly given in (75)-(77) and (79)-(82).

Thus, we have derived a microscopic based effective Hamiltonian of the RPM given in terms of the collective variables $\rho_{\mathbf{k},N}$ connected with the order parameter. Taking into account the charge-charge correlations through integration over the charge subsystem (the collective variables $\rho_{\mathbf{k},Q}$) we get the contribution to the coefficients at the second order which describes the effective attraction of short-range character. The resulting Hamiltonian has a structure of the GLW Hamiltonian of an Ising model in an external magnetic field. We conclude that the form of the effective Hamiltonian of the RPM confirms the fact that its critical behaviour near the gas-liquid critical point belongs to the universal class of a three-dimensional Ising model.

7.2. *Crossover behaviour in fluids with Coulomb interactions*

A challenging problem for theory, simulations and experiments in the field of critical phenomena of ionic fluids is the crossover from the mean-field-like behaviour to the Ising model criticality when approaching the critical temperature. In particular, a sharp crossover was reported for the systems $\mathrm{Na} - \mathrm{NH_3}$[7] (see also[10,12,14]).

The analysis of experimental data for various ionic solutions confirmed that such systems generally exhibit crossover or, at least a tendency to crossover, from the Ising behaviour asymptotically close to the critical point to the mean-field behaviour upon increasing distance from the critical point.[4] Moreover, the systematic experimental investigations of the ionic systems such as tetra-n-butylammonium picrate, $\mathrm{Bu_4NPic}$, (for tetra-n-butylammonium picrate we will follow the notations from[19,20]) in long chain n-alkanols with dielectric constant ranging from 3.6 for 1-tetradecanol to 16.8 for 2-propanol suggest an increasing tendency for crossover to the mean-field behaviour when the Coulomb contribution becomes essential.[17,19,20] They also indicate that the "Coulomb limit" reduced temperature of the RPM $T_c \simeq 0.05$ is valid for the almost non-polar long chain alkanols.[17,20] It has been stressed[17] that for solutions of Bu_4NPic in 1-alkanols, the upper critical solution points are found to increase linearly with the chain length of the alcohols (it corresponds to the decrease of dielectric constant of the solvent). The experimental data for the critical points and the dielectric permittivities for solutions of Bu_4NPic in 1-alkanols are given in Table 3.[17]

Table 3. The experimental parameters of the critical points (critical temperature T_c, critical mass fraction w_c) and the corresponding dielectric constants ϵ for solutions of Bu_4NPic in 1-alkanols.[17]

Solvent	$\epsilon(T_c)$	T_c/K	w_c
1-oktanol	9.5	298.55	0.336
1-nonanol	7.9	308.64	0.325
1-decanol	6.4	318.29	0.3152
1-undecanol	5.4	326.98	0.303
1-dodecanol	4.7	335.91	0.2951
1-tridecanol	4.3	342.35	0.284
1-tetradecanol	3.6	351.09	0.2721

The assumption that the *Coulombic systems* demonstrate a much smaller asymptotic region than the *solvophobic systems* may be checked using the Ginzburg criterion.[104] According to this criterion, one can consider two limiting types of near-critical behaviour, namely, Ising-like behaviour for $\tau \ll \tau_G$ and classical mean-field behaviour for $\tau \gg \tau_G$, and a crossover region for τ and τ_G of the same order of magnitude. It is worth noting that for nonionic fluids the Ginzburg criterion gives a reasonable estimate of the temperature region where a crossover is observed.

In order to quantify the Ginzburg criterion one should estimate the dimensionless ratio[5]

$$E_{LG} = \frac{|\int_V \mathrm{d^d}\mathbf{r}\ G(r)|}{\int_V \mathrm{d^d}\mathbf{r}\ [m(r)]^2},$$

where the integrals are taken over a correlation volume $V = [\xi(T)]^{\mathrm{d}}$, $\xi(T)$ is the correlation length, and $G(r)$ is the density-density connected correlation function. The Landau theory is valid if E_{LG} is small. From the condition $E_{LG} \ll 1$ one can get the estimate of the Ginzburg temperature in terms of coefficients of the Landau expansion, namely:[99,100]

$$t_G \simeq \frac{9u_4^2}{8\pi^2 c_2}\left(\frac{\sigma}{b_2}\right)^6, \tag{83}$$

where c_2 and u_4 are the coefficients of the second and fourth orders of the Landau expansion, b_2^2 is the coefficient of the square gradient term, and σ is the hard sphere diameter. The numerical factor in (83) is to some extent arbitrary and its different values are known in the literature (see Refs.[5,101]). The Ginzburg temperature marks a lower bound of the temperature region where (along the critical isochore) the Landau theory is valid

Theoretically, the crossover behaviour in ionic systems was studied for the RPM by Fisher and Levin,[98] Leote de Carvalho and Evans,[99] Fisher and Lee[100] and Schröer and Weiss.[101,102] The results obtained for the Ginzburg temperature were similar to those found for simple fluids in comparable fashion which is in variance to what is expected from the experiments.[19,20] Nearly at the same time in[105] the crossover behaviour of the lattice version of a fluid exhibiting the Ising behaviour was studied as additional symmetrical electrostatic interactions were turned on. Based on the microscopic ground, the effective Hamiltonian in terms of the fluctuating field conjugate to the number density was derived in this work. Then, the crossover between the mean-field and Ising-like behaviour was estimated using the Ginzburg criterion. The resulting crossover temperature calculated as function of the ionicity

$$\mathcal{I} = \frac{1}{\beta^*} = \frac{|q_+q_-|}{k_BT\epsilon\sigma}, \tag{84}$$

which defines the strength of the Coulomb interaction relative to the short-range interaction, indicates its weak dependence but with the trends correlating with those observed experimentally.

Following the ideas of Ref.,[105] the effect of the interplay of short- and long-range interactions on the crossover behaviour has been studied more recently but for *a continuous version* of the charge-asymmetric model.[106] Below we briefly sketch the main results obtained in this work.

Let us start with a classical two-component charge-asymmetric system. The pair interaction potential of the model is given in Eq. (19) under conditions $v^S_{++}(r) = v^S_{--}(r) = v^S_{+-}(r)$. Then, the model without the Coulomb interactions is characterized by the dielectric constant ϵ and exhibits a gas-liquid critical point belonging to the Ising class of criticality. In this work the effective Ginzburg-Landau-Wilson Hamiltonian is derived in terms of the fluctuating field conjugate to the order parameter associated with the gas-liquid phase separation of the full model. As in Ref.,[105] the coefficients of this Hamiltonian are of the form of an expansion in powers of ionicity $\mathcal{I}$ (see (84)). Based on these coefficients the Ginzburg temperature t_G depending on the ionicity (in the approximation $\mathcal{I}^2$) is calculated for a specific model which consists of charged hard spheres of the same diameter interacting through the additional square-well potentials. To study the effect of the interplay between short- and long-range interactions, the range of the square-well potential λ is also changed. In Fig. 9 the ratio of reduced Ginzburg temperatures, $t_G(\mathcal{I})/t_G(0)$ where $t_G(0) = t_G(\mathcal{I} = 0)$ is the Ginzburg temperature of the model system without electrostatic interactions, is shown at different values of λ. It is worth noting that, as in,[105] $t_G(\mathcal{I})$ demonstrates the non-monotonous behaviour which becomes more pronounced as λ increases.

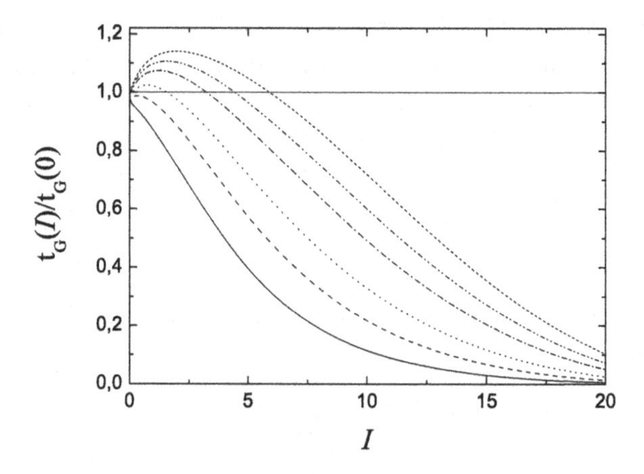

Fig. 9. The ratio $t_G(\mathcal{I})/t_G(0)$ as a function of the ionicity at different values of λ ($\rho = \rho_c$). Lines from the bottom to the top: $\lambda = 1.3$, $\lambda = 1.5$, $\lambda = 1.7$, $\lambda = 2.0$, $\lambda = 2.2$, and $\lambda = 2.5$.

In Table 4 we compare the results for the ionicity dependence of the Ginzburg temperature obtained for the continuous model (at $\lambda = 1.5$)[106] with the results of[105] for the lattice model as well as with experimental data for the crossover temperatures $t_\times$ (data for $\mathcal{I}$ and $t_\times$ are taken from Ref.[105]). The systems (a)-(c) correspond to the same ionic species Bu_4NPic within solvents of different dielectric constant. As is seen, the results obtained for the systems (a)-(c) in Ref. [106] are in good agreement (qualitative and quantitative) with the experimental findings. This confirms the experimental observations that an interplay between the solvophobic and Coulomb interactions alters the temperature region of the crossover regime, namely: the increase of the ionicity that can be related to the decrease of dielectric constant leads to the decrease of the crossover region. In Table 4 we also present (see case "d") the experimental and theoretical results obtained for the sodium solution in ammonia. For this system, the sharp crossover behaviour was observed experimentally. This system, of course, might be described by the potential $v^S(r)$ with the different attraction range λ. For instance, for $\lambda = 2$ we obtain $t_G(\mathcal{I} = 6.97) = 0.8 \times 10^{-2}$ (see Fig. 7 in[106]) which correlates with the experimental value $t_\times = 0.6 \times 10^{-2}$. Case "e" corresponds to the ionic system tetra-n-pentylammonium bromide in water and case (f) corresponds to the simple fluid; the both systems display the Ising-type critical behaviour.

Table 4. Experimentally assessed crossover temperature, $t_\times$, taken from:[105] (a) tetra-n-butylammonium picrate (Bu_4NPic) in 1-tridecanol; (b) Bu_4NPic in 1-dodecanol; (c) Bu_4NPic in 75% 1-dodecanol plus 25% 1,4-butanediol; (d) Na in NH_3; (e) tetra-n-pentylammonium bromide in water and the reduced Ginzburg temperature, t_G, found theoretically in Ref.[105] ($t_G^\dagger$) and in Ref.[106] ($t_G^\diamond$). System (f) corresponds to the uncharged fluid.

System	Ionicity,$\mathcal{I}$	$t_\times$	$t_G^\dagger$	$t_G^\diamond$
(a)	17.9	$\sim 10^{-3}$	~ 0.712	2.7×10^{-3}
(b)	16.8	$\sim 0.9 \times 10^{-2}$	~ 0.717	0.38×10^{-2}
(c)	8.9	$\sim 3 \times 10^{-2}$	~ 0.777	2.5×10^{-2}
(d)	6.97	$\sim 0.6 \times 10^{-2}$	~ 0.807	3.7×10^{-2}
(e)	~ 1.4	$\mathcal{O}(\mathcal{I})$	1	~ 0.09
(f)	0	$\mathcal{O}(\mathcal{I})$	1	~ 0.09

As a result, we have obtained a similar tendency for the reduced Ginzburg temperature as in[105] when the region of short-range attraction increases. However, our results demonstrate a much faster decrease of the Ginzburg temperature when the ionicity increases. In Ref.[106] a good qualitative and sufficient quantitative agreement with the experimental findings for Bu_4NPic in n-alkanols is found. This confirms the experimental observations that an interplay between the solvophobic and Coulomb interactions alters the temperature region of the crossover regime i.e., the increase of the ionicity that can be related to the decrease of dielectric constant leads to the decrease of the crossover region.

7.3. *Charge-charge correlations in the models of ionic fluids*

In the context of criticality in ionic systems, the relevant issue concerns the behaviour of the two-body charge-charge correlation function $G_{QQ}(r)$ near the critical point where the density fluctuations strongly diverge. The renormalization group tells us that near the criticality, the Fourier transform of the density-density correlation function diverges as[5]

$$\widetilde{G}_{NN}(k=0;T,\rho_c) \sim \tau^{-\gamma} \qquad \widetilde{G}_{NN}(k;T_c,\rho_c) \sim \frac{1}{k^{2-\eta}}$$

But what is the behaviour of the charge-charge correlation function at the critical point? This issue, in turn, is tightly connected with the validity of the SL rules[29] for $G_{QQ}(r)$.

The behaviour of the charge-charge correlation function was addressed in a series of works.[35,107–109] In particular, Aqua and Fisher[108] using a class of exactly soluble spherical models studied both symmetric and nonsymmetric version of 1:1 ionic lattice models. They showed that in the former case the two-point charge correlations remain of short range and obey the SL rule near and at the critical point. Otherwise they found the divergence of the charge correlation length to be precisely the same as the divergence of the density correlation length. They also found that the SL rule fails at criticality. Nearly at the same time, a continuous charge- and size-asymmetric model of ionic fluids with the interaction potential (19) was considered in Ref.[109] Here the following expression for $G_{QQ}(r)$ along the critical isochore was obtained in RPA

$$\begin{aligned}\widetilde{G}_{QQ}(k) &= \frac{k^2}{4\pi\beta q^2}\left(1-\frac{k^2}{\overline{\kappa}_D^2}\right)+\frac{(\widetilde{V}_{NQ}(0))^2k^4}{16\pi^2q^4}\\ &\quad\times\frac{1}{a(k^2+\xi^{-2})}+\mathcal{O}(k^6),\end{aligned} \tag{85}$$

where ξ is the density-density correlation length, $\overline{\kappa}_D^2$ is the effective squared Debye number, $\overline{\kappa}_D^2 = \kappa_D^2/(1+\beta\widetilde{V}_{QQ}(0)\rho z)$, a is the quantity dependent on the model parameters, the density and the temperature, $\widetilde{V}_{NQ}(k)$ and $\widetilde{V}_{QQ}(k)$ have the form:

$$\widetilde{V}_{NQ}(k) = \frac{1}{(1+z)^2}\left(\widetilde{V}_{++}(k)+(z-1)\widetilde{V}_{+-}(k)-z\widetilde{V}_{--}(k)\right),$$

$$\widetilde{V}_{QQ}(k) = \frac{1}{(1+z)^2}\left(\widetilde{V}_{++}(k)-2\widetilde{V}_{+-}(k)+\widetilde{V}_{--}(k)\right).$$

Equation (85) is valid far from and at the critical point at small k.

Away from the critical point ($\beta\neq\beta_c$) one has $\widetilde{G}_{QQ}(k)\sim k^2/4\pi\beta q^2$ which means that the both SL rules are satisfied. This is in agreement with the results obtained in[110] in RPA. In the general case, at the critical point ($\beta=\beta_c, \rho=\rho_c$), $\xi^{-1}=0$ and we have for $\widetilde{G}_{QQ}(k)$

$$\widetilde{G}_{QQ}(k) = k^2\left(\frac{1}{4\pi\beta q^2}+\frac{(\widetilde{V}_{NQ}(0))^2}{16\pi^2q^4}\frac{1}{a}\right)+\mathcal{O}(k^4). \tag{86}$$

As is seen from Eq. (86), the both SL rules are satisfied for a symmetrical model ($\widetilde{V}_{NQ}=0$). However, for an asymmetric model, the second SL rule is violated at the critical point which is confirmed by the results obtained within the mean-field theories.

It is obvious that in order to treat the fluctuations in a proper way one needs to go beyond the mean-field approximation taking into account the fluctuations of the higher-order than the second order. The phenomenological generalization of these results on the basis of the scaling relations yields the following expression for the charge-charge correlation function at the critical point

$$\widetilde{G}_{QQ}(k) = \frac{k^2}{4\pi\beta q^2} + \mathcal{O}(k^{2+\eta}), \tag{87}$$

where η is a small critical exponent. This preliminary analysis allows one to make an assumption that the second-moment SL rule is verified for the asymmetric ionic model even at the critical point if the fluctuation effects are taken properly into account.

8. Conclusions

In this chapter we have reviewed the theoretical results directed towards understanding the nature of the gas-liquid phase equilibrium in ionic fluids with dominant Coulomb interactions. Since 1993, several review papers devoted to the subject have been published (see Refs. [9,20,32,35,111]). Because of this, we have endeavored to avoid a comprehensive analysis of early studies focusing mainly on the recent research works carried out by theoretical methods. Recent reviews of simulation studies are also available in Refs.[48,49]

It should be emphasized that a major part of the work on the phase and critical behaviour of ionic systems has been performed on the so-called *primitive models* (PMs) that contain no solvent. These models belong to the class of strong coupled systems. For example, for the RPM near the gas-liquid critical point, the thermal energy is about $1/20$th of the energy of two ions at contact. As a result, such systems tend to form clusters (dimers, trimers, etc.) at small distances which together with the long-range character of the Coulomb potential leads to significant difficulties in computer simulations. As concerns the theory, several theoretical methods of the mean-field type have been proposed in which the ion association is explicitly taken into account. The major of them are the generalized DH (GDH) theory and the associating MSA. These theories are based on the addition of the chemical association model of Bjerrum or Ebeling and Grigo. The GDH theory (solvated ion-cluster theory with hard-core term) yields the following estimations for the critical parameters of the RPM: $T_c^* = 0.0557$, $\rho_c^* = 0.0261$.[87] On the other hand, the theoretical approaches which use the functional integration methods have been proposed more recently. They allow one to obtain, on the microscopic grounds, the functional effective Hamiltonians that take into account the contributions from higher-order correlations between the positive and negative ions. In particular, the CVs based theory is the statistical field theory of this kind. As was shown in Sec. 5, the well-known approximations for the free energy, in particular DH limiting law and MSA, can be reproduced within the framework of this theory. Links between this approach and the other theories that use the functional methods are also established (see Refs. [42,112,113]). The recent results obtained using this theory are presented in Sec 6 and Sec 7. These include the best theoretical quantitative estimates for the critical parameters of the RPM ($T_c^* = 0.0503$, $\rho_c^* = 0.042$[90]), the trends of the critical parameters of the PMs with size and charge asymmetry that qualitatively agree with the Monte Carlo simulation findings, a microscopically based effective GLW Hamiltonian that belongs to the universal class of a three-dimensional Ising model. As is also shown in Sec 7, the recent results obtained for the Ginzburg temperature confirm the experimental observations that the interplay between solvophobic and Coulomb interactions alters the temperature region of the the crossover regime. It should be noted that the results listed above are obtained by taking into account the higher-order correlation effects without a direct inclusion of ionic association on a semi-phenomenological level.

Despite the significant progress attained in understanding the nature of the gas-liquid phase equilibrium for PMs, a number of problems still remain. In particular, further studies are needed to get the quantitative description of phase diagrams of asymmetric PMs. The results obtained within the framework of the CVs based theory allow us to expect that the quantitative agreement between the theory and simulations can be reached by taking into account the higher-order correlations using the scheme outlined in Sec 6. On the other hand, it would be of great interest to study the systems with strong asymmetry in order to understand the phase behaviour of highly asymmetric ionic fluids. Fist steps in this direction have been made in[114] where the mean-field phase diagram of the PM is obtained in the limit of strong asymmetry ($\lambda \to \infty$, $z \to \infty$). Such models can be used for the description of charged colloid systems.

Another interesting issue is a global phase diagram of the PM. In 1968, Stillinger and Lovett[29] sketched a phase diagram of the RPM including fluid-solid and solid-solid equilibria in addition to the gas-liquid phase transition. This prediction was confirmed by computer simulations.[115–119] The global phase diagram of the RPM constructed by the recent simulations[118] has a fluid phase in the low density region, a fluid-face-centered cubic crystal phase coexistence in the high temperature-high density region, a fluid-CsCl crystal phase coexistence in the low temperature region, a sequence of CsCl, CuAu and tetragonal phases with an increase of density. As was shown in Ref.,[118] the phase behaviour of the RPM is qualitatively similar to that of oppositely charged colloids. The fluid-solid phase coexistence in the RPM was studied theoretically using the cell theory[116] and the density functional theory of freezing.[120] Phase coexistence with an ionic crystal was found within the field-theoretical framework in Refs. [121,122] by applying the Brazovskii-type approximation. In Ref.,[40] a mean-field stability analysis of the effect of size and charge asymmetry on the fluid-solid transition was performed. Nevertheless, further theoretical and simulation studies are required in order to gain a greater insight into fluid-solid and solid-solid phase equilibriia of symmetric and asymmetric PMs.

Summarizing, one should note that the PMs are the "coarse-grained" systems which do not take into account the structure of a solvent. Even for these simplified models, both theories and simulations encounter significant difficulties. Therefore, the description of more realistic models of ionic fluids still remains a relevant problem.

Acknowledgement

We are grateful to A. Ciach, J.-M. Caillol and T. Patsahan who have collaborated with us on topics covered in this review.

References

1. T. Welton, Chem. Rev. **99**, 2071-2083 (1999).
2. M. J. Earle and K. R. Seddon, Pure Appl. Chem. **72**, 1391-1398 (2000).
3. F. Wegner, Phys. Rev.B **5** (1972) 4529-4536.
4. K. Gutkowski, M.A. Anisimov, and J.V. Sengers, J. Chem. Phys. **114**, 3133-3148 (2001).
5. N. Goldenfeld, *Lectures on phase transitions and the renormalization group* (Addison-Wesley Publishing Company), 389 p.
6. M. Buback and E.U. Franck, Ber. Bunsenges. Phys. Chem. **76**,350-354 (1972).
7. P. Chieux and M.J. Sienko, J.Chem.Phys. **53**, 566-570 (1970).
8. R.R. Singh and K.S. Pitzer, J. Chem. Phys. **92**, 6775-6778 (1990).
9. J.M.H. Levelt Sengers and J.A. Given, Mol. Phys. **80**, 899–913 (1993).
10. T. Narayanan and K.S. Pitzer, J. Phys. Chem. **98**, 9170-9174 (1994).
11. T. Narayanan and K.S. Pitzer, Phys. Rev. Lett. **73**, 3002-3005 (1994).
12. T. Narayanan and K.S. Pitzer, J. Phys. Chem. **102**, 8118-8130 (1995).
13. M.L. Japas and J.M.H. Levelt Sengers, J. Phys. Chem. **94**, 5361-5368 (1990).
14. M.A. Anisimov, J. Jacob, A. Kumar, V.A. Agayan, and J.V. Sengers, Phys. Rev. Lett. **85**, 2336-2339 (2000).
15. H. Weingärtner, M. Kleemeier, S. Wiegand, and W. Schröer, J.Stat. Phys. **78**, 169-198 (1995).

16. S. Wiegand, R.F.Berg, and J.M.H. Levelt Sengers, J. Chem. Phys. **109**, 4533-4545 (1998).
17. M.Kleemeier, S. Wiegand, W. Schröer, and H. Weingärtner, J. Chem. Phys. **110**, 3085-3099 (1999).
18. H.Weingärtner, Pure Appl. Chem. **73**, 1733-1748 (2001).
19. W. Schröer and H. Weingärtner, Pure Appl. Chem. **76**, 19-27 (2004).
20. W. Schröer, in *Ionic Soft Matter: Modern Trends and Applications*, (D. Henderson, M.Holovko, A. Trokhymchuk, eds., Dordrecht: NATO ASI Series II, Springer, 2004) P. 143-180.
21. J.-M. Caillol, D. Levesque, and J.-J. Weis, J. Chem. Phys. **116**, 10794-10800 (2002).
22. E. Luijten, M.E. Fisher, and A.Z. Panagiotopoulos, Phys. Rev. Lett. **88**, 185701-4 (2002).
23. Y.C. Kim, M.E. Fisher, and A.Z. Panagiotopoulos, Phys. Rev. Lett. **95**, 195703-195706 (2005).
24. A. Ciach, Phys. Rev.E **73**, 066110-28 (2006).
25. O.V. Patsahan and I.M. Mryglod, J. Phys.: Condens. Matter. **16**, L235-L241 (2004).
26. W. Schröer, W. Wagner, and O. Stanga, J. Mol. Liq. **127**, 2-9 (2006).
27. A. Butka, V.R. Vale, D. Saracsan, C. Rybarsch, V.C. Weiss, and W. Schröer, Pure Appl. Chem. **80**, 1613-1630 (2008).
28. W. Schröer and V.R. Vale, J.Phy.: Condens. Matter **21**, 424119-21 (2009).
29. F.H. Stillinger, Jr. and R. Lovett, J. Chem. Phys. **48**, 3858-3868 (1968).
30. P.N. Vorontsov-Veliaminov and V.P. Chasovskikh, Teplofiz. Vys. Temp. **13**, 1153-1159 (1975) (in Russian).
31. G.Stell, K.C. Wu, and B. Larsen, Phys. Rev. Lett. **37**, 1369-1372 (1976).
32. M.E. Fisher, J. Stat. Phys. **75**, 1-36 (1994).
33. Y. Levin and M.E. Fisher, Physica A. **225**, 164-220 (1996).
34. M.E. Fisher, J. Phys.: Condens. Matter. **8**, 9103-9109 (1996).
35. G. Stell, J. Stat. Phys. **78**, 197-238 (1995).
36. G. Stell, J. Phys.: Condens. Matter. **8**, 9329-9333 (1996).
37. J.-M. Caillol, J. Stat. Phys. **115**, 1461-1504 (2004).
38. J.-M. Caillol, Mol. Phys. **103**, 1271-1283 (2005).
39. A. Ciach and G. Stell, J. Mol. Liq.**87**, 255-273 (2000).
40. A. Ciach, W.T. Góźdź, and G. Stell, Phys. Rev. E **75**, 051505-19 (2007).
41. O. Patsahan and I. Mryglod, Condens. Matter Phys. **9**, 659-668 (2006).
42. O.V. Patsahan, I.M. Mryglod, and T.M. Patsahan, J.Phys.: Condens. Matter.**18**, 10223-10235 (2006).
43. O.V. Patsahan and T.M. Patsahan, Phys. Rev. E 81 (2010) 031110.
44. J. Valleau, J. Chem. Phys. **95**, 584-589 (1991).
45. A.Z. Panagiotopoulos, Fluid Phase Equilib. **76**, 97-112 (1992).
46. G. Orkoulas and A.Z. Panagiotopoulos, J. Chem. Phys. **101**, 1452-1459 (1994).
47. J.M. Caillol, J. Chem. Phys. **100**, 2161-2169 (1994).
48. A.Z. Panagiotopoulos, J. Phys.:Condens. Matter. **175**, S3205-S3213 (2005).
49. A.-P. Hynninen and A.Z. Panagiotopoulos, Mol. Phys. **106**, 2039-2051 (2008).
50. A.D. Bruce and N.B. Wilding, Phys. Rev. Lett. **68** 193-196 (1992).
51. J.M. Caillol, D. Levesque, and J.J. Weis, Phys. Rev. Lett. **77**, 4039-4042 (1996).
52. J.M. Caillol, D. Levesque, and J.J. Weis,J. Chem. Phys. **107**, 1565-1575 (1997).

53. G. Orkoulas and A.Z. Panagiotopoulos, J. Chem. Phys. **110**, 1581-1590 (1999).
54. Q. Yan and J.J. de Pablo, J. Chem. Phys. **111**, 9509-9516 (1999).
55. A.Z. Panagiotopoulos, J. Chem. Phys. **116**, 3007-3011 (2002).
56. Y.C. Kim and M.E. Fisher, Phys. Rev. Lett. **92**, 185703-4 (2004).
57. S. Arrhenius, Z. Phys. Chem. **1**, 631-648 (1887).
58. P. Debye and E. Hückel, Phys. Z. **24**, 185-206 (1923).
59. D.A. McQuarrie, Statistical Mechanics (Harper and Row, New York, 1976).
60. N.B. Bjerrum, Kgl. Danske Videnskab. Selskab. Mat.-Fys. Medd. **7**, 1-48 (1926).
61. J.P. Hansen and I.R. McDonald Theory of simple liquids (Academic Press, 1986).
62. E. Waisman and J.L. Lebowitz, J. Chem. Phys. **56**, 3086-3093 (1972).
63. W. Ebeling, Z. Phys. Chem. (Leipzig), **249**, 140-142 (1972).
64. Y. Q. Zhou, S. Yeh, and G. Stell, J. Chem. Phys. **102**, 5785-5795 (1995).
65. S. Yeh, Y. Q. Zhou, and G. Stell, J. Phys. Chem. 100, 1415-1419 (1996).
66. Yu.V. Kalyuzhnyj and P.T. Cummings, J. Chem. Phys. **115**, 540-551 (2001).
67. J. Jiang, L. Blum, O. Bernard, J.M. Prausnitz and S.I. Sandler, J. Chem. Phys. **116**, 7977-7982 (2002).
68. Yu.V. Kalyuzhnyi, G. Kahl and P.T. Cummings, J. Chem. Phys. **123**, 124501-15 (2001).
69. J.D. Weeks, D. Chandler, and H.C. Andersen, J. Chem. Phys. **54**, 5237-5247 (1971).
70. P.J. Camp and G.N. Patey, J. Chem. Phys. **111**, 9000-9008 (1999).
71. Q. Yan, J.J and de Pablo, Phys. Rev. Lett. **88**, 095504-4 (2002).
72. A.Z. Panagiotopoulos and M.E. Fisher, Phys. Rev. Lett. **88**, 045701-4 (2002).
73. D.N. Zubarev , Dokl. Acad. Nauk SSSR, **95**, 757-760 (1954) (in Russian).
74. I.R. Yukhnovsky , Zh. Eksp. Ter. Fiz., **34**, 379-389 (1958) (in Russian).
75. I. R. Yukhnovskii and M.F. Holovko, *Statistical Theory of Classical Equilibrium Systems*, (Naukova Dumka, Kiev, 1980) (in Russian).
76. I.R. Yukhnovskii, *Phase Transitions of the Second Order: Collective Variables Method* (Singapore, World Scientific, 1987).
77. I. Nezbeda, R. Melnyk and A. Trokhymchuk, J. of Supercritical Fluids, **55**, 448-454 (2010).
78. J. W. Negele and H. Orland, *Quantum many-particle systems* (Frontiers in Physics, Addison-Wesley, 1988).
79. R. R. Netz and H. Orland, Europhys. Lett. **45**, 726-732 (1999).
80. J.M. Romero-Enrique, G. Orkoulas, A. Z. Panagiotopoulos, and M. E. Fisher, Phys. Rev. Lett. **85**, 4558-4561 (2000).
81. Q. Yan and J.J. de Pablo, Phys. Rev. Lett. **8**6, 2054-2057 (2001).
82. Q. Yan and J.J. de Pablo, J. Chem. Phys. **116**, 2967-2972 (2002).
83. D.W. Cheong, and A.Z. Panagiotopoulos, J. Chem. Phys. **1**19, 8526-8536 (2003).
84. E. González-Tovar, Mol. Phys. **9**7, 1203-1206 (1999).
85. D.M. Zuckerman, M.E. Fisher, and S. Bekiranov, Phys. Rev. E **6**4, 011206-13 (2001).
86. M.N. Artyomov, V. Kobelev, and A.B. Kolomeisky, J.Chem.Phys. 118, 6394-6402 (2003).
87. M.E. Fisher, J.-N. Aqua, and S. Banerjee, Phys. Rev. Lett. **95**, 135701-4 (2005).
88. Y.V. Kalyuzhnyi, M.F. Holovko, and V. Vlachy, J. Stat. Phys. 100, 243-265 (2000).
89. D. di Caprio, J.P. Badiali, and M. Holovko, J. Phys. A: Math. Theor. **42**, 214038-214041 (2009).

90. O.V. Patsahan, Condens. Matter Phys. **7**, 35-52 (2004).
91. J.L. Lebowitz, Phys. Rev. **133**, 895-899 (1964).
92. J.L. Lebowitz and J.S. Rowlinson, J. Chem. Phys. **41**, 133-138 (1964).
93. O.V. Patsahan and T.M. Patsahan, Condens. Matter Phys. **13**, 23004-10 (2010) 23004.
94. G. Stell, Phys. Rev. B **1**, 2265-2270 (1970).
95. M.E. Fisher, S.-K. Ma, and B.G. Nickel, Phys. Rev. Lett. **29**, 917-920 (1972).
96. J. Sak, Phys. Rev. B **8**, 281-285 (1973).
97. R.F. Kayser and H.J. Raveché, Phys. Rev. A **29**, 1013-1015 (1984).
98. M.E. Fisher and Y. Levin, Phys. Rev. Lett. **71**, 3826-3829 (1993).
99. J.F. Leote de Carvalho and R. Evans, J. Phys.: Condens. Matter. **7**, L57-L65 (1995).
100. M.E. Fisher and B.P. Lee, Phys. Rev. Lett. **77**, 3561-3564 (1996).
101. V.C. Weiss and W. Schröer, J. Chem. Phys. **106**, 1930-1940 (1997).
102. W. Schröer and V.C. Weiss, J. Chem. Phys. **109**, 8504-8513 (1998).
103. N.V. Brilliantov, C. Bagnuls, and C. Bervillier, Phys. Lett. A **245**, 274-278 (1998).
104. V.L. Ginzburg, Solid State Physics **2**, 2031-2043 (1960) (in Russian).
105. A.G. Moreira, M.M. Telo de Gama, and M.E. Fisher, J. Chem. Phys. **110**, 10058-10066 (1999).
106. O.V. Patsahan, J.-M. Caillol, and I.M. Mryglod, Europ. Phys. Journal, **B58**, 449-459 (2007).
107. J. Stafiej and J. P. Badiali, J. Chem. Phys. **106**, 8579-8586 (1997).
108. J.-N. Aqua and M. E. Fisher, Phys. Rev. Lett. **92**, 135702-4 (2004).
109. O.V. Patsahan, I.M. Mryglod, and J.-M. Caillol, J. Phys.:Condens.Matter. **17**, L251-L256 (2005).
110. R. Evans and T.J. Sluckin, Mol. Phys. **40**, 413-435 (1980).
111. B. Guillot and I. Guissani, Mol. Phys. **87**, 37-86 (1996).
112. J.-M. Caillol, O. Patsahan, and I. Mryglod, Condens. Matter Phys. **8**, 665-684 (2005).
113. O.V. Patsahan, I.M. Mryglod, J.Phys. A: Math. Gen. **39**, L583-L588 (2006).
114. A. Ciach, W.T. Góźdź, and G. Stell, J. Phys.: Condens. Matter. **18**, 1629-1648 (2006).
115. B. Smit, K. Esselink and D. Frenkel, Mol. Phys. **87**, 159-166 (1996).
116. C. J. Vega, F. Bresme and J.L.F. Abascal, Phys. Rev. E. **54**, 2746-2760 (1996).
117. F. Bresme, C. J. Vega and J.L.F. Abascal, Phys. Rev. Lett. **85**, 3217-3220 (2000).
118. A.P. Hynninen, M. E. Leunissen, A. van Blaaderen and M. Dijkstra, Phys. Rev. Lett. **96**, 018303-4 (2006).
119. C. Vega, E. Sanz, J.L.F. Abascal, E.G. Noya, J. Phys. : Condes. Matter. **20**, 153101 (2008).
120. J.-L. Barrat, J. Phys.: Solid State Phys. **20**, 1031–1041 (1987).
121. A. Ciach and O. Patsahan, Phys. Rev. E, **74**, 021508-13 (2006).
122. O. Patsahan and A. Ciach, J. Phys.: Condens. Matter. **19**, 236203-20 (2007).

Chapter 3

Monte Carlo Simulations in Statistical Physics — From Basic Principles to Advanced Applications

Wolfhard Janke

Institut für Theoretische Physik and Centre for Theoretical Sciences (NTZ), Universität Leipzig, Postfach 100 920, 04009 Leipzig, Germany
wolfhard.janke@itp.uni-leipzig.de

This chapter starts with an overview of Monte Carlo computer simulation methodologies which are illustrated for the simple case of the Ising model. After reviewing importance sampling schemes based on Markov chains and standard local update rules (Metropolis, Glauber, heat-bath), nonlocal cluster-update algorithms are explained which drastically reduce the problem of critical slowing down at second-order phase transitions and thus improve the performance of simulations. How this can be quantified is explained in the section on statistical error analyses of simulation data including the effect of temporal correlations and autocorrelation times. Histogram reweighting methods are explained in the next section. Eventually, more advanced generalized ensemble methods (simulated and parallel tempering, multicanonical ensemble, Wang-Landau method) are discussed which are particularly important for simulations of first-order phase transitions and, in general, of systems with rare-event states. The setup of scaling and finite-size scaling analyses is the content of the following section. The chapter concludes with two advanced applications to complex physical systems. The first example deals with a quenched, diluted ferromagnet, and in the second application we consider the adsorption properties of macromolecules such as polymers and proteins to solid substrates. Such systems often require especially tailored algorithms for their efficient and successful simulation.

Contents

1. Introduction

Classical statistical physics is conceptually a well understood subject which poses, however, many difficult problems when specific properties of interacting systems are considered. In almost all non-trivial applications, analytical methods can only provide approximate answers. Experiments, on the other hand, are often plagued by side effects which are difficult to control. Numerical computer simulations are, therefore, an important third complementary method of modern physics. The relationship between theory, experiment, and computer simulation is sketched in Fig. 1. On the one hand a computer simulation allows one to assess the range of validity of approximate analytical work for generic models and on the other hand it can bridge the gap to experiments for real systems with typically fairly complicated interactions. Computer simulations are thus helpful on our way to a deeper understanding of complex physical systems such as disordered magnets and (spin) glasses or of biologically motivated problems such as protein folding and adsorption of macromolecules to solid substrates, to mention only a few. Quantum statistical problems in condensed matter or the broad field of elementary particle physics and quantum gravity are other major applications which, after suitable mappings, basically rely on the same simulation techniques.

This chapter provides an overview of computer simulations employing Monte Carlo methods based on Markov chain importance sampling. Most methods can be illustrated with the simple Ising spin model. Not all aspects can be discussed

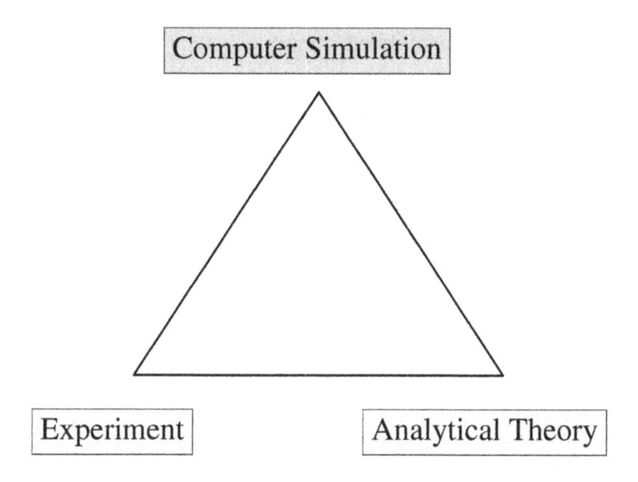

Fig. 1. Sketch of the relationship between theory, experiment and computer simulation.

in detail and for further study the reader is referred to recent textbooks,[1–4] where some of the material presented here is discussed in more depth. The rest of this chapter is organized as follows. In the next Sect. 2, first the definition of the standard Ising model is briefly recalled. Then the basic method underlying all importance sampling Monte Carlo simulations is described and some properties of local update algorithms (Metropolis, Glauber, heat-bath) are discussed. The following subsection is devoted to non-local cluster algorithms which in some cases can dramatically speed up the simulations. A fairly detailed account of statistical error analyses is given in Sect. 3. Here temporal correlation effects and autocorrelation times are discussed, which explain the problems with critical slowing down at a continuous phase transition and exponentially large flipping times at a first-order transition. Reweighting techniques are discussed in Sect. 4 which are particularly important for finite-size scaling studies. More advanced generalized ensemble simulation methods are briefly outlined in Sect. 5, focusing on simulated and parallel tempering, the multicanonical ensemble and the Wang-Landau method. In Sect. 6 suitable observables for scaling analyses (specific heat, magnetization, susceptibility, correlation functions, ...) are briefly discussed. Some characteristic properties of phase transitions, scaling laws, the definition of critical exponents and the method of finite-size scaling are summarized. In order to illustrate how all these techniques can be put to good use, in Sect. 7 two concrete applications are discussed: The phase diagram of a quenched, diluted ferromagnet and the adsorption properties of polymers to solid substrates. Finally, in Sect. 8 this chapter closes with a few concluding remarks.

2. The Monte Carlo Method

The goal of Monte Carlo simulations is to estimate expectation values

$$\langle \mathcal{O} \rangle \equiv \sum_{\text{states } \sigma} \mathcal{O}(\sigma) e^{-\beta \mathcal{H}(\sigma)} / \mathcal{Z} \ , \tag{1}$$

where $\mathcal{O}$ stands for any quantity of the system defined by its Hamiltonian $\mathcal{H}$ and

$$\mathcal{Z} = e^{-\beta \mathcal{F}} = \sum_{\text{states } \sigma} e^{-\beta \mathcal{H}(\sigma)} = \sum_{E} \Omega(E) e^{-\beta E} \tag{2}$$

is the (canonical) partition function. The first sum runs over all possible microstates of the system and the second sum runs over all energies, where the density of states $\Omega(E)$ counts the number of microstates contributing to a given energy E. The state space may be discrete or continuous (where sums become integrals etc.). As usual $\beta \equiv 1/k_B T$ denotes the inverse temperature fixed by an external heat bath and k_B is Boltzmann's constant.

In the following most simulation methods will be illustrated for the minimalistic Ising model[5] where

$$\mathcal{H}(\sigma) = -J \sum_{\langle ij \rangle} \sigma_i \sigma_j - h \sum_i \sigma_i \ , \qquad \sigma_i = \pm 1 \ . \tag{3}$$

Here J is a coupling constant which is positive for a ferromagnet ($J > 0$) and negative for an anti-ferromagnet ($J < 0$), h is an external magnetic field, and the symbol $\langle ij \rangle$ indicates that the lattice sum is restricted to all nearest-neighbor pairs of spins living at the lattice sites i. In the examples discussed below, usually D-dimensional simple-cubic lattices with $V = L^D$ spins subject to periodic boundary conditions are considered. From now on we will always assume natural units in which $k_B = 1$ and $J = 1$.

For any realistic number of degrees of freedom, complete enumeration of all microstates contributing to (1) or (2) is impossible. For the Ising model with only two states per site, enumeration still works up to a, say, 6×6 square lattice where $2^{36} \approx 6.9 \times 10^{10}$ microstates contribute. Since this yields the exact expectation value of *any* quantity, enumeration for very small systems is a useful exercise for comparison with the numerical methods discussed here. However, already for a moderate 10^3 lattice, the number of terms would be astronomically large:[a] $2^{1000} \approx 10^{300}$.

[a]For comparison, a standard estimate for the number of protons in the Universe is 10^{80}.

2.1. *Random sampling*

One way out is stochastic sampling of the huge state space. Simple random sampling, however, does not work for statistical systems with many degrees of freedom. Here the problem is that the region of state space that contributes significantly to canonical expectation values at a given temperature $T \ll \infty$ is extremely narrow and hence far too rarely hit by random sampling. In fact, random sampling corresponds to setting $\beta = 1/T = 0$, i.e., exploring mainly the typical microstates at infinite temperature. Of course, the low-energy states in the tails of this distribution contain theoretically (that is, for infinite statistics) all information about the system's properties at finite temperature, too, but this is of very little practical relevance since the probability to hit this tail in random sampling is by far too small. With finite statistics consisting of typically $10^9 - 10^{12}$ randomly drawn microstates, this tail region is virtually not sampled at all.

2.2. *Importance sampling*

The solution to this problem has been known since long as *importance sampling*[6,7] where a Markov chain[8–10] is set up to draw a microstate σ_i not at random but according to the given equilibrium distribution

$$\mathcal{P}_i^{\mathrm{eq}} \equiv \mathcal{P}^{\mathrm{eq}}(\sigma_i) = e^{-\beta\mathcal{H}(\sigma_i)}/\mathcal{Z} \ . \tag{4}$$

For definiteness, on the r.h.s. a canonical ensemble governed by the *Boltzmann weight* $e^{-\beta\mathcal{H}(\sigma_i)}$ was assumed, but this is not essential for most of the following.

A Markov chain is defined by the transition probability $W_{ij} \equiv W(\sigma_i \to \sigma_j)$ for a given microstate σ_i to "evolve" into another microstate σ_j (which may be again σ_i) subject to the condition that this probability *only* depends on the preceding state σ_i but *not* on the history of the whole trajectory in state space, i.e., the stochastic process is almost local in time. Mnemonically this can be depicted as

$$\dots \xrightarrow{W} \sigma^{(k)} \xrightarrow{W} \sigma^{(k+1)} \xrightarrow{W} \sigma^{(k+2)} \xrightarrow{W} \dots \ ,$$

where $\sigma^{(k)}$ is the current state of the system after the kth step of the Markov chain. To ensure that, after an initial transient or equilibration period, microstates occur with the given probability (4), the transition probability W_{ij} has to satisfy three conditions:

$$i)\ W_{ij} \geq 0 \qquad \forall i, j \ , \tag{5}$$

$$ii)\ \sum_j W_{ij} = 1 \qquad \forall i \ , \tag{6}$$

$$iii)\ \sum_i W_{ij} \mathcal{P}_i^{\rm eq} = \mathcal{P}_j^{\rm eq} \qquad \forall j \ . \tag{7}$$

The first two conditions merely formalize that, for any initial state σ_i, W_{ij} should be a properly normalized probability distribution. The equal sign in (5) may occur and, in fact, does so for almost all pairs of microstates i, j in any realistic implementation of the Markov process. To ensure *ergodicity* one additionally has to require that starting from *any* given microstate σ_i *any* other σ_j can be reached in a *finite* number of steps, i.e., an integer $n < \infty$ must exist such that $(W^{n+1})_{ij} = \sum_{k_1,k_2,\dots,k_n} W_{ik_1} W_{k_1 k_2} \dots W_{k_n j} > 0$. In other words, at least one (finite) path connecting σ_i and σ_j must exist in state space that can be realized with non-zero probability.[b]

The *balance condition* (7) implies that the transition probability W has to be chosen such that the desired equilibrium distribution (4) is a fixed point of W, i.e., an eigenvector of W with unit eigenvalue. The usually employed *detailed balance* is a stronger, sufficient condition:

$$W_{ij}\, \mathcal{P}_i^{\rm eq} = W_{ji}\, \mathcal{P}_j^{\rm eq} \ . \tag{8}$$

By summing over i and using the normalization condition (6), one easily proves the more general balance condition (7).

After an initial equilibration period, expectation values can be estimated as arithmetic mean over the Markov chain,

$$\langle \mathcal{O} \rangle = \sum_\sigma \mathcal{O}(\sigma) \mathcal{P}^{\rm eq}(\sigma) \approx \overline{\mathcal{O}} \equiv \frac{1}{N} \sum_{k=1}^{N} \mathcal{O}(\sigma^{(k)}) \ , \tag{9}$$

where $\sigma^{(k)}$ stands for a microstate at "time" k.[c] Since in equilibrium $\langle \mathcal{O}(\sigma^{(k)}) \rangle = \langle \mathcal{O} \rangle$ at any "time" k, one immediately sees that $\langle \overline{\mathcal{O}} \rangle = \langle \mathcal{O} \rangle$, showing that the mean value $\overline{\mathcal{O}}$ is a so-called *unbiased estimator* of the expectation value $\langle \mathcal{O} \rangle$. A more detailed exposition of the mathematical concepts underlying any Markov chain Monte Carlo algorithm can be found in many textbooks and reviews.[1–4,11–13]

[b] In practice, one may nevertheless observe "effective" ergodicity breaking when $(W^{n+1})_{ij}$ is so small that this event will typically not happen in finite simulation time.

[c] In Monte Carlo simulations, "time" refers to the stochastic evolution in state space and is *not* directly related to physical time as for instance in molecular dynamics simulations where the trajectories are determined by Newton's deterministic equation.

2.3. *Local update algorithms*

The Markov chain conditions (5)–(7) are rather general and can be satisfied with many different transition probabilities. A very flexible prescription is the original Metropolis algorithm,[14] which is applicable in practically all cases (lattice/off-lattice, discrete/continuous, short-range/long-range interactions, ...). Here one first proposes with *selection probability*

$$f_{ij} = f(\sigma_i \longrightarrow \sigma_j) \ , \quad f_{ij} \geq 0 \ , \quad \sum_j f_{ij} = 1 \ , \tag{10}$$

a potential update from the current "old" microstate $\sigma_{\mathrm{o}} = \sigma_i$ to some microstate σ_j. The proposed microstate σ_j is then accepted as the "new" state $\sigma_{\mathrm{n}} = \sigma_j$ with an *acceptance probability*

$$w_{ij} = w(\sigma_i \longrightarrow \sigma_j) = \min\left(1, \frac{f_{ji}}{f_{ij}} \frac{\mathcal{P}_j^{\mathrm{eq}}}{\mathcal{P}_i^{\mathrm{eq}}}\right) \ , \tag{11}$$

where $\mathcal{P}^{\mathrm{eq}}$ is the desired equilibrium distribution specified in (4). Otherwise the system remains in the old microstate, $\sigma_{\mathrm{n}} = \sigma_{\mathrm{o}}$, which may also trivially happen when $f_{ii} \neq 0$.

Keeping this in mind, one readily sees that the transition probability W_{ij} is given as

$$W_{ij} = \begin{cases} f_{ij} w_{ij} & j \neq i \\ f_{ii} + \sum_{j \neq i} f_{ij}(1 - w_{ij}) & j = i \end{cases} \ . \tag{12}$$

Since $f_{ij} \geq 0$ and $0 \leq w_{ij} \leq 1$, the first Markov condition $W_{ij} \geq 0$ follows immediately. Also the second condition (6) is easy to prove:

$$\begin{aligned} \sum_j W_{ij} &= W_{ii} + \sum_{j \neq i} W_{ij} \\ &= f_{ii} + \sum_{j \neq i} f_{ij}(1 - w_{ij}) + \sum_{j \neq i} f_{ij} w_{ij} = \sum_j f_{ij} = 1 \ . \end{aligned} \tag{13}$$

Finally we show that W_{ij} satisfies the detailed balance condition (8). We first consider the case $f_{ji}\mathcal{P}_j^{\mathrm{eq}} > f_{ij}\mathcal{P}_i^{\mathrm{eq}}$. Then, from (11), one immediately finds $W_{ij}\mathcal{P}_i^{\mathrm{eq}} = f_{ij}\mathcal{P}_i^{\mathrm{eq}}$ for the l.h.s. of (8). Since $W_{ji} = f_{ji}\min\left(1, \frac{f_{ij}}{f_{ji}} \frac{\mathcal{P}_i^{\mathrm{eq}}}{\mathcal{P}_j^{\mathrm{eq}}}\right)$, the r.h.s. of (8) becomes

$$W_{ji}\mathcal{P}_j^{\mathrm{eq}} = f_{ji} \frac{f_{ij}}{f_{ji}} \frac{\mathcal{P}_i^{\mathrm{eq}}}{\mathcal{P}_j^{\mathrm{eq}}} \mathcal{P}_j^{\mathrm{eq}} = f_{ij}\mathcal{P}_i^{\mathrm{eq}} \ , \tag{14}$$

which completes the proof. For the second case $f_{ji}\mathcal{P}_j^{\mathrm{eq}} < f_{ij}\mathcal{P}_i^{\mathrm{eq}}$, the proof proceeds precisely along the same lines.

The update prescription (10), (11) is still very general: (a) The selection probability may be asymmetric ($f_{ij} \neq f_{ji}$), (b) it has not yet been specified how to pick the trial state σ_j given σ_o, and (c) $\mathcal{P}^\text{eq}$ could be "some" arbitrary probability distribution. The last point (c) is obviously trivial, but the resulting formulas simplify when a Boltzmann weight as in (4) is assumed. Then

$$\frac{\mathcal{P}_j^\text{eq}}{\mathcal{P}_i^\text{eq}} = e^{-\beta\Delta E} \tag{15}$$

where $\Delta E = E_j - E_i = E_\text{n} - E_\text{o}$ is the energy difference between the proposed new and the old microstate. The second point (b), on the other hand, is of great practical relevance since an arbitrary proposal for σ_n would typically lead to a large ΔE and hence a high rejection rate if $\beta > 0$. One therefore commonly tries to update only one degree of freedom at a time. Then σ_n differs only *locally* from σ_o. For short-range interactions this automatically has the additional advantage that only the local neighborhood of the selected degree of freedom contributes to ΔE, so that there is no need to compute the total energies in each update step. These two specializations are usually employed, but the selection probabilities may still be chosen asymmetrically. If this is the case, one refers to this update prescription as the *Metropolis-Hastings*[15] update algorithm. For a recent example with asymmetric f_{ij} in the context of polymer simulations see, e.g., Ref. 16.

2.3.1. *Metropolis algorithm*

In generic applications, however, the f_{ij} are symmetric. For instance, if we pick one of the V Ising spins at random and propose to flip it, then $f_{ij} = 1/V$ does not depend on i and j and hence is trivially symmetric. In this case the acceptance probability simplifies to

$$\begin{aligned} w_{ij} &= \min\left(1, \frac{\mathcal{P}_j^\text{eq}}{\mathcal{P}_i^\text{eq}}\right) = \min\left(1, e^{-\beta\Delta E}\right) \\ &= \begin{cases} 1 & E_\text{n} < E_\text{o} \\ \exp\left[-\beta(E_\text{n} - E_\text{o})\right] & E_\text{n} \geq E_\text{o} \end{cases} . \end{aligned} \tag{16}$$

This is the standard *Metropolis* update algorithm, which is very easy to implement.

If the proposed update lowers the energy, it is always accepted. On the other hand, when the new microstate has a higher energy, the update has still to be accepted with probability (16) in order to ensure the proper treatment of entropic contributions – in thermal equilibrium, it is the *free* energy $F = U - TS$ which has to be minimized and not the energy. Only in the limit of zero temperature, $\beta \to \infty$, the acceptance probability for new states with higher energy tends to zero and the Metropolis method degenerates to a minimization algorithm for the energy functional. With some additional refinements, this is the basis for the *simulated annealing* technique,[17] which is often applied to hard optimization and minimization problems.

For the Ising model with only two states per spin, a spin flip is the only admissible local update proposal. Hence in this simple example there is no parameter available by which one could tune the *acceptance ratio*, which is defined as the fraction of trial moves that are accepted. For models with many states per spin (e.g., q-state Potts or Z_n clock models) or in continuous systems (e.g., Heisenberg spin model or off-lattice molecular systems), however, it is in the most cases not recommendable to propose the new state uniformly out of all available possibilities. Rather, one usually restricts the trial states to a neighborhood of the current "old" state. For example, in a continuous atomic system, a trial move may consist of displacing a randomly chosen atom by a random step size up to some maximum S_{max} in each Cartesian direction. If S_{max} is small, almost all attempted moves will be accepted and the acceptance ratio is close to unity, but the configuration space is explored slowly. On the other hand, if S_{max} is large, a successful move would make a large step in configuration space, but many trial moves would be rejected because configurations with low Boltzmann weight are very likely, yielding an acceptance ratio close to zero. As a compromise of these two extreme situations, one often applies the common rule of thumb that S_{max} is adjusted to achieve an acceptance ratio of 0.5.[18,19]

Empirically this value proves to be a reasonable but at best heuristically justified choice. In principle, one should measure the statistical error bars as a function of S_{max} for otherwise identical simulation conditions and then choose that S_{max} which minimises the statistical error. In general the optimal S_{max} depends on the model at hand and even on the considered observable, so finally some "best average" would have to be used. At any rate, the corresponding acceptance ratio would certainly not coincide with 0.5. Example computations of this type reported values in the range $0.4 - 0.6$ (Refs. 18,20) but for certain models also much smaller (or larger) values may be favourable. Incidentally, there appeared recently a proof in the mathematical literature[21] claiming an optimal acceptance ratio of 0.234 which, however, relies on assumptions[22] not met in a typical statistical physics simulation.[d]

Whether relying on the rule of thumb value 0.5 or trying to optimise S_{max}, this should be done *before* the actual simulation run. Trying to maintain a given acceptance ratio automatically during the run by periodically updating S_{max} is at

[d]Thanks are due to Yuko Okamoto who pointed to this paper and to Bob Swendsen who immediately commented on it during the CompPhys11 Workshop in November 2011 in Leipzig.

least potentially dangerous.[19] The reason is that the accumulated average of the acceptance ratio and hence the updated $S_{\rm max}$ are dependent on the recent history of the Monte Carlo trajectory – and *not* only on the current configuration – what violates the Markovian requirement. Consequently the balance condition is no longer fulfilled which may lead to more or less severe systematic deviations (bias). As claimed already a while ago in Ref. 18 and reemphasized recently in Ref. 20, by following a carefully determined schedule for the adjustments of $S_{\rm max}$, the systematic error may be kept smaller than the statistical error in a controlled way, but to be on the safe side one should be very cautious with this type of refinements.

Finally a few remarks on the practical implementation of the Metropolis method. To decide whether a proposed update should be accepted or not, one draws a uniformly distributed random number $r \in [0, 1)$, and if $r \leq w_{ij}$, the new state is accepted. Otherwise one keeps the old configuration and continues with the next spin. In computer simulations, random numbers are generated by means of "pseudo-random number generators" (RNGs), which produce – according to some deterministic rule – (more or less) uniformly distributed numbers whose values are "very hard" to predict.[23] In other words, given a finite sequence of subsequent pseudo-random numbers, it should be (almost) impossible to predict the next one or to even uncover the deterministic rule underlying their generation. The "goodness" of a RNG is thus assessed by the difficulty to derive its underlying deterministic rule. Related requirements are the absence of correlations and a very long period, what can be particularly important in high-statistics simulations. Furthermore, a RNG should be portable among different computer platforms and, very importantly, it should yield reproducible results for testing purposes. The design of RNGs is a science in itself, and many things can go wrong with them.[e] As a recommendation one should better not experiment too much with some fancy RNG picked up somewhere from the WWW, say, but rely on well-documented and well-tested subroutines.

2.3.2. *Glauber algorithm*

As indicated earlier the Markov chain conditions (5)–(7) are rather general and the Metropolis rule (11) or (16) for the acceptance probability w_{ij} is not the only possible choice. For instance, when flipping a spin at site i_0 in the Ising model, w_{ij} can also be taken as[25]

$$w_{ij} = w(\sigma_{i_0} \to -\sigma_{i_0}) = \frac{1}{2}\left[1 - \sigma_{i_0} \tanh\left(\beta S_{i_0}\right)\right] \,, \tag{17}$$

[e] A prominent example is the failure of the by then very prominent and apparently well-tested R250 generator when applied to the single-cluster algorithm.[24]

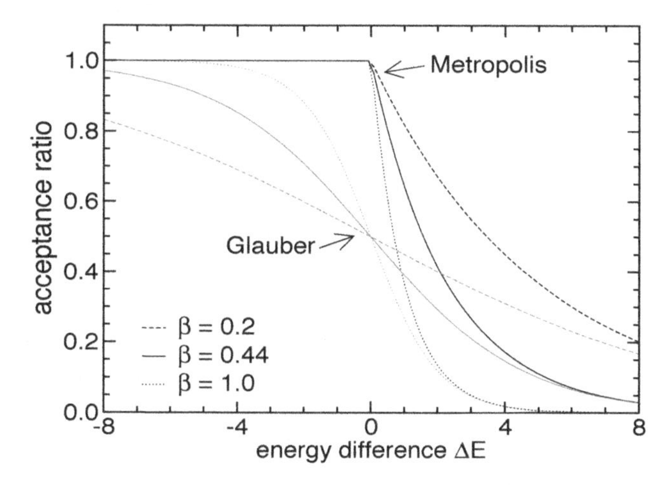

Fig. 2. Comparison of the acceptance ratio for a spin flip in the two-dimensional Ising model with the Glauber (or equivalently heat-bath) and Metropolis update algorithm for three different inverse temperatures β.

where $S_{i_0} = \sum_k \sigma_k + h$ is an effective spin or field collecting all neighboring spins (in their "old" states) interacting with the spin at site i_0 and h is the external magnetic field. This is the *Glauber* update algorithm. Detailed balance is straightforward to prove. Rewriting $\sigma_{i_0} \tanh(\beta S_{i_0}) = \tanh(\beta \sigma_{i_0} S_{i_0})$ (making use of $\sigma_{i_0} = \pm 1$ and the point symmetry of the hyperbolic tangent) and noting that $\Delta E = E_{\rm n} - E_{\rm o} = 2\sigma_{i_0} S_{i_0}$ (where σ_{i_0} is the "old" spin value and $(-\sigma_{i_0})$ the "new" one), Eq. (17) becomes

$$w(\sigma_{i_0} \to -\sigma_{i_0}) = \frac{1}{2}\left[1 - \tanh(\beta \Delta E/2)\right] = \frac{e^{-\beta \Delta E/2}}{e^{\beta \Delta E/2} + e^{-\beta \Delta E/2}} \,, \qquad (18)$$

showing explicitly that the acceptance probability of the Glauber algorithm also only depends on the total energy change as in the Metropolis case. In this form it is thus possible to generalize the Glauber update rule from the Ising model with only two states per spin to any general model that can be simulated with the Metropolis procedure. The acceptance probability (18) is plotted in Fig. 2 as a function of ΔE for various (inverse) temperatures and compared with the corresponding probability (16) of the Metropolis algorithm. Note that for all values of ΔE and temperature, the Metropolis acceptance probability is higher than that of the Glauber algorithm. As we shall see in the next paragraph, for the Ising model, the Glauber and heat-bath algorithms are identical.

The Glauber update algorithm for the Ising model is also theoretically of interest since for the one-dimensional case the dynamics of the Markov chain can be calculated analytically. For the relaxation time of the magnetisation one finds the remarkably simple result[25] $m(t) = m(0)\exp(-t/\tau_{\rm relax})$ with $\tau_{\rm relax} = 1/[1 - \tanh(2\beta)]$. For two and higher dimensions, however, no exact solutions are known.

2.3.3. *Heat-bath algorithm*

The heat-bath algorithm is different from the two previous update algorithms in that it does not follow the previous scheme "update proposal plus accept/reject step". Rather, the new value of σ_{i_0} at a randomly selected site i_0 is determined by testing all its possible states in the "heat bath" of its (fixed) neighbors (e.g., 4 on a square lattice and 6 on a simple-cubic lattice with nearest-neighbor interactions). For models with a finite number of states per degree of freedom the transition probability reads

$$w(\sigma_{\rm o} \to \sigma_{\rm n}) = \frac{e^{-\beta\mathcal{H}(\sigma_{\rm n})}}{\sum_{\sigma_{i_0}} e^{-\beta\mathcal{H}(\sigma_{\rm o})}} = \frac{e^{-\beta\sum_k H_{i_0k}}}{\sum_{\sigma_{i_0}} e^{-\beta\sum_k H_{i_0k}}} \,, \tag{19}$$

where $\sum_k H_{i_0k}$ collect all terms involving the spin σ_{i_0}. All other contributions to the energy not involving σ_{i_0} cancel due to the ratio in (19), so that for the update at each site i_0 only a small number of computations is necessary (e.g, about 4 for a square and 6 for a simple-cubic lattice of arbitrary size). Detailed balance (8) is obviously satisfied since

$$e^{-\beta\mathcal{H}(\sigma_{\rm o})}\frac{e^{-\beta\mathcal{H}(\sigma_{\rm n})}}{\sum_{\sigma_{i_0}} e^{-\beta\mathcal{H}(\sigma_{\rm n})}} = e^{-\beta\mathcal{H}(\sigma_{\rm n})}\frac{e^{-\beta\mathcal{H}(\sigma_{\rm o})}}{\sum_{\sigma_{i_0}} e^{-\beta\mathcal{H}(\sigma_{\rm o})}} \,. \tag{20}$$

How is the probability (19) realized in practice? Due to the summation over all local states, special tricks are necessary when each degree of freedom can take many different states, and only in special cases the heat-bath method can be efficiently generalized to continuous degrees of freedom. In many applications, however, the admissible local states of σ_{i_0} can be labeled by a small number of integers, say $n = 1, \dots, N$, which occur with probabilities p_n according to (19). Since this probability distribution is normalized to unity, the sequence $(p_1, p_2, \dots, p_n, \dots, p_N)$ decomposes the unit interval into segments of length $\propto p_n$. If one now draws a random number $R \in [0, 1)$ and compares the accumulated probabilities $\sum_{k=1}^{n} p_k$ with R, then the new state n is the smallest upper bound that satisfies $\sum_{k=1}^{n} p_k \geq R$. Clearly, for a large number of possible local states, the determination of n can become quite time-consuming (in particular, if many small p_n are at the beginning of the sequence, in which case a clever permutation of the p_n by relabeling the admissible local states can improve the performance).

In the special case of the Ising model with only two states per spin, $\sigma_i = \pm 1$, (19) simplifies to

$$w(\sigma_{\mathrm{o}} \to \sigma_{\mathrm{n}}) = \frac{e^{\beta \sigma_{i_0} S_{i_0}}}{e^{\beta S_{i_0}} + e^{-\beta S_{i_0}}} , \tag{21}$$

where σ_{i_0} is the *new* spin value and $S_{i_0} = \sum_k \sigma_k + h$ represents the effective spin interacting with σ_{i_0} as defined already below (17). And since $\Delta E = E_{\mathrm{n}} - E_{\mathrm{o}} = -(\sigma_{i_0} - (-\sigma_{i_0}))S_{i_0} = -2\sigma_{i_0} S_{i_0}$, the probability for a spin flip becomes[26]

$$w(-\sigma_{i_0} \to \sigma_{i_0}) = \frac{e^{-\beta \Delta E/2}}{e^{\beta \Delta E/2} + e^{-\beta \Delta E/2}} . \tag{22}$$

This is *identical* to the acceptance probability (18) for a spin flip in the Glauber update algorithm, that is, for the Ising model, the Glauber and heat-bath update rules give precisely the same results.

2.4. *Temporal correlations*

Data generated with a Markov chain method always exhibit temporal correlations which can be estimated from the autocorrelation function

$$A(k) = \frac{\langle \mathcal{O}_i \mathcal{O}_{i+k} \rangle - \langle \mathcal{O}_i \rangle \langle \mathcal{O}_i \rangle}{\langle \mathcal{O}_i^2 \rangle - \langle \mathcal{O}_i \rangle \langle \mathcal{O}_i \rangle} , \tag{23}$$

where $\mathcal{O}$ denotes any measurable quantity, for example the energy or magnetization (technical issues and the way in which temporal correlations enter statistical error estimates will be discussed in more detail in Sect. 3.1.3). For large time separations k, $A(k)$ decays exponentially ($a =$ const),

$$A(k) \stackrel{k \to \infty}{\longrightarrow} a e^{-k/\tau_{\mathcal{O},\mathrm{exp}}} , \tag{24}$$

which defines the *exponential* autocorrelation time $\tau_{\mathcal{O},\mathrm{exp}}$. At smaller distances usually also other modes contribute and $A(k)$ behaves no longer purely exponentially.

This is illustrated in Fig. 3 for the 2D Ising model on a rather small 16×16 square lattice with periodic boundary conditions at the infinite-volume critical point $\beta_c = \ln(1 + \sqrt{2})/2 = 0.440\,686\,793\ldots$. The spins were updated in sequential order by proposing always to flip a spin and accepting or rejecting this proposal according to (16). The raw data of the simulation are collected in a time-series file, storing 1 000 000 measurements of the energy and magnetization taken after each sweep over the lattice, after discarding (quite generously) the first 200 000 sweeps for equilibrating the system from a disordered start configuration.

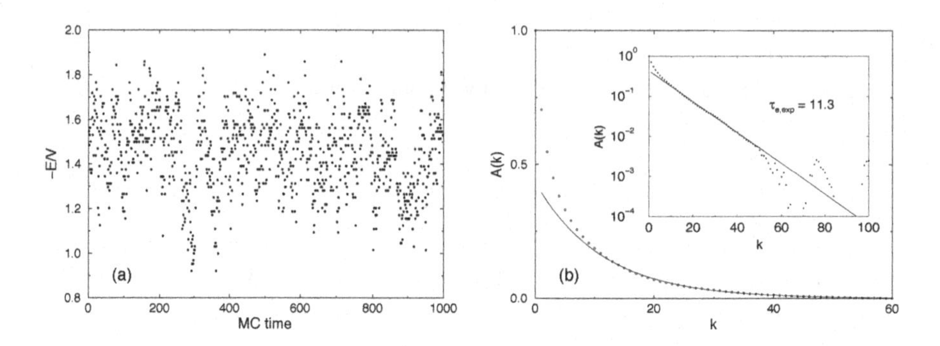

Fig. 3. (a) Part of the time evolution of the energy $e = E/V$ for the 2D Ising model on a 16×16 lattice at $\beta_c = \ln(1+\sqrt{2})/2 = 0.440\,686\,793\ldots$ and (b) the resulting autocorrelation function. In the inset the same data are plotted on a logarithmic scale, revealing a fast initial drop for very small k and noisy behaviour for large k. The solid lines show a fit to the ansatz $A(k) = a\exp(-k/\tau_{e,\mathrm{exp}})$ in the range $10 \leq k \leq 40$ with $\tau_{e,\mathrm{exp}} = 11.3$ and $a = 0.432$.

The last 1000 sweeps of the time evolution of the energy are shown in Fig. 3(a). Using the complete time series the autocorrelation function was computed according to (23) which is shown in Fig. 3(b). On the linear-log scale of the inset we clearly see the asymptotic linear behaviour of $\ln A(k)$. A linear fit of the form (24), $\ln A(k) = \ln a - k/\tau_{e,\mathrm{exp}}$, in the range $10 \leq k \leq 40$ yields an estimate for the exponential autocorrelation time of $\tau_{e,\mathrm{exp}} \approx 11.3$. In the small k behaviour of $A(k)$ we observe an initial fast drop, corresponding to faster relaxing modes, before the asymptotic behaviour sets in. This is the generic behaviour of autocorrelation functions in realistic models where the small-k deviations are, in fact, often much more pronounced than for the 2D Ising model.

The influence of autocorrelation times is particular pronounced for phase transitions and critical phenomena.[27–30] For instance, close to a critical point, the autocorrelation time typically scales in the infinite-volume limit as

$$\tau_{\mathcal{O},\mathrm{exp}} \propto \xi^{z} \ , \tag{25}$$

where $z \geq 0$ is the so-called *dynamical critical exponent*. Since the spatial correlation length $\xi \propto |T - T_c|^{-\nu} \to \infty$ when $T \to T_c$, also the autocorrelation time $\tau_{\mathcal{O},\mathrm{exp}}$ diverges when the critical point is approached, $\tau_{\mathcal{O},\mathrm{exp}} \propto |T - T_c|^{-\nu z}$. This leads to the phenomenon of *critical slowing down* at a continuous phase transition which can be observed experimentally for instance in critical opalescence.[31] The reason is that local spin-flip Monte Carlo dynamics (or diffusion dynamics in a lattice-gas picture) describes at least qualitatively the true physical dynamics of a system in contact with a heat bath. In a finite system, the correlation length ξ is limited by the linear system size L, so that the characteristic length scale is then L and the scaling law (25) is replaced by

$$\tau_{\mathcal{O},\mathrm{exp}} \propto L^z \ . \tag{26}$$

For local dynamics, the critical slowing down effect is quite pronounced since the dynamical critical exponent takes a rather large value around

$$z \approx 2 \ , \tag{27}$$

which is only weakly dependent on the dimensionality and can be understood by a simple random-walk or diffusion argument in energy space. Non-local update algorithms such as multigrid schemes[32–36] or in particular the cluster methods discussed in the next section can reduce the value of the dynamical critical exponent z significantly, albeit in a strongly model-dependent fashion.

At a first-order phase transition, a completely different mechanism leads to an even more severe "slowing-down" problem.[37,38] Here, the keyword is "phase coexistence". A finite system close to the (pseudo-) transition point can flip between the coexisting pure phases by crossing a two-phase region. Relative to the weight of the pure phases, this region of state space is strongly suppressed by an additional Boltzmann factor $\exp(-2\sigma L^{d-1})$, where σ denotes the interface tension between the coexisting phases, L^{d-1} is the (projected) "area" of the interface and the factor 2 accounts for periodic boundary conditions, which enforce always an even number of interfaces for simple topological reasons. The time spent for crossing this highly suppressed rare-event region scales proportional to the inverse of this interfacial Boltzmann factor, implying that the autocorrelation time increases exponentially with the system size,

$$\tau_{\mathcal{O},\mathrm{exp}} \propto e^{2\sigma L^{d-1}} \ . \tag{28}$$

In the literature, this behaviour is sometimes termed *supercritical slowing down*, even though, strictly speaking, nothing is "critical" at a first-order phase transition. Since this type of slowing-down problem is directly related to the shape of the probability distribution, it appears for all types of update algorithms, i.e., in contrast to the situation at a second-order transition, here it cannot be cured by employing multigrid or cluster techniques. It can be overcome, however, at least in part by means of multicanonical methods which are briefly discussed at the end of this chapter in Sect. 5.

2.5. *Cluster algorithms*

The critical slowing down at a second-order phase transition reflects that excitations on all length scales become important, leading to diverging spatial correlations. This suggests that some sort of non-local update rules should be able to alleviate this problem. Natural candidates are rules where whole clusters or droplets of spins are flipped at a time. Still, it took until 1987 before Swendsen and Wang[39] proposed the first legitimate cluster update procedure satisfying detailed balance. For the Ising model this follows from the identity

$$Z = \sum_{\{\sigma_i\}} \exp\left(\beta \sum_{\langle ij\rangle} \sigma_i \sigma_j\right) \tag{29}$$

$$= \sum_{\{\sigma_i\}} \prod_{\langle ij\rangle} e^{\beta} \left[(1-p) + p\delta_{\sigma_i,\sigma_j}\right] \tag{30}$$

$$= \sum_{\{\sigma_i\}} \sum_{\{n_{ij}\}} \prod_{\langle ij\rangle} e^{\beta} \left[(1-p)\delta_{n_{ij},0} + p\delta_{\sigma_i,\sigma_j}\delta_{n_{ij},1}\right] \;, \tag{31}$$

where

$$p = 1 - e^{-2\beta} \;. \tag{32}$$

Here the n_{ij} are bond occupation variables which can take the values $n_{ij} = 0$ or 1, interpreted as "deleted" or "active" bonds. The representation (30) follows from the observation that the product $\sigma_i\sigma_j$ of two Ising spins can only take the two values ± 1, so that $\exp(\beta\sigma_i\sigma_j) = x + y\delta_{\sigma_i\sigma_j}$ can easily be solved for x and y. And in the third line (31) we made use of the trivial (but clever) identity $a+b = \sum_{n=0}^{1}\left(a\delta_{n,0} + b\delta_{n,1}\right)$. Going one step further and performing in (31) the summation over spins, one arrives at the so-called Fortuin-Kasteleyn representation.[40–43]

2.5.1. *Swendsen-Wang multiple-cluster algorithm*

According to (31) a cluster update sweep consists of two alternating steps. One first updates the bond variables n_{ij} for given spins and then updates the spins σ_i for a given bond configuration:

(1) If $\sigma_i \neq \sigma_j$, set $n_{ij} = 0$, or if $\sigma_i = \sigma_j$, assign values $n_{ij} = 1$ and 0 with probability p and $1 - p$, respectively, cf. Fig. 4.
(2) Identify *stochastic* clusters of spins that are connected by "active" bonds ($n_{ij} = 1$).
(3) Draw a random value ± 1 independently for each cluster (including one-site clusters), which is then assigned to all spins in a cluster.

Technically the cluster identification part is the most complicated step, but there are efficient algorithms from percolation theory available for this task.[44–47]

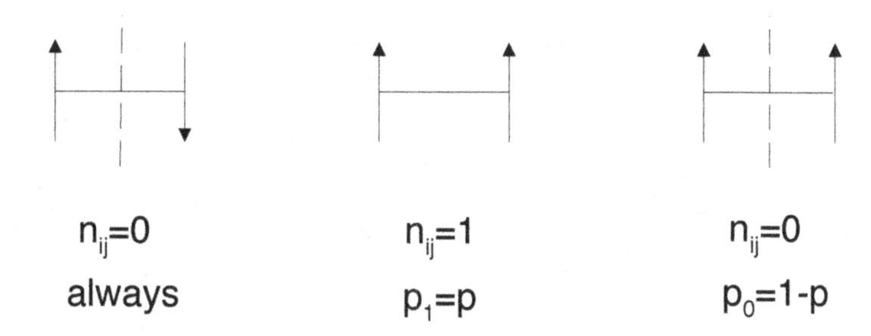

Fig. 4. Illustration of the bond variable update. The bond between unlike spins is always "deleted" as indicated by the dashed line. A bond between like spins is only "active" with probability $p = 1 - \exp(-2\beta)$. Only at zero temperature ($\beta \longrightarrow \infty$) stochastic and geometrical clusters coincide.

Notice the difference between the just defined stochastic clusters and *geometrical* clusters whose boundaries are defined by drawing lines through bonds between unlike spins. In fact, since in the stochastic cluster definition bonds between like spins are "deleted" with probability $p_0 = 1 - p = \exp(-2\beta)$, stochastic clusters are on the average smaller than geometrical clusters. Only at zero temperature ($\beta \longrightarrow \infty$) p_0 approaches zero and the two cluster definitions coincide. It is worth pointing out that at least for the 2D Ising and more generally 2D Potts models the *geometrical* clusters also do encode critical properties – albeit those of different but related (tricritical) models.[48]

As described above, the cluster algorithm is referred to as Swendsen-Wang (SW) or multiple-cluster update.[39] The distinguishing point is that the *whole* lattice is decomposed into stochastic clusters whose spins are assigned a random value $+1$ or -1. In one sweep one thus attempts to update all spins of the lattice.

2.5.2. *Wolff single-cluster algorithm*

In the single-cluster algorithm of Wolff[49] one constructs only the one cluster connected with a randomly chosen site and then flips all spins of this cluster. Typical configuration plots before and after the cluster flip are shown in Fig. 5, which also illustrates the difference between *stochastic* and *geometrical* clusters mentioned in the last paragraph: The upper right plot clearly shows that, due to the randomly distributed inactive bonds between like spins, the stochastic cluster is much smaller than the underlying black geometrical cluster which connects *all* neighboring like spins.

In the single-cluster variant some care is necessary with the definition of the unit of "time" since the number of flipped spins varies from cluster to cluster. It

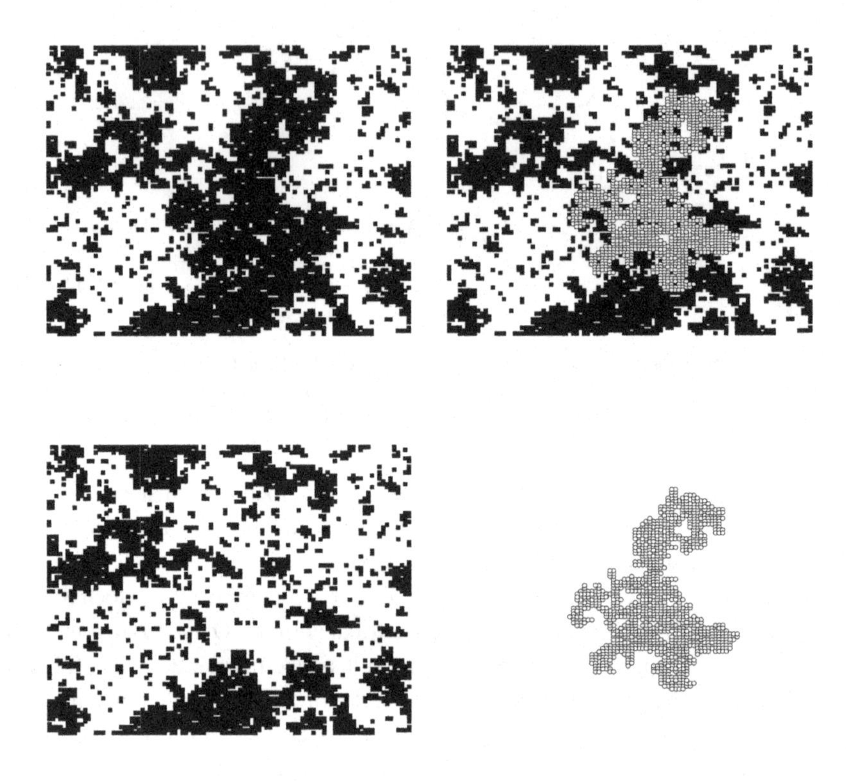

Fig. 5. Illustration of the Wolff single-cluster update for the 2D Ising model on a 100×100 square lattice at $0.97 \times \beta_c$. *Upper left:* Initial configuration. *Upper right:* The stochastic cluster is marked. Note how it is embedded into the larger geometric cluster connecting *all* neighboring like (black) spins. *Lower left:* Final configuration after flipping the spins in the cluster. *Lower right:* The flipped cluster.

also depends crucially on temperature since the average cluster size automatically adapts to the correlation length. With $\langle |C| \rangle$ denoting the average cluster size, a sweep is usually defined to consist of $V/\langle |C| \rangle$ single cluster steps, assuring that on the average V spins are flipped in one sweep. With this definition, autocorrelation times are directly comparable with results from the Swendsen-Wang or Metropolis algorithm. Apart from being somewhat easier to program, Wolff's single-cluster variant is usually more efficient than the Swendsen-Wang multiple-cluster algorithm, especially in 3D. The reason is that with the single-cluster method, on the average, larger clusters are flipped.

2.5.3. *Embedded cluster algorithm*

While for q-state Potts models[50] with Hamiltonian $\mathcal{H}_{\text{Potts}} = -\sum_{\langle ij \rangle} \delta_{\sigma_i \sigma_j}$, $\sigma_i = 1, \dots, q$, the generalization of (29)–(32) is straightforward (because also the Potts spin-spin interaction $\delta_{\sigma_i \sigma_j}$ contributes only two possible values to the energy, as in the Ising model), for O(n) spin models with $n \geq 2$ defined by the Hamiltonian

$$\mathcal{H}_{O(n)} = -J \sum_{\langle ij \rangle} \vec{\sigma}_i \cdot \vec{\sigma}_j \ , \quad \vec{\sigma}_i = (\sigma_{i,1}, \sigma_{i,2}, \dots, \sigma_{i,n}) \ , \quad |\vec{\sigma}_i| = 1 \ , \tag{33}$$

one needs a new strategy.[49,51–53] The basic idea is to isolate Ising degrees of freedom by projecting the spins $\vec{\sigma}_i$ onto a randomly chosen unit vector $\vec{r}$,

$$\vec{\sigma}_i = \vec{\sigma}_i^{\parallel} + \vec{\sigma}_i^{\perp} \ , \quad \vec{\sigma}_i^{\parallel} = \epsilon_i \, |\vec{\sigma}_i \cdot \vec{r}| \, \vec{r} \ , \quad \epsilon_i = \text{sign}(\vec{\sigma}_i \cdot \vec{r}) \ . \tag{34}$$

Inserting this in (33) one ends up with an effective Hamiltonian

$$\mathcal{H}_{O(n)} = -\sum_{\langle ij \rangle} J_{ij} \epsilon_i \epsilon_j + \text{const} \ , \tag{35}$$

with positive random couplings $J_{ij} = J|\vec{\sigma}_i \cdot \vec{r}||\vec{\sigma}_j \cdot \vec{r}| \geq 0$, whose Ising degrees of freedom ϵ_i can be updated with a cluster algorithm as described above.

2.5.4. *Performance of cluster algorithms*

Beside the generalization to O(n)-symmetric spin models, cluster update algorithms have also been constructed for many other models.[36] Close to criticality, they clearly outperform local algorithms with dynamical critical exponent $z \approx 2$, that is, for both cluster variants much smaller values of z have been obtained in 2D and 3D.[36,54–59] For a rigorous lower bound for the autocorrelation time of the Swenden-Wang algorithm, see Ref. 60. In 2D, the efficiencies of Swendsen-Wang and Wolff cluster updates are comparable, whereas in 3D, the Wolff update is favourable.

2.5.5. *Improved estimators*

The intimate relationship of cluster algorithms with the correlated percolation representation of Fortuin and Kasteleyn[40–43] leads to another quite important improvement which is not directly related with the dynamical properties discussed so far. Within the percolation picture, it is quite natural to introduce alternative estimators ("measurement prescriptions") for most standard quantities which turn out to be so-called "improved estimators". By this one means measurement prescriptions that yield the same expectation value as the standard ones but have a smaller statistical variance which helps to reduce the statistical errors.

Suppose we want to estimate the expectation value $\langle \mathcal{O} \rangle$ of an observable $\mathcal{O}$. Then any estimator $\hat{\mathcal{O}}$ satisfying $\langle \hat{\mathcal{O}} \rangle = \langle \mathcal{O} \rangle$ is permissible. This does not determine $\hat{\mathcal{O}}$ uniquely since there are infinitely many other possible choices, $\hat{\mathcal{O}}' = \hat{\mathcal{O}} + \hat{\mathcal{X}}$, as long as the added estimator $\hat{\mathcal{X}}$ has zero expectation, $\langle \hat{\mathcal{X}} \rangle = 0$. The variance of the estimator $\hat{\mathcal{O}}'$, however, can be quite different and is not necessarily related to any physical quantity (contrary to the standard mean-value estimator of the energy, for instance, whose variance is proportional to the specific heat). It is exactly this freedom in the choice of $\hat{\mathcal{O}}$ which allows the construction of improved estimators.

For the single-cluster algorithm an improved "cluster estimator" for the spin-spin correlation function in the high-temperature phase, $G(\vec{x}_i - \vec{x}_j) \equiv \langle \vec{\sigma}_i \cdot \vec{\sigma}_j \rangle$, is given by[53]

$$\hat{G}(\vec{x}_i - \vec{x}_j) = n \frac{V}{|C|} \vec{r} \cdot \vec{\sigma}_i \, \vec{r} \cdot \vec{\sigma}_j \, \Theta_C(\vec{x}_i) \Theta_C(\vec{x}_j) \ , \tag{36}$$

where $\vec{r}$ is the normal of the mirror plane used in the construction of the cluster of size $|C|$ and $\Theta_C(\vec{x})$ is its characteristic function (=1 if $\vec{x} \in C$ and 0 otherwise). In the Ising case ($n = 1$), this simplifies to

$$\hat{G}(\vec{x}_i - \vec{x}_j) = \frac{V}{|C|} \Theta_C(\vec{x}_i) \Theta_C(\vec{x}_j) \ , \tag{37}$$

i.e., to the test whether the two sites $\vec{x}_i$ and $\vec{x}_j$ belong to same stochastic cluster or not. Only in the former case, the average over clusters is incremented by one, otherwise nothing is added. This implies that $\hat{G}(\vec{x}_i - \vec{x}_j)$ is strictly positive which is not the case for the standard estimator $\vec{\sigma}_i \cdot \vec{\sigma}_j$, where ± 1 contributions have to average to a positive value. It is therefore at least intuitively clear that the cluster (or percolation) estimator has a smaller variance and is thus indeed an improved estimator, in particular for large separations $|\vec{x}_i - \vec{x}_j|$. For the Fourier transform, $\tilde{G}(\vec{k}) = \sum_{\vec{x}} G(\vec{x}) \exp(-i\vec{k} \cdot \vec{x})$, Eq. (36) implies the improved estimator

$$\hat{\tilde{G}}(\vec{k}) = \frac{n}{|C|} \left[\left(\sum_{i \in C} \vec{r} \cdot \vec{\sigma}_i \cos \vec{k}\vec{x}_i \right)^2 + \left(\sum_{i \in C} \vec{r} \cdot \vec{\sigma}_i \sin \vec{k}\vec{x}_i \right)^2 \right] \ , \tag{38}$$

which, for $\vec{k} = \vec{0}$, reduces to an improved estimator for the susceptibility $\chi' = \beta V \langle m^2 \rangle$ in the high-temperature phase,

$$\hat{\tilde{G}}(\vec{0}) = \hat{\chi}'/\beta = \frac{n}{|C|} \left(\sum_{i \in C} \vec{r} \cdot \vec{\sigma}_i \right)^2 \ . \tag{39}$$

For the Ising model ($n = 1$) this reduces to $\chi'/\beta = \langle |C| \rangle$, i.e., the improved estimator of the susceptibility is just the average cluster size of the single-cluster update algorithm. For the XY ($n = 2$) and Heisenberg ($n = 3$) models one finds empirically that in two as well as in three dimensions $\langle |C| \rangle \approx 0.81\chi'/\beta$ for $n = 2$ (Refs. 51,58) and $\langle |C| \rangle \approx 0.75\chi'/\beta$ for $n = 3$ (Refs. 53,59), respectively.

Close to criticality, the average cluster size becomes large, growing in a finite system of linear length L (cf. Sect. 6) as $\chi' \propto L^{\gamma/\nu} \simeq L^2$, since $\gamma/\nu = 2 - \eta$ with η usually small, and the advantage of cluster estimators diminishes. In fact, in particular for short-range quantities such as the energy (the next-neighbor correlation) it may even degenerate into a "deproved" or "deteriorated" estimator, while long-range quantities such as $G(\vec{x}_i - \vec{x}_j)$ for large distances $|\vec{x}_i - \vec{x}_j|$ usually still profit from it. A significant reduction of variance by means of the estimators (36)–(39) can, however, always be expected outside the critical region where the average cluster size is small compared to the volume of the system.

3. Statistical Analysis of Monte Carlo Data

3.1. *Statistical errors and autocorrelation times*

3.1.1. *Estimators*

When discussing the importance sampling idea in Sect. 2.2 we already saw in Eq. (9) that within Markov chain Monte Carlo simulations, the expectation value $\langle \mathcal{O} \rangle$ of some quantity $\mathcal{O}$, for instance the energy, can be estimated as arithmetic mean,

$$\langle \mathcal{O} \rangle = \sum_{\sigma} \mathcal{O}(\sigma) P^{\text{eq}}(\sigma) \approx \overline{\mathcal{O}} = \frac{1}{N} \sum_{k=1}^{N} \mathcal{O}_k \ , \tag{40}$$

where the "measurement" $\mathcal{O}_k = \mathcal{O}(\sigma^{(k)})$ is obtained from the kth microstate $\sigma^{(k)}$ and N is the number of measurement sweeps. Of course, this is only valid after a sufficiently long thermalization period without measurements, which is needed to equilibrate the system after starting the Markov chain in an arbitrarily chosen initial configuration.

Conceptually it is important to distinguish between the expectation value $\langle \mathcal{O} \rangle$, an ordinary number representing the exact result (which is usually unknown, of course), and the mean value $\overline{\mathcal{O}}$, which is a so-called *estimator* of the former. In contrast to $\langle \mathcal{O} \rangle$, the estimator $\overline{\mathcal{O}}$ is a *random* variable which for finite N fluctuates around the theoretically expected value. Certainly, from a single Monte Carlo simulation with N measurements, we obtain only a single number for $\overline{\mathcal{O}}$ at the end of the day. For estimating the statistical uncertainty due to the fluctuations, i.e., the statistical error, it seems at first sight that one would have to repeat the whole simulation many times. Fortunately, this is not so because one can express the variance of $\overline{\mathcal{O}}$,

$$\sigma^2_{\overline{\mathcal{O}}} = \langle [\overline{\mathcal{O}} - \langle \overline{\mathcal{O}} \rangle]^2 \rangle = \langle \overline{\mathcal{O}}^2 \rangle - \langle \overline{\mathcal{O}} \rangle^2 \ , \tag{41}$$

in terms of the statistical properties of the individual measurements $\mathcal{O}_k, k = 1, \ldots, N$, of a single Monte Carlo run.

3.1.2. *Uncorrelated measurements*

Inserting (40) into (41) gives

$$\begin{aligned}\sigma^2_{\overline{\mathcal{O}}} &= \langle \overline{\mathcal{O}}^2 \rangle - \langle \overline{\mathcal{O}} \rangle^2 \\ &= \frac{1}{N^2} \sum_{k=1}^{N} \left(\langle \mathcal{O}_k^2 \rangle - \langle \mathcal{O}_k \rangle^2 \right) + \frac{1}{N^2} \sum_{k \neq l}^{N} \left(\langle \mathcal{O}_k \mathcal{O}_l \rangle - \langle \mathcal{O}_k \rangle \langle \mathcal{O}_l \rangle \right) \ ,\end{aligned} \tag{42}$$

where we have collected diagonal and off-diagonal terms. The second, off-diagonal term encodes the "temporal" correlations between measurements at "times" k and l and thus vanishes for completely uncorrelated data (which is, of course, never really the case for importance sampling Monte Carlo simulations). Assuming equilibrium, the variances $\sigma^2_{\mathcal{O}_k} = \langle \mathcal{O}_k^2 \rangle - \langle \mathcal{O}_k \rangle^2$ of individual measurements appearing in the first, diagonal term do not depend on "time" k, such that $\sigma^2_{\mathcal{O}_k} = \sigma^2_{\mathcal{O}}$ and (42) simplifies to

$$\sigma^2_{\overline{\mathcal{O}}} = \sigma^2_{\mathcal{O}} / N \ . \tag{43}$$

Whatever form the distribution $\mathcal{P}(\mathcal{O}_k)$ assumes (which, in fact, is often close to Gaussian because the $\mathcal{O}_k$ are usually already lattice averages over many degrees of freedom), by the central limit theorem the distribution of the mean value is Gaussian, at least for weakly correlated data in the asymptotic limit of large N. The variance of the mean, $\sigma^2_{\overline{\mathcal{O}}}$, is the squared width of this (N dependent) distribution which is usually taken as the "one-sigma" squared error, $\epsilon^2_{\overline{\mathcal{O}}} \equiv \sigma^2_{\overline{\mathcal{O}}}$, and quoted together with the mean value $\overline{\mathcal{O}}$. Under the assumption of a Gaussian distribution for the mean, the interpretation is that about 68% of all simulations under the same conditions would yield a mean value in the range $[\langle \mathcal{O} \rangle - \sigma_{\overline{\mathcal{O}}}, \langle \mathcal{O} \rangle + \sigma_{\overline{\mathcal{O}}}]$.[61] For a "two-sigma" interval which also is sometimes used, this percentage goes up to about 95.4%, and for a "three-sigma" interval which is rarely quoted, the confidence level is higher than 99.7%.

3.1.3. *Correlated measurements and autocorrelation times*

For correlated data the off-diagonal term in (42) does not vanish and things become more involved.[62–65] Using the symmetry $k \leftrightarrow l$ to rewrite the summation

$\sum_{k\neq l}^{N}$ as $2\sum_{k=1}^{N}\sum_{l=k+1}^{N}$, reordering the summation, and using time-translation invariance in equilibrium, one obtains[66]

$$\sigma_{\overline{\mathcal{O}}}^2 = \frac{1}{N}\left[\sigma_{\mathcal{O}}^2 + 2\sum_{k=1}^{N}\Big(\langle\mathcal{O}_1\mathcal{O}_{1+k}\rangle - \langle\mathcal{O}_1\rangle\langle\mathcal{O}_{1+k}\rangle\Big)\left(1-\frac{k}{N}\right)\right] , \tag{44}$$

where, due to the last factor $(1-k/N)$, the $k = N$ term may be trivially kept in the summation. Factoring out $\sigma_{\mathcal{O}}^2$, this can be written as

$$\sigma_{\overline{\mathcal{O}}}^2 = \frac{\sigma_{\mathcal{O}}^2}{N}\,2\tau_{\mathcal{O},\mathrm{int}} , \tag{45}$$

where we have introduced the *integrated* autocorrelation time

$$\tau_{\mathcal{O},\mathrm{int}} = \frac{1}{2} + \sum_{k=1}^{N} A(k)\left(1-\frac{k}{N}\right) , \tag{46}$$

with

$$A(k) \equiv \frac{\langle\mathcal{O}_1\mathcal{O}_{1+k}\rangle - \langle\mathcal{O}_1\rangle\langle\mathcal{O}_{1+k}\rangle}{\sigma_{\mathcal{O}}^2} \overset{k\to\infty}{\longrightarrow} a e^{-k/\tau_{\mathcal{O},\mathrm{exp}}} \tag{47}$$

being the normalized autocorrelation function ($A(0) = 1$). In any meaningful simulation study one chooses $N \gg \tau_{\mathcal{O},\mathrm{exp}}$, so that $A(k)$ is already exponentially small before the correction term $(1-k/N)$ in (46) becomes important. It is therefore often omitted for simplicity.

As far as the accuracy of Monte Carlo data is concerned, the important point of Eq. (45) is that due to temporal correlations of the measurements the statistical error $\epsilon_{\overline{\mathcal{O}}} \equiv \sqrt{\sigma_{\overline{\mathcal{O}}}^2}$ on the Monte Carlo estimator $\overline{\mathcal{O}}$ is enhanced by a factor of $\sqrt{2\tau_{\mathcal{O},\mathrm{int}}}$. This can be rephrased by writing the statistical error similar to the uncorrelated case as $\epsilon_{\overline{\mathcal{O}}} = \sqrt{\sigma_{\mathcal{O}}^2/N_{\mathrm{eff}}}$, but now with a parameter

$$N_{\mathrm{eff}} = N/2\tau_{\mathcal{O},\mathrm{int}} \leq N , \tag{48}$$

describing the *effective* statistics. This shows more clearly that only every $2\tau_{\mathcal{O},\mathrm{int}}$ iterations the measurements are approximately uncorrelated and gives a better idea of the relevant effective size of the statistical sample. In view of the scaling behaviour of the autocorrelation time in (25), (26) or (28), it is obvious that without extra care this effective sample size may become very small close to a continuous or first-order phase transition, respectively.

3.1.4. *Bias*

A too small effective sample size does not only affect the error bars, but for some quantities even the mean values can be severely underestimated. This happens for so-called *biased* estimators, as is for instance the case for the specific heat and susceptibility. The specific heat can be computed as $C = \beta^2 V \left(\langle e^2\rangle - \langle e\rangle^2\right) = \beta^2 V \sigma_e^2$, with the standard estimator for the variance

$$\hat{\sigma}^2_{\mathcal{O}} = \overline{\mathcal{O}^2} - \overline{\mathcal{O}}^2 = \overline{(\mathcal{O} - \overline{\mathcal{O}})^2} = \frac{1}{N}\sum_{k=1}^{N}\left(\mathcal{O}_k - \overline{\mathcal{O}}\right)^2 . \tag{49}$$

Subtracting and adding $\langle\overline{\mathcal{O}}\rangle^2$, one finds for the *expected* value of $\hat{\sigma}^2_{\mathcal{O}}$,

$$\langle\hat{\sigma}^2_{\mathcal{O}}\rangle = \langle\overline{\mathcal{O}^2} - \overline{\mathcal{O}}^2\rangle = \left(\langle\overline{\mathcal{O}^2}\rangle - \langle\overline{\mathcal{O}}\rangle^2\right) - \left(\langle\overline{\mathcal{O}}^2\rangle - \langle\overline{\mathcal{O}}\rangle^2\right) = \sigma^2_{\mathcal{O}} + \sigma^2_{\overline{\mathcal{O}}} . \tag{50}$$

Using (45) this gives

$$\langle\hat{\sigma}^2_{\mathcal{O}}\rangle = \sigma^2_{\mathcal{O}}\left(1 - \frac{2\tau_{\mathcal{O},\mathrm{int}}}{N}\right) = \sigma^2_{\mathcal{O}}\left(1 - \frac{1}{N_{\mathrm{eff}}}\right) \neq \sigma^2_{\mathcal{O}} . \tag{51}$$

The estimator $\hat{\sigma}^2_{\mathcal{O}}$ in (49) thus systematically underestimates the true value by a term of the order of $\tau_{\mathcal{O},\mathrm{int}}/N$. Such an estimator is called *weakly biased* ("weakly" because the statistical error $\propto 1/\sqrt{N}$ is asymptotically larger than the systematic bias; for medium or small N, however, also prefactors need to be carefully considered).

We thus see that for large autocorrelation times, the bias may be quite large. Since for local update algorithms $\tau_{\mathcal{O},\mathrm{int}}$ scales quite strongly with the system size, some care is necessary when choosing the run time N. Otherwise the system-size dependence of the specific heat or susceptibility may be systematically influenced by temporal correlations.[67] Any serious simulation should therefore provide at least a rough order-of-magnitude estimate of autocorrelation times.

3.1.5. *Numerical estimation of autocorrelation times*

The above considerations show that not only for the error estimation but also for the computation of static quantities themselves, it is important to have control over autocorrelations. Unfortunately, it is very difficult to give reliable *a priori* estimates, and an accurate numerical analysis is often too time consuming. As a rough estimate it is about ten times harder to get precise information on dynamic quantities than on static quantities like critical exponents. Similar to the estimator (49) for the variance a (weakly biased) estimator $\hat{A}(k)$ for the autocorrelation function is obtained by replacing in (47) the expectation values (ordinary

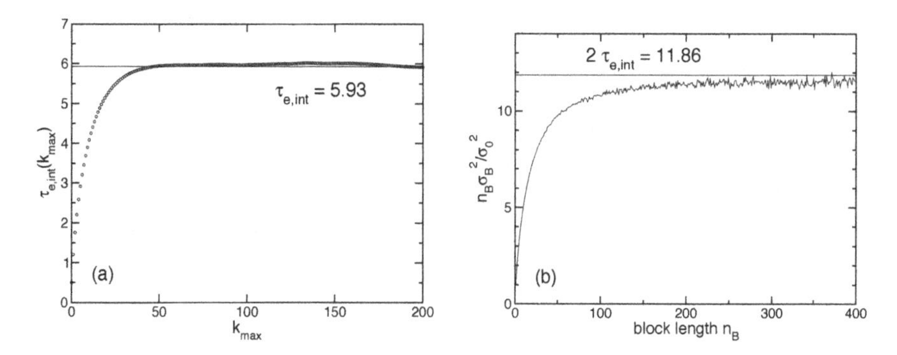

Fig. 6. (a) Integrated autocorrelation time approaching $\tau_{e,\mathrm{int}} \approx 5.93$ for large upper cutoff k_{max} and (b) binning analysis for the energy of the 2D Ising model on a 16×16 lattice at β_c, using the same data as in Fig. 3. The horizontal line in (b) shows $2\tau_{e,\mathrm{int}}$ with $\tau_{e,\mathrm{int}}$ read off from (a).

numbers) by mean values (random variables), e.g., $\langle \mathcal{O}_1 \mathcal{O}_{1+k} \rangle$ by $\overline{\mathcal{O}_1 \mathcal{O}_{1+k}}$. With increasing separation k the relative variance of $\hat{A}(k)$ diverges rapidly. To get at least an idea of the order of magnitude of $\tau_{\mathcal{O},\mathrm{int}}$ and thus the correct error estimate (45), it is useful to record the "running" autocorrelation time estimator

$$\hat{\tau}_{\mathcal{O},\mathrm{int}}(k_{\mathrm{max}}) = \frac{1}{2} + \sum_{k=1}^{k_{\mathrm{max}}} \hat{A}(k) \ , \tag{52}$$

which approaches $\tau_{\mathcal{O},\mathrm{int}}$ in the limit of large k_{max} where, however, the statistical error rapidly increases. As an example, Fig. 6(a) shows results for the 2D Ising model from an analysis of the same raw data as in Fig. 3.

As a compromise between systematic and statistical errors, an often employed procedure is to determine the upper limit k_{max} self-consistently by cutting off the summation once $k_{\mathrm{max}} \geq 6\hat{\tau}_{\mathcal{O},\mathrm{int}}(k_{\mathrm{max}})$, where $A(k) \approx e^{-6} \approx 10^{-3}$. In this case an *a priori* error estimate is available,[34,35,63]

$$\epsilon_{\tau_{\mathcal{O},\mathrm{int}}} = \tau_{\mathcal{O},\mathrm{int}} \sqrt{\frac{2(2k_{\mathrm{max}}+1)}{N}} \approx \tau_{\mathcal{O},\mathrm{int}} \sqrt{\frac{12}{N_{\mathrm{eff}}}} \ . \tag{53}$$

For a 5% relative accuracy one thus needs at least $N_{\mathrm{eff}} \approx 5\,000$ or $N \approx 10\,000\,\tau_{\mathcal{O},\mathrm{int}}$ measurements. For an order of magnitude estimate consider the 2D Ising model on a square lattice with $L = 100$ simulated with a local update algorithm. Close to criticality, the integrated autocorrelation time for this example is of the order of $L^z \approx L^2 \approx 100^2$ (ignoring an unknown prefactor of "order unity" which depends on the considered quantity), implying $N \approx 10^8$. Since in each sweep L^2 spins have to be updated and assuming that each spin update takes about 0.1 μsec, we end up with a total time estimate of about 10^5 seconds ≈ 1 CPU-day to achieve this accuracy.

An alternative is to approximate the tail end of $A(k)$ by a single exponential as in (24). Summing up the small k part exactly, one finds[68]

$$\tau_{\mathcal{O},\text{int}}(k_{\text{max}}) = \tau_{\mathcal{O},\text{int}} - ce^{-k_{\text{max}}/\tau_{\mathcal{O},\text{exp}}} \;, \tag{54}$$

where c is a constant. The latter expression may be used for a numerical estimate of both the exponential and integrated autocorrelation times.[68]

3.2. *Binning analysis*

It should be clear by now that ignoring autocorrelation effects can lead to severe underestimates of statistical errors. Applying the full machinery of autocorrelation analyses discussed above, however, is often too cumbersome. On a day by day basis the following binning analysis is much more convenient (though somewhat less accurate). By grouping the N original time-series data into N_B non-overlapping bins or blocks of length n_B (such that[f] $N = N_B n_B$), one forms a new, shorter time series of block averages,

$$\mathcal{O}_j^{(B)} \equiv \frac{1}{n_B} \sum_{i=1}^{n_B} \mathcal{O}_{(j-1)n_B+i} \;, \qquad j = 1, \ldots, N_B \;, \tag{55}$$

which by choosing the block length $n_B \gg \tau$ are almost uncorrelated and can thus be analyzed by standard means. The mean value over all block averages obviously satisfies $\overline{\mathcal{O}^{(B)}} = \overline{\mathcal{O}}$ and their variance can be computed according to the standard (unbiased) estimator, leading to the squared statistical error of the mean value,

$$\epsilon^2_{\overline{\mathcal{O}}} \equiv \sigma^2_{\overline{\mathcal{O}}} = \sigma^2_B/N_B = \frac{1}{N_B(N_B-1)} \sum_{j=1}^{N_B} (\mathcal{O}_j^{(B)} - \overline{\mathcal{O}^{(B)}})^2 \;. \tag{56}$$

By comparing with (45) we see that $\sigma^2_B/N_B = 2\tau_{\mathcal{O},\text{int}}\sigma^2_{\mathcal{O}}/N$. Recalling the definition of the block length $n_B = N/N_B$, this shows that one may also use

$$2\tau_{\mathcal{O},\text{int}} = n_B \sigma^2_B/\sigma^2_{\mathcal{O}} \tag{57}$$

for the estimation of $\tau_{\mathcal{O},\text{int}}$. This is demonstrated in Fig. 6(b). Estimates of $\tau_{\mathcal{O},\text{int}}$ obtained in this way are often referred to as "blocking τ" or "binning τ".

A simple toy model (bivariate time series), where the behaviour of the "blocking τ" and also of $\tau_{\mathcal{O},\text{int}}(k_{\text{max}})$ for finite n_B resp. k_{max} can be worked out exactly, is discussed in Ref. 26. These analytic formulas are very useful for validating the computer implementations.

[f] Here we assume that N was chosen cleverly. Otherwise one has to discard some of the data and redefine N.

3.3. *Jackknife analysis*

Even if the data are completely uncorrelated in time, one still has to handle the problem of error estimation for quantities that are not "directly" measured in the simulation but are computed as a non-linear combination of "basic" observables such as $\langle \mathcal{O} \rangle^2$ or $\langle \mathcal{O}_1 \rangle / \langle \mathcal{O}_2 \rangle$. This problem can either be solved by error propagation or by using the Jackknife method,[69,70] where instead of considering rather small blocks of length n_B and their fluctuations as in the binning analysis, one forms N_B large Jackknife blocks $\mathcal{O}_j^{(J)}$ containing all data but the jth block of the previous binning method,

$$\mathcal{O}_j^{(J)} = \frac{N\overline{\mathcal{O}} - n_B \mathcal{O}_j^{(B)}}{N - n_B} \; , \qquad j = 1, \ldots, N_B \; , \tag{58}$$

cf. the schematic sketch in Fig. 7. Each of the Jackknife blocks thus consists of $N - n_B = N(1 - 1/N_B)$ data, i.e., it contains almost as many data as the original time series. When non-linear combinations of basic variables are estimated, the bias is hence comparable to that of the total data set (typically $1/(N - n_B)$ compared to $1/N$). The N_B Jackknife blocks are, of course, trivially correlated because one and the same original data is re-used in $N_B - 1$ different Jackknife blocks. This trivial correlation caused by re-using the original data over and over again has nothing to do with temporal correlations. As a consequence, the Jacknife block variance σ_J^2 will be much smaller than the variance estimated in the binning method. Because of the trivial nature of the correlations, however, this reduction can be corrected by multiplying σ_J^2 with a factor $(N_B - 1)^2$, leading to

$$\epsilon_{\overline{\mathcal{O}}}^2 \equiv \sigma_{\overline{\mathcal{O}}}^2 = \frac{N_B - 1}{N_B} \sum_{j=1}^{N_B} (\mathcal{O}_j^{(J)} - \overline{\mathcal{O}^{(J)}})^2 \; . \tag{59}$$

To summarize this section, any realization of a Markov chain Monte Carlo update algorithm is characterised by autocorrelation times which enter directly into the statistical errors of Monte Carlo estimates. Since temporal correlations always increase the statistical errors, it is thus a very important issue to develop Monte Carlo update algorithms that keep autocorrelation times as small as possible. This is the reason why cluster and other non-local algorithms are so important.

4. Reweighting Techniques

The physics underlying reweighting techniques[71,72] is extremely simple and the basic idea has been known since long (see the list of references in Ref. 72), but

Fig. 7. Sketch of the organization of Jackknife blocks. The grey part of the N data points is used for calculating the total and the Jackknife block averages. The white blocks enter into the more conventional binning analysis using non-overlapping blocks.

their power in practice has been realized only relatively late in 1988. The important observation by Ferrenberg and Swendsen[71,72] was that the best performance is achieved *near* criticality where histograms are usually broad. In this sense reweighting techniques are complementary to improved estimators, which usually perform best *off* criticality.

4.1. *Single-histogram technique*

The single-histogram reweighting technique[71] is based on the following very simple observation. Denoting the number of states (spin configurations) that have the same energy $e = E/V$ by $\Omega(e)$, the partition function at the simulation point $\beta_0 = 1/k_B T_0$ can always be written as[g]

$$Z(\beta_0) = \sum_{\sigma} e^{-\beta_0 \mathcal{H}(\sigma)} = \sum_{e} \Omega(e) e^{-\beta_0 E} \propto \sum_{e} P_{\beta_0}(e) \ , \tag{60}$$

where we have introduced the unnormalized energy histogram (density)

$$P_{\beta_0}(e) \propto \Omega(e) e^{-\beta_0 E} \ . \tag{61}$$

If we would normalize $P_{\beta_0}(e)$ to unit area, the r.h.s. would have to be divided by $\sum_e P_{\beta_0}(e) = Z(\beta_0)$, but the normalization will be unimportant in what follows. Let us assume we have performed a Monte Carlo simulation at inverse temperature

[g]For simplicity we consider here only models with *discrete* energies. If the energy varies continuously, sums have to be replaced by integrals, etc. Also lattice size dependences are suppressed to keep the notation short.

β_0 and thus know $P_{\beta_0}(e)$. It is then easy to see that

$$P_\beta(e) \propto \Omega(e)e^{-\beta E} = \Omega(e)e^{-\beta_0 E}e^{-(\beta-\beta_0)E} \propto P_{\beta_0}(e)e^{-(\beta-\beta_0)E} , \quad (62)$$

i.e., the histogram at any point β can be derived, in principle, by *reweighting* the simulated histogram at β_0 with the exponential factor $\exp[-(\beta - \beta_0)E]$. Notice that in reweighted expectation values,

$$\langle f(e)\rangle(\beta) = \sum_e f(e)P_\beta(e) / \sum_e P_\beta(e) , \quad (63)$$

the normalization of $P_\beta(e)$ indeed cancels. This gives for instance the energy $\langle e\rangle(\beta)$ and the specific heat $C(\beta) = \beta^2 V[\langle e^2\rangle(\beta) - \langle e\rangle(\beta)^2]$, in principle, as a continuous function of β from a single Monte Carlo simulation at β_0, where $V = L^D$ is the system size.

As an example of this reweighting procedure, using actual Swendsen-Wang cluster simulation data (with 5000 sweeps for equilibration and 50 000 sweeps for measurements) of the 2D Ising model at $\beta_0 = \beta_c = \ln(1+\sqrt{2})/2 = 0.440\,686\ldots$ on a 16×16 lattice with periodic boundary conditions, the reweighted data points for the specific heat $C(\beta)$ are shown in Fig. 8(a) and compared with the continuous curve obtained from the exact Kaufman solution[73,74] for finite $L_x \times L_y$ lattices. Note that the location of the peak maximum is slightly displaced from the infinite-volume transition point β_c due to the rounding and shifting of $C(\beta)$ caused by finite-size effects discussed in more detail in Sect. 6. This comparison clearly demonstrates that, in practice, the β-range over which reweighting can be trusted is limited. The reason for this limitation are unavoidable statistical errors in the numerical determination of P_{β_0} using a Monte Carlo simulation. In the tails of the histograms the relative statistical errors are largest, and the tails are exactly the regions that contribute most when multiplying $P_{\beta_0}(e)$ with the exponential reweighting factor to obtain $P_\beta(e)$ for β-values far off the simulation point β_0. This is illustrated in Fig. 8(b) where the simulated histogram at $\beta_0 = \beta_c$ is shown together with the reweighted histograms at $\beta = 0.375 \approx \beta_0 - 0.065$ and $\beta = 0.475 \approx \beta_0 + 0.035$, respectively. For the 2D Ising model the quality of the reweighted histograms can be judged by comparing with the curves obtained from Beale's[75] exact expression for $\Omega(e)$.

4.1.1. *Reweighting range*

As a rule of thumb, the range over which reweighting should produce accurate results can be estimated by requiring that the peak location of the reweighted histogram should not exceed the energy value at which the input histogram had decreased to about one half or one third of its maximum value. In most applications

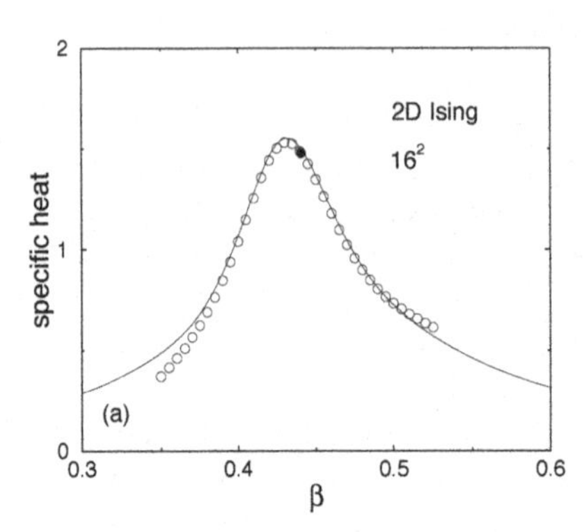

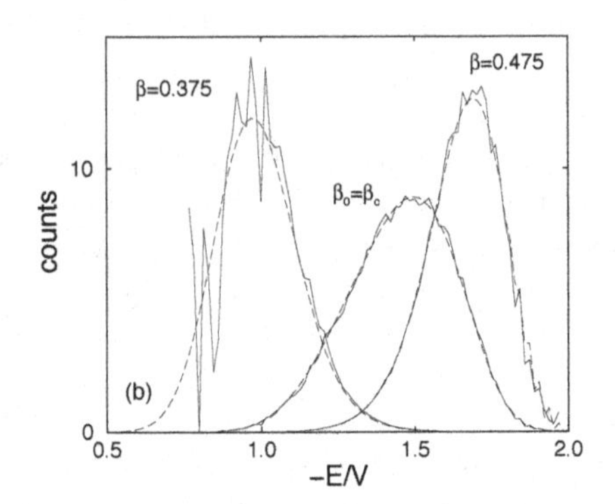

Fig. 8. (a) The specific heat of the 2D Ising model on a 16×16 square lattice computed by reweighting from a single Monte Carlo simulation at $\beta_0 = \beta_c$, marked by the filled data symbol. The continuous line shows for comparison the exact solution of Kaufman.[73,74] (b) The corresponding energy histogram at β_0, and reweighted to $\beta = 0.375$ and $\beta = 0.475$. The dashed lines show for comparison the exact histograms obtained from Beale's expression.[75]

this range is wide enough to locate from a single simulation, e.g., the specific-heat maximum by employing a standard maximization subroutine to the continuous function $C(\beta)$. This is by far more convenient, accurate and faster than the traditional way of performing many simulations close to the peak of $C(\beta)$ and trying to determine the maximum by splines or least-squares fits.

For an analytical estimate of the reweighting range we now require that the peak of the reweighted histogram is within the width $\langle e\rangle(T_0)\pm\Delta e(T_0)$ of the input histogram (where a Gaussian histogram would have decreased to $\exp(-1/2) \approx 0.61$ of its maximum value),

$$|\langle e\rangle(T) - \langle e\rangle(T_0)| \leq \Delta e(T_0) \ , \tag{64}$$

where we assumed that for a not too asymmetric histogram $P_{\beta_0}(e)$ the maximum location approximately coincides with $\langle e\rangle(T_0)$. Recalling that the half width Δe of a histogram is related to the specific heat via $(\Delta e)^2 \equiv \langle (e - \langle e\rangle)^2\rangle = \langle e^2\rangle - \langle e\rangle^2 = C(\beta_0)/\beta_0^2 V$ and using the Taylor expansion $\langle e\rangle(T) = \langle e\rangle(T_0) + C(T_0)(T - T_0) + \ldots$, this can be written as $C(T_0)|T - T_0| \leq T_0\sqrt{C(T_0)/V}$ or

$$\frac{|T - T_0|}{T_0} \leq \frac{1}{\sqrt{V}} \frac{1}{\sqrt{C(T_0)}} \ . \tag{65}$$

Since $C(T_0)$ is known from the input histogram this is quite a general estimate of the reweighting range. For the example in Fig. 8 with $V = 16 \times 16$, $\beta_0 = \beta_c \approx 0.44$ and $C(T_0) \approx 1.5$, this estimate yields $|\beta - \beta_0|/\beta_0 \approx |T - T_0|/T_0 \leq 0.05$, i.e., $|\beta - \beta_0| \leq 0.02$ or $0.42 \leq \beta \leq 0.46$. By comparison with the exact solution we see that this is indeed a fairly conservative estimate of the reliable reweighting range.

If we only want to know the scaling behaviour with system size $V = L^D$, we can go one step further by considering three generic cases:

i) *Off-critical*, where $C(T_0) \approx$ const, such that

$$\frac{|T - T_0|}{T_0} \propto V^{-1/2} = L^{-D/2} \ . \tag{66}$$

ii) *Critical*, where $C(T_0) \simeq a_1 + a_2 L^{\alpha/\nu}$, with a_1 and a_2 being constants, and α and ν denoting the standard critical exponents of the specific heat and correlation length, respectively. For $\alpha > 0$, the leading scaling behaviour becomes $|T - T_0|/T_0 \propto L^{-D/2} L^{-\alpha/2\nu}$. Assuming hyperscaling ($\alpha = 2 - D\nu$) to be valid, this simplifies to

$$\frac{|T - T_0|}{T_0} \propto L^{-1/\nu} \ , \tag{67}$$

i.e., the typical scaling behaviour of pseudo-transition temperatures in the finite-size scaling regime of a second-order phase transition.[76] For $\alpha < 0$, $C(T_0)$ approaches asymptotically a constant and the leading scaling behaviour of the reweighting range is as in the off-critical case.

iii) *First-order transitions*, where $C(T_0) \propto V = L^D$. This yields

$$\frac{|T - T_0|}{T_0} \propto V^{-1} = L^{-D} \ , \tag{68}$$

which is again the typical finite-size scaling behaviour of pseudo-transition temperatures close to a first-order phase transition.[38]

4.1.2. *Reweighting of non-conjugate observables*

If we also want to reweight other quantities such as the magnetization $m = \langle \mu \rangle$ we have to go one step further. The conceptually simplest way would be to store two-dimensional histograms $P_{\beta_0}(e, \mu)$ where $e = E/V$ is the energy and $\mu = \sum_i \sigma_i / V$ the magnetization. We could then proceed in close analogy to the preceding case, and even reweighting to non-zero magnetic field h would be possible, which enters via the Boltzmann factor $\exp(\beta h \sum_i \sigma_i) = \exp(\beta V h \mu)$. However, the storage requirements may be quite high (of the order of V^2), and it is often preferable to proceed in the following way. For any function $g(\mu)$, e.g.,

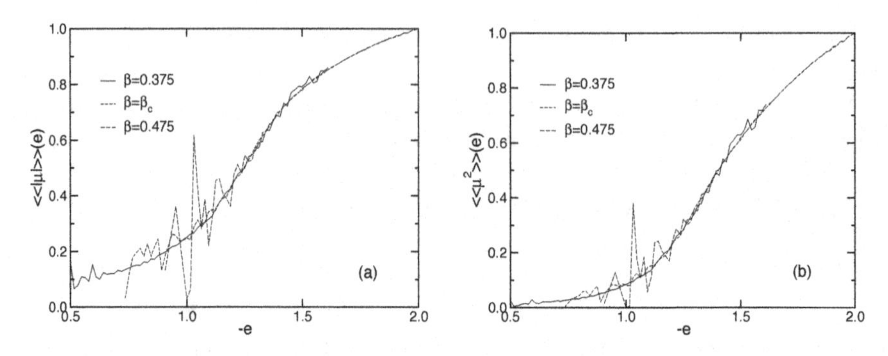

Fig. 9. Microcanonical expectation values for (a) the absolute magnetization and (b) the magnetization squared obtained from the 2D Ising model simulations shown in Fig. 8.

$g(\mu) = \mu^k$, we can write

$$\langle g(\mu)\rangle = \sum_{\sigma} g(\mu(\sigma))e^{-\beta_0 \mathcal{H}(\sigma)}/Z(\beta_0) = \sum_{e,\mu} \Omega(e,\mu)g(\mu)e^{-\beta_0 E}/Z(\beta_0)$$
$$= \sum_{e} \frac{\sum_{\mu} \Omega(e,\mu)g(\mu)}{\sum_{\mu} \Omega(e,\mu)} \sum_{\mu} \Omega(e,\mu)e^{-\beta_0 E}/Z(\beta_0) \ . \tag{69}$$

Recalling that $\sum_{\mu} \Omega(e,\mu)e^{-\beta_0 E}/Z(\beta_0) = \Omega(e)e^{-\beta_0 E}/Z(\beta_0) = P_{\beta_0}(e)$ and defining the *microcanonical* expectation value of $g(\mu)$ at fixed energy e (sometimes denoted as a "list"),

$$\langle\langle g(\mu)\rangle\rangle(e) \equiv \frac{\sum_{\mu} \Omega(e,\mu)g(\mu)}{\sum_{\mu} \Omega(e,\mu)} \ , \tag{70}$$

we arrive at

$$\langle g(\mu)\rangle = \sum_{e} \langle\langle g(\mu)\rangle\rangle(e) P_{\beta_0}(e) \ . \tag{71}$$

Identifying $\langle\langle g(\mu)\rangle\rangle(e)$ with $f(e)$ in Eq. (63), the actual reweighting procedure is precisely as before. An example for computing $\langle\langle|\mu|\rangle\rangle(e)$ and $\langle\langle\mu^2\rangle\rangle(e)$ using the data of Fig. 8 is shown in Fig. 9. Mixed quantities, e.g. $\langle e^k \mu^l\rangle$, can be treated similarly. One caveat of this method is that one has to decide beforehand which "lists" $\langle\langle g(\mu)\rangle\rangle(e)$ one wants to store during the simulation, e.g., which powers k in $\langle\langle\mu^k\rangle\rangle(e)$ are relevant.

An alternative and more flexible method is based on time series. Suppose we have performed a Monte Carlo simulation at β_0 and stored the time series of N measurements $e_1, e_2, \ldots, e_N$ and $\mu_1, \mu_2, \ldots, \mu_N$. Then the most general expectation values at another inverse temperature β can simply be obtained from

$$\langle f(e,\mu)\rangle = \sum_{i=1}^{N} f(e_i,\mu_i)e^{-(\beta-\beta_0)E_i} / \sum_{i=1}^{N} e^{-(\beta-\beta_0)E_i} \quad , \tag{72}$$

i.e., in particular all moments $\langle e^k \mu^l \rangle$ can be computed. Notice that this can also be written as

$$\langle f(e,\mu)\rangle = \langle f(e,\mu)e^{-(\beta-\beta_0)E}\rangle_0 / \langle e^{-(\beta-\beta_0)E}\rangle_0 \quad , \tag{73}$$

where the subscript 0 refers to expectation values taken at β_0. Another very important advantage of the last formulation is that it works without any systematic discretization error also for continuously distributed energies and magnetizations.

As nowadays hard-disk space is no real limitation anymore, it is advisable to store time series in any case. This guarantees the greatest flexibility in the data analysis. As far as the memory requirement of the actual reweighting code is concerned, however, the method of choice is sometimes not so clear. Using directly histograms and lists, one typically has to store about $(6-8)V$ data, while working directly with the time series one needs $2N$ computer words. The cheaper solution (also in terms of CPU time) thus obviously depends on both, the system size V and the run length N. It is hence sometimes faster to generate from the time series first histograms and the required lists and then proceed with reweighting the latter quantities.

4.2. *Multi-histogram technique*

The basic idea of the multi-histogram technique[77] can be summarized as follows:

i) Perform m Monte Carlo simulations at $\beta_1, \beta_2, \ldots, \beta_m$ with N_i, $i = 1, \ldots, m$, measurements,
ii) reweight all runs to a common reference point β_0,
iii) combine at β_0 all information by computing error weighted averages,
iv) reweight the "combined histogram" to any other β.

Since a weighted combination of several histograms enters this method it is also referred to as "weighted histogram analysis method" or "WHAM".[78,79] In fact, in chemistry and biochemistry the multi-histogram method is basically only known under this acronym.

To proceed we first note that the exact normalized energy distribution at $\beta = \beta_i$ can we written as

$$P_i(e) \equiv P_{\beta_i}(e) = \frac{\Omega(e)e^{-\beta_i E}}{Z_i} \quad , \tag{74}$$

where $Z_i \equiv Z(\beta_i)$ so that $\sum_e P_i(e) = 1$. This can be estimated by the empirical histogram $H_i(e)$ obtained from the simulation at β_i,

$$\hat{P}_i(e) = \frac{H_i(e)}{N_i} \ , \tag{75}$$

which also satisfies the normalization constraint $\sum_e \hat{P}_i(e) = 1$. Rearranging (74) and replacing the exact $P_i(e)$ by its estimator $\hat{P}_i(e)$ yields an estimator for the density of states (this corresponds to choosing the common reference point as $\beta_0 = 0$):

$$\hat{\Omega}_i(e) = Z_i e^{\beta_i E} \frac{H_i(e)}{N_i} \ . \tag{76}$$

Notice that we have introduced a subscript i to label the m estimators $\hat{\Omega}_i(e)$. The expectation value of each $\hat{\Omega}_i(e)$ should be the exact $\Omega(e)$, but being random variables their statistical properties are different as can be quantified by estimating their variance. This is simplest done by interpreting the histogram entries $H_i(e)$ as result of measuring $\mathcal{O} = \delta_{e_t,e}$ where e_t denotes the energy after the t's sweep of the simulation at β_i:

$$\frac{H_i(e)}{N_i} = \overline{\delta_{e_t,e}} = \frac{1}{N_i} \sum_{t=1}^{N_i} \delta_{e_t,e} \ . \tag{77}$$

As in (40) and (41) the expected value is $\langle H_i(e)/N_i \rangle = (1/N_i) \sum_{t=1}^{N_i} \langle \delta_{e_t,e} \rangle = P_i(e)$ and, neglecting temporal correlations for the moment,

$$\begin{aligned} \left\langle \left(\frac{H_i(e)}{N_i} \right)^2 \right\rangle &= \left\langle \frac{1}{N_i^2} \sum_{t,t'=1}^{N_i} \delta_{e_t,e} \delta_{e_{t'},e} \right\rangle \\ &= \frac{1}{N_i^2} \left[N_i(N_i - 1) \langle \delta_{e_t,e} \rangle \langle \delta_{e_{t'},e} \rangle + N_i \langle \delta_{e_t,e} \delta_{e_{t'},e} \rangle \right] \\ &= P_i(e)^2 + \frac{1}{N_i} P_i(e)[1 - P_i(e)] \ , \end{aligned} \tag{78}$$

such that

$$\sigma^2_{H_i(e)/N_i} = \left\langle \left(\frac{H_i(e)}{N_i} - \left\langle \frac{H_i(e)}{N_i} \right\rangle \right)^2 \right\rangle = \frac{1}{N_i} P_i(e)[1 - P_i(e)] \ . \tag{79}$$

For sufficiently many energy bins, the normalized probabilities $P_i(e)$ are much smaller than unity, such that the second term $[1 - P_i(e)]$ can usually be neglected. Taking autocorrelations into account, as in (45) the variance (79) would be enhanced by a factor $2\tau_{\text{int},i}(e)$. Recall that the subscript i of $\tau_{\text{int},i}(e)$ refers to the simulation point and the argument e to the energy bin. Note that the autocorrelation times of the histogram bins are usually much smaller than the autocorrelation time $\tau_{\mathrm{int},e}$ of the mean energy. For the following it is useful to define the effective statistics parameter $N_{\mathrm{eff},i}(e) = N_i/2\tau_{\mathrm{int},i}(e)$. Recalling (76), the variance of the m estimators $\hat{\Omega}_i(e)$ can then be written as

$$\sigma^2_{\hat{\Omega}_i(e)} = \frac{Z_i^2 e^{2\beta_i E}}{N_{\mathrm{eff},i}(e)} P_i(e) = \frac{Z_i e^{\beta_i E}}{N_{\mathrm{eff},i}(e)} \Omega(e) \; . \tag{80}$$

As usual the error weighted average

$$\hat{\Omega}_{\mathrm{opt}}(e) = \frac{\sum_{i=1}^m w_i(e)\hat{\Omega}_i(e)}{\sum_{i=1}^m w_i(e)} \tag{81}$$

with $w_i(e) = 1/\sigma^2_{\hat{\Omega}_i(e)}$ is an optimised estimator with minimal variance $\sigma^2_{\hat{\Omega}_{\mathrm{opt}}(e)} = 1/\sum_{i=1}^m w_i(e)$. This can be simplified to

$$\hat{\Omega}_{\mathrm{opt}}(e) = \frac{\sum_{i=1}^m H_i(e)/2\tau_{\mathrm{int},i}(e)}{\sum_{i=1}^m N_{\mathrm{eff},i}(e) Z_i^{-1} e^{-\beta_i E}} \tag{82}$$

and

$$\sigma^2_{\hat{\Omega}_{\mathrm{opt}}(e)}/\Omega^2(e) = \frac{1}{\sum_{i=1}^m \langle H_i(e)\rangle/2\tau_{\mathrm{int},i}(e)} \; . \tag{83}$$

So far the partition function values $Z_i \equiv Z(\beta_i)$ have been assumed to be exact (albeit usually unknown) parameters which are now self-consistently determined from

$$Z_j = \sum_e \hat{\Omega}_{\mathrm{opt}}(e) e^{-\beta_j E} = \sum_e \frac{\sum_{i=1}^m H_i(e)/2\tau_{\mathrm{int},i}(e)}{\sum_{i=1}^m (N_i/2\tau_{\mathrm{int},i}(e)) Z_i^{-1} e^{-\beta_i E}} e^{-\beta_j E} \; , \tag{84}$$

up to an unimportant overall constant. A good starting point for the recursion is to fix, say, $Z_1 = 1$ and use single histogram reweighting to get an estimate of $Z_2/Z_1 = \exp[-(\hat{F}_2 - \hat{F}_1)]$, where $\hat{F}_i = \beta_i F(\beta_i)$. Once Z_2 is determined, the same procedure can be applied to estimate Z_3 and so on. In the limit of infinite statistics, this would already yield the solution of (84). In realistic simulations the statistics is of course limited and the remaining recursions average this uncertainty to get a self-consistent set of Z_i. In order to work in practice, the histograms at neighboring β-values must have sufficient overlap, i.e., the spacings of the simulation points must be chosen according to the estimates (66)–(68). The issue of optimal convergence of the WHAM equations (84) has recently been discussed in detail in Ref. 80.

Multiple-histogram reweighting has been employed in a wide spectrum of applications. In many applications the influence of autocorrelations has been neglected since it is quite cumbersome to estimate the $\tau_{\mathrm{int},i}(e)$ for each of the m

simulations and *all* energy bins. For work dealing with autocorrelations in this context see, e.g., Refs. 81,82. Note that, even when ignoring the $\tau_{\mathrm{int},i}(e)$, the error weighted average in (81) does still give a correct estimator for $\Omega(e)$ – it is only no longer properly optimised. Moreover, since for each energy bin typically only the histograms at neighboring simulation points contribute significantly, the two or three $\tau_{\mathrm{int},i}(e)$ values relevant for each energy bin e are close to each other. And since an overall constant drops out of the WHAM equation (84), the influence of autocorrelations on the final result turns out to be very minor anyway.

Alternatively[59] one may also compute from each of the m independent simulations by reweighting all quantities of interest as a function of β, together with their proper statistical errors including autocorrelation effects as discussed in Sect. 3.1.3. As a result one obtains, at each β-value, m estimates, e.g. $e_1(\beta) \pm \Delta e_1, e_2(\beta) \pm \Delta e_2, \ldots, e_m(\beta) \pm \Delta e_m$, which may be optimally combined according to their error bars to give $e(\beta) \pm \Delta e$, where

$$e(\beta) = \left(\frac{e_1(\beta)}{(\Delta e_1)^2} + \frac{e_2(\beta)}{(\Delta e_2)^2} + \cdots + \frac{e_m(\beta)}{(\Delta e_m)^2} \right) (\Delta e)^2 \,, \tag{85}$$

and

$$\frac{1}{(\Delta e)^2} = \frac{1}{(\Delta e_1)^2} + \frac{1}{(\Delta e_2)^2} + \cdots + \frac{1}{(\Delta e_m)^2} \,. \tag{86}$$

Notice that by this method the average for each quantity can be individually optimised.

5. Generalized Ensemble Methods

All Monte Carlo methods described so far assumed a conventional canonical ensemble where the probability distribution of microstates is governed by a Boltzmann factor $\propto \exp(-\beta E)$. A simulation at some inverse temperature β_0 then covers a certain range of the state space but not all (recall the discussion of the reweighting range). In principle a broader range can be achieved by patching several simulations at different temperatures using the multi-histogram method. Loosely speaking generalized ensemble methods aim at replacing this "static" patching by a single simulation in an appropriately defined "generalized ensemble". The purpose of this section is to give at least a brief survey of the available methods.

5.1. *Simulated tempering*

One approach are tempering methods which may be characterized as "dynamical" multi-histogramming. Similarly to the static reweighting approach, in "simulated" as well as in "parallel" tempering one considers m simulation points $\beta_1 < \beta_2 < \cdots < \beta_m$ which here, however, are connected already during the simulation in a specific, dynamical way.

In *simulated* tempering simulations[83,84] one starts from a joint partition function (*expanded* ensemble)

$$\mathcal{Z}_{\rm ST} = \sum_{i=1}^{m} e^{g_i} \sum_{\sigma} e^{-\beta_i \mathcal{H}(\sigma)} \ , \tag{87}$$

where $g_i = \beta_i f(\beta_i)$ and the inverse temperature β is treated as an additional dynamical degree of freedom that can take the values $\beta_1, \ldots, \beta_m$. Employing a Metropolis update algorithm, a proposed move from $\beta = \beta_i$ to β_j with σ fixed is accepted with probability

$$w = \min\left\{1, \exp[-(\beta_j - \beta_i)\mathcal{H}(\sigma) + g_j - g_i]\right\} \ . \tag{88}$$

Similar to multi-histogram reweighting (and also to multicanonical simulations discussed below), the free-energy parameters g_i are *a priori* unknown and have to be adjusted iteratively. To assure a reasonable acceptance rate for the β-update moves (usually between neighboring β_i-values), the histograms at β_i and β_{i+1}, $i = 1, \ldots, m-1$, must overlap. An estimate for a suitable spacing $\Delta\beta = \beta_{i+1} - \beta_i$ of the simulation points β_i is hence immediately given by the results (66)–(68) for the reweighting range,

$$\Delta\beta \propto \begin{cases} L^{-D/2} & \text{off-critical} \ , \\ L^{-1/\nu} & \text{critical} \ , \\ L^{-D} & \text{first-order} \ . \end{cases} \tag{89}$$

Overall the simulated tempering method shows some similarities to the "avoiding rare events" variant of multicanonical simulations briefly discussed in subsection 5.3.

5.2. *Parallel tempering*

In *parallel* tempering or *replica exchange* or *multiple Markov chain* Monte Carlo simulations,[85–88] the starting point is a product of partition functions (*extended* ensemble),

$$\mathcal{Z}_{\rm PT} = \prod_{i=1}^{m} \mathcal{Z}(\beta_i) = \prod_{i=1}^{m} \sum_{\sigma_i} e^{-\beta_i \mathcal{H}(\sigma_i)} \ , \tag{90}$$

and all m systems at different simulation points $\beta_1 < \beta_2 < \cdots < \beta_m$ are simulated in parallel, using any legitimate update algorithm (Metropolis, cluster,...). This freedom in the choice of update algorithm is a big advantage of a parallel tempering simulation[88] which is a special case of the earlier replica exchange Monte Carlo method[85] proposed in the context of spin-glass simulations (to some extent the focus on this special application hides the general aspects of the method as becomes clearer in Ref. 86). After a certain number of sweeps, exchanges of the current configurations σ_i and σ_j are attempted (equivalently, the β_i may be exchanged, as is done in most implementations). Adapting the Metropolis criterion (16) to the present situation, the proposed exchange will be accepted with probability

$$w = \min\{1, \exp[(\beta_j - \beta_i)(E_j - E_i)]\} \tag{91}$$

where $E_i \equiv E(\sigma_i)$. To assure a reasonable acceptance rate, usually only "nearest-neighbor" exchanges ($j = i \pm 1$) are attempted and, as a first rough guess, the β_i could again be spaced by $\Delta\beta$ given in (89). By carefully monitoring the dynamics of the algorithm, recently much more refined prescriptions for the optimal choice of the simulation points β_i have been proposed.[89,90] In most applications, the smallest inverse temperature β_1 is chosen in the high-temperature phase where the autocorrelation time is expected to be very short and the system decorrelates rapidly. Conceptually this approach follows again the "avoiding rare events" strategy.

Notice that in parallel tempering no free-energy parameters have to be adjusted. The method is thus very robust and moreover can be almost trivially parallelized. For instance it it straightforward to implement this algorithm on a graphics card and perform "parallel tempering GPU computations".[91]

5.3. *Multicanonical ensembles*

To conclude this introduction to simulation techniques, at least a very brief outline of multicanonical ensembles[92,93] shall be given. For more details, in particular on practical implementations, see the earlier reviews[94–97] and the textbook by Berg.[4] Similarly to the tempering methods of the last section, multicanonical simulations may also be interpreted as a dynamical multi-histogram reweighting method. This interpretation is stressed by the notation used in the original papers by Berg and Neuhaus[92,93] and explains the name "*multi*canonical". At the same time, this method may also be viewed as a specific realization of non-Boltzmann sampling[98] which has been known since long to be a legitimate alternative to the more standard Monte Carlo approaches.[99] The practical significance of non-Boltzmann sampling was first realized in the so-called "umbrella sampling" method,[100] but it took many years before the introduction of the multicanonical ensemble turned non-Boltzmann sampling into a widely appreciated practical tool in computer simulation studies of phase transitions. Once the feasibility of such a generalized ensemble approach was realized, many related methods and further refinements were developed. By now the applications of the method range from physics and chemistry to biophysics, biochemistry and biology to engineering problems.

Conceptually the method can be divided into two main strategies. The first strategy can be best described as "avoiding rare events" which is close in spirit to the alternative tempering methods. In this variant one tries to connect the important parts of phase space by "easy paths" which go around suppressed rare-event regions which hence cannot be studied directly. The second approach is based on "enhancing the probability of rare event states", which is for example the typical strategy for dealing with the highly suppressed mixed-phase region of first-order phase transitions[38,97] and the very rugged free-energy landscapes of spin glasses.[101–104] This allows a direct study of properties of the rare-event states such as, e.g., interface tensions or more generally free energy barriers, which would be very difficult (or practically impossible) with canonical simulations and also with the tempering methods described in Sects. 5.1 and 5.2.

In general the idea goes as follows. With σ representing generically the degrees of freedom (discrete spins or continuous field variables), the canonical Boltzmann distribution

$$\mathcal{P}_{\rm can}(\sigma) \propto e^{-\beta\mathcal{H}(\sigma)} \tag{92}$$

is replaced by an auxiliary multicanonical distribution

$$\mathcal{P}_{\rm muca}(\sigma) \propto W(Q(\sigma))e^{-\beta\mathcal{H}(\sigma)} \equiv e^{-\beta\mathcal{H}_{\rm muca}(\sigma)} \ , \tag{93}$$

introducing a multicanonical weight factor $W(Q)$ where Q stands for any macroscopic observable such as the energy or magnetization. This defines formally $\mathcal{H}_{\rm muca} = \mathcal{H} - (1/\beta)\ln W(Q)$ which may be interpreted as an effective "multicanonical" Hamiltonian. The Monte Carlo sampling can then be implemented as usual by comparing $\mathcal{H}_{\rm muca}$ before and after a proposed update of σ, and canonical expectation values can be recovered exactly by inverse reweighting,

$$\langle\mathcal{O}\rangle_{\rm can} = \langle\mathcal{O}W^{-1}(Q)\rangle_{\rm muca}/\langle W^{-1}(Q)\rangle_{\rm muca} \ , \tag{94}$$

similarly to Eq. (73). The goal is now to find a suitable weight factor W such that the dynamics of the multicanonical simulation profits most.

To be specific, let us assume in the following that the relevant macroscopic observable is the energy E itself. This is for instance the case at a temperature driven

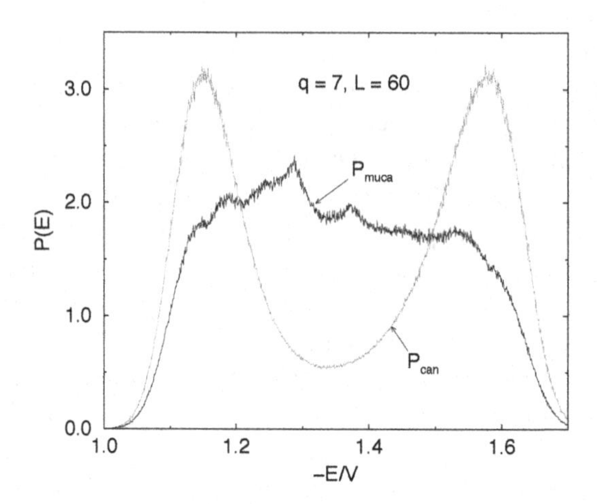

Fig. 10. The canonical energy density $P_{\rm can}(E)$ of the 2D 7-state Potts model on a 60×60 lattice at inverse temperature $\beta_{{\rm eqh},L}$, where the two peaks are of equal height, together with the multicanonical energy density $P_{\rm muca}(E)$, which is approximately constant between the two peaks.

first-order phase transition, where the canonical energy distribution $P_{\rm can}(E)$ develops a characteristic double-peak structure.[38] As an illustration, simulation data for the 2D 7-state Potts model[105] are shown in Fig. 10. With increasing system size, the region between the two peaks becomes more and more suppressed by the interfacial Boltzmann factor $\propto \exp(-2\sigma_{od}L^{D-1})$, where σ_{od} is the (reduced) interface tension, L^{D-1} the cross-section of a D-dimensional system, and the factor 2 accounts for the fact that with the usually employed periodic boundary condition at least two interfaces are present due to topological reasons. The time needed to cross this strongly suppressed rare-event two-phase region thus grows *exponentially* with the system size L, i.e., the autocorrelation time scales as $\tau \propto \exp(+2\sigma_{od}L^{D-1})$. In the literature, this is sometimes termed "supercritical slowing down" (even though nothing is "critical" here). Given such a situation, one usually adjusts $W = W(E)$ such that the multicanonical distribution $P_{\rm muca}(E)$ is approximately constant between the two peaks of $P_{\rm can}(E)$, thus aiming at a random-walk (pseudo-) dynamics of the Monte Carlo process,[106,107] cf. Fig. 10.

The crucial non-trivial point is, of course, *how* this can be achieved. On a piece of paper, $W(E) \propto 1/P_{\rm can}(E)$ – but we do not know $P_{\rm can}(E)$ (otherwise there would be little need for the simulation ...). The solution of this problem is a recursive computation. Starting with the canonical distribution, or some initial guess based on results for already simulated smaller systems together with finite-size scaling extrapolations, one performs a relatively short simulation to get an improved estimate of the canonical distribution. When this is inverted one obtains a new estimate of the multicanonical weight factor, which then is used in the next iteration and so on. In this naive variant only the simulation data of the last iteration are used in the construction of the improved weight factor.

A more sophisticated recursion, in which the updated weight factor, or more conveniently the ratio $R(E) = W(E + \Delta E)/W(E)$, is computed from *all* available data accumulated so far, works as follows:[97,108–110]

1. Perform a simulation with $R_n(E)$ to obtain the nth histogram $H_n(E)$.
2. Compute the statistical weight of the nth run:

$$p(E) = H_n(E)H_n(E + \Delta E)/[H_n(E) + H_n(E + \Delta E)] \ . \tag{95}$$

3. Accumulate statistics:

$$p_{n+1}(E) = p_n(E) + p(E) \ , \tag{96}$$

$$\kappa(E) = p(E)/p_{n+1}(E) \ . \tag{97}$$

4. Update weight ratios:

$$R_{n+1}(E) = R_n(E)\,[H_n(E)/H_n(E + \Delta E)]^{\kappa(E)} \ . \tag{98}$$

Goto 1.

The recursion is initialized with $p_0(E) = 0$. To derive this recursion one assumes that (unnormalized) histogram entries $H_n(E)$ have an *a priori* statistical error $\sqrt{H_n(E)}$ and (quite crudely) that all data are uncorrelated. Due to the accumulation of statistics, this procedure is rather insensitive to the length of the nth run in the first step and has proved to be rather stable and efficient in practice.

In most applications local update algorithms have been employed, but for certain classes of models also non-local multigrid methods[34,35,111] are applicable.[68,112] A combination with non-local cluster update algorithms, on the other hand, is not straightforward. Only by making direct use of the random-cluster representation as a starting point, a multibondic variant[113–115] has been developed. For a recent application to improved finite-size scaling studies of second-order phase transitions, see Ref. 116. If P_{muca} was completely flat and the Monte Carlo update moves would perform an ideal random walk, one would expect that after V^2 local updates the system has travelled on average a distance V in total energy. Since one lattice sweep consists of V local updates, the autocorrelation time should scale in this idealized picture as $\tau \propto V$. Numerical tests for various models with a first-order phase transition have shown that in practice the data are at best consistent with a behaviour $\tau \propto V^\alpha$, with $\alpha \geq 1$. While for the temperature-driven transitions of 2D Potts models the multibondic variant seems to saturate the bound,[113–115] employing local update algorithms, typical fit results are $\alpha \approx 1.1 - 1.3$, and due to the limited accuracy of the data even a weak exponential growth law cannot be excluded.

In fact, at least for the field-driven first-order transition of the 2D Ising model below T_c, where one works with the magnetization instead of the energy (sometimes called "multimagnetical" simulations), it has been demonstrated recently[117] that even for a perfectly flat multicanonical distribution there are two "hidden" free energy barriers (in directions "orthogonal" to the magnetization) which lead to an exponential growth of τ with lattice size, which is albeit much weaker than the leading "supercritical slowing down" of the canonical simulation. Physically the two barriers are related to the nucleation of a large droplet of the "wrong phase" (say "$-$" spins in the background of "$+$" spins)[118–123] and the transition of this large, more or less spherical droplet to the strip phase (coexisting strips of "$-$" and "$+$" spins, separated by two straight interfaces) around $m = 0$.[124]

5.4. *Wang-Landau method*

Another more recently proposed method deals directly with estimators $\Omega(E)$ of the density of states.[125] By flipping spins randomly, the transition probability from energy level E_1 to E_2 is

$$w(E_1 \to E_2) = \min\left[1, \frac{\Omega(E_1)}{\Omega(E_2)}\right] . \tag{99}$$

Each time an energy level is visited, the estimator is multiplicatively updated,

$$\Omega(E) \to f\,\Omega(E) , \tag{100}$$

where initially $\Omega(E) = 1$ and $f = f_0 = e^1$. Once the accumulated energy histogram is sufficiently flat, the factor f is refined,

$$f_{n+1} = \sqrt{f_n} , \qquad n = 0, 1, \ldots , \tag{101}$$

and the energy histogram reset to zero until some small value such as $f = e^{10^{-8}} \approx 1.00000001$ is reached.

For the 2D Ising model this procedure converges very rapidly towards the exactly known density of states, and also for other applications a fast convergence has been reported. Since the procedure violates the Markovian requirement and hence does not satisfy the balance condition (7), some care is necessary in setting up a proper protocol for the recursion (this is similar in spirit to the automatic updating of the optimal step size $S_{\max}$ in the Metropolis update algorithm discussed in Sect. 2.3.1). Most authors who employ the obtained density of states directly to extract canonical expectation values by standard reweighting argue that, once

f is close enough to unity, systematic deviations become negligible. While this claim can be verified empirically for the 2D Ising model (where exact results are available for judgement), possible systematic deviations are difficult to assess in the general case. A safe way would be to consider the recursion (99)–(101) as an alternative method to determine the multicanonical weights, and then to perform a usual multicanonical simulation employing these *fixed* weights. As emphasized earlier, any deviations of multicanonical weights from their optimal shape do not show up in the final canonical expectation values; they rather only influence the dynamics of the multicanonical simulations.

6. Scaling Analyses

Equipped with the various technical tools discussed above, the purpose of this section is to outline typical scaling and finite-size scaling (FSS) analyses of Monte Carlo simulations of second-order phase transitions. The described procedure is generally applicable but to keep the notation short, all formulas are formulated for Ising like systems. For instance for O(n) symmetric models, m should be replaced by $\vec{m}$ etc. The main results of such studies are usually estimates of the critical temperature and the critical exponents characterising the universality class of the transition.

Basic observables are the internal energy per site, $u = U/V$, with $U = -\mathrm{d}\ln\mathcal{Z}/\mathrm{d}\beta = \langle\mathcal{H}\rangle \equiv \langle E\rangle$, and the specific heat,

$$C = \frac{\mathrm{d}u}{\mathrm{d}T} = \beta^2\left(\langle E^2\rangle - \langle E\rangle^2\right)/V = \beta^2 V\left(\langle e^2\rangle - \langle e\rangle^2\right) \ , \tag{102}$$

where we have set $\mathcal{H} \equiv E = eV$ with V denoting the number of lattice sites, i.e., the "lattice volume". In simulations one usually employs the variance definition (since any discretized numerical differentiation would introduce some systematic error). The magnetization per site $m = M/V$ and the susceptibility χ are defined as[h]

$$m = \langle|\mu|\rangle \ , \qquad \mu = \frac{1}{V}\sum_i \sigma_i \ , \tag{103}$$

[h]Notice that here and in the following formulas, $|\mu|$ is used instead of μ as would follow from the formal definition of the zero-field magnetization $m(\beta) = (1/V\beta)\lim_{h\to 0}\partial\ln\mathcal{Z}(\beta,h)/\partial h$. The reason is that for a symmetric model on finite lattices one obtains $\langle\mu\rangle(\beta) = 0$ for all temperatures due to symmetry. Only in the proper infinite-volume limit, that is $\lim_{h\to 0}\lim_{V\to\infty}$, spontaneous symmetry breaking can occur below T_c. In a simulation on finite lattices, this is reflected by a symmetric double-peak structure of the magnetization distribution (provided the runs are long enough). By averaging μ one thus gets zero by symmetry, while the peak locations $\pm m_0(L)$ are close to the spontaneous magnetization so that the average of $|\mu|$ is a good estimator. Things become more involved for slightly asymmetric models, where this recipe would produce a systematic error and thus cannot be employed. For strongly asymmetric models, on the other hand, one peak clearly dominates and the average of μ can usually be measured without too many problems.

and

$$\chi = \beta V \left(\langle \mu^2 \rangle - \langle |\mu| \rangle^2 \right) \ . \tag{104}$$

In the disordered phase for $T > T_c$, where $m = \langle \mu \rangle = 0$ by symmetry, one often works with the definition

$$\chi' = \beta V \langle \mu^2 \rangle \ . \tag{105}$$

The correlation between spins σ_i and σ_j at sites labeled by i and j can be measured by considering correlation functions like the two-point spin-spin correlation

$$G(\vec{r}) = G(i,j) = \langle \sigma_i \sigma_j \rangle - \langle \sigma_i \rangle \langle \sigma_j \rangle \ , \tag{106}$$

where $\vec{r} = \vec{r}_j - \vec{r}_i$ (assuming translational invariance). Away from criticality and at large distances $|\vec{r}| \gg 1$ (assuming a lattice spacing $a = 1$), $G(\vec{r})$ decays exponentially,

$$G(\vec{r}) \sim |\vec{r}|^{-\kappa} e^{-|\vec{r}|/\xi} \ , \tag{107}$$

where ξ is the spatial correlation length and the exponent κ of the power-law prefactor depends in general on the dimension and on whether one studies the ordered or disordered phase. Strictly speaking ξ depends on the direction of $\vec{r}$.

6.1. *Critical exponents and scaling relations*

The most characteristic feature of a second-order phase transition is the divergence of the correlation length at T_c. As a consequence thermal fluctuations are equally important on *all* length scales, and one therefore expects power-law singularities in thermodynamic functions. The leading divergence of the correlation length is usually parameterized in the high-temperature phase as

$$\xi = \xi_{0+} |1 - T/T_c|^{-\nu} + \ldots \quad (T \geq T_c) \ , \tag{108}$$

where the ... indicate sub-leading analytical as well as confluent corrections. This defines the critical exponent $\nu > 0$ and the critical amplitude ξ_{0+} on the high-temperature side of the transition. In the low-temperature phase one expects a similar behaviour,

$$\xi = \xi_{0-} (1 - T/T_c)^{-\nu} + \ldots \quad (T \leq T_c) \ , \tag{109}$$

with the same critical exponent ν but a different critical amplitude $\xi_{0-} \neq \xi_{0+}$.

The singularities of the specific heat, magnetization (for $T < T_c$), and susceptibility are similarly parameterized by the critical exponents α, β, and γ, respectively,

$$C = C_{\text{reg}} + C_0|1 - T/T_c|^{-\alpha} + \dots \ , \tag{110}$$

$$m = m_0(1 - T/T_c)^{\beta} + \dots \ , \tag{111}$$

$$\chi = \chi_0|1 - T/T_c|^{-\gamma} + \dots \ , \tag{112}$$

where C_{reg} is a regular background term, and the amplitudes are again in general different on the two sides of the transition. Right at the critical temperature T_c, two further exponents δ and η are defined through

$$m \propto h^{1/\delta} \qquad (T = T_c) \ , \tag{113}$$

$$G(\vec{r}) \propto r^{-D+2-\eta} \qquad (T = T_c) \ . \tag{114}$$

An important consequence of the divergence of the correlation length is that qualitative properties of second-order phase transitions should not depend on short-distance details of the Hamiltonian. This is the basis of the *universality* hypothesis[126] which means that all (short-ranged) systems with the same symmetries and same dimensionality should exhibit similar singularities governed by one and the same set of critical exponents. For the amplitudes this is not true, but certain amplitude ratios such as ξ_{0+}/ξ_{0-} or χ_{0+}/χ_{0-} are also universal.

In the 1960s, Rushbrooke,[127] Griffiths,[128] Josephson,[129] and Fisher[130] showed that the six critical exponents defined above are related via four inequalities. Subsequent experimental evidence indicated that these scaling relations were in fact equalities which are now firmly established by renormalization group (RG) considerations and fundamentally important in the theory of critical phenomena:

$$2\beta + \gamma = 2 - \alpha \quad \text{(Rushbrooke's law)} \ , \tag{115}$$

$$\beta(\delta - 1) = \gamma \quad \text{(Griffiths' law)} \ , \tag{116}$$

$$\nu(2 - \eta) = \gamma \quad \text{(Fisher's law)} \ . \tag{117}$$

The fourth equality involves the dimension D. It is therefore a (somewhat weaker) so-called hyperscaling relation:

$$D\nu = 2 - \alpha \quad \text{(Josephson's law)} \ . \tag{118}$$

In the conventional scaling scenario, Rushbrooke's and Griffiths' laws can be deduced from the Widom scaling hypothesis that the Helmholtz free energy is a

Table 1. Critical exponents of the Ising model. All 2D exponents are exactly known.[144,145] For the 3D Ising model the "world-average" for ν and γ calculated in Ref. 146 is quoted. The other exponents follow from hyperscaling ($\alpha = 2 - D\nu$) and scaling ($\beta = (2 - \alpha - \gamma)/2$, $\delta = \gamma/\beta + 1$, $\eta = 2 - \gamma/\nu$) relations. For all $D \geq D_u = 4$ the mean-field exponents are valid (in 4D up to multiplicative logarithmic corrections).

	ν	α	β	γ	δ	η
$D = 2$	1	0 (log)	1/8	7/4	15	1/4
$D = 3$	0.630 05(18)	0.109 85	0.326 48	1.237 17(28)	4.7894	0.036 39
$D \geq 4$	1/2	0 (disc)	1/2	1	3	0

homogeneous function.[131] Widom scaling and the remaining two laws can in turn be derived from the Kadanoff block-spin construction[132] and ultimately from RG considerations.[133] Josephson's law can also be derived from the hyperscaling hypothesis, namely that the free-energy density behaves near criticality as the inverse correlation volume: $f \sim \xi^{-D}$. Twice differentiating this relation and inserting the scaling law (110) for the specific heat gives immediately (118).

The paradigm model for systems exhibiting a continuous (or, roughly speaking, second-order) phase transition is the Ising model. When the temperature is varied the system passes at T_c from an ordered low-temperature to a disordered high-temperature phase. In two dimensions (2D), the thermodynamic limit of this model in zero external field has been solved exactly by Onsager,[134] and even for finite $L_x \times L_y$ lattices the exact partition function is known.[73,74] Also the exact density of states can be calculated by means of computer algebra up to reasonably large lattice sizes.[75] This provides a very useful testing ground for any new algorithmic idea in computer simulations. For infinite lattices, the correlation length has been calculated in arbitrary lattice directions.[135,136] The exact magnetization for $h = 0$, apparently already known to Onsager,[137] was first derived by Yang[138] and later generalized by Chang.[139] The only quantity which up to date is not truly exactly known is the susceptibility. However, its properties have been characterized to very high precision[140–142] (for both, low- and high-temperature series expansions, 2000 terms are known exactly[141]). In three dimensions (3D) *no* exact solutions are available, but analytical and numerical results from various methods give a consistent and very precise picture. In four dimensions (4D) the so-called upper critical dimension D_u is reached and for $D \geq D_u = 4$ the critical exponents take their mean-field values (in 4D up to multiplicative logarithmic corrections[143]). The critical exponents of the Ising model are collected in Table 1.[144–146]

6.2. *Finite-size scaling (FSS)*

In computer simulation studies, the (linear) system size L is always necessarily finite. The correlation length may hence become large (of the order of L) but never diverges in a mathematical sense. For the divergences in other quantities this implies that they are also rounded and shifted.[11,147–149] How this happens is described by finite-size scaling (FSS) theory, which in a nut-shell may be explained as follows: Near T_c the role of ξ is taken over by the linear size L of the system. By rewriting (108) or (109) and replacing ξ by L, it is easy to see that

$$|1 - T/T_c| \propto \xi^{-1/\nu} \longrightarrow L^{-1/\nu} \ . \tag{119}$$

It follows that the scaling laws (110)–(112) have to be replaced by the *finite-size scaling* (FSS) ansatz,

$$C = C_{\rm reg} + aL^{\alpha/\nu} + \ldots \ , \tag{120}$$

$$m \propto L^{-\beta/\nu} + \ldots \ , \tag{121}$$

$$\chi \propto L^{\gamma/\nu} + \ldots \ , \tag{122}$$

where $C_{\rm reg}$ is a regular, smooth background term and a a constant. As a mnemonic rule, a critical exponent x in a temperature scaling law is replaced by $-x/\nu$ in the corresponding FSS law. This describes the rounding of the singularities quantitatively.

In general these scaling laws are valid in a vicinity of T_c as long as the scaling variable

$$x = (1 - T/T_c)L^{1/\nu} \tag{123}$$

is kept fixed.[11,147–149] In this more general formulation the scaling law for, e.g., the susceptibility reads

$$\chi(T, L) = L^{\gamma/\nu} f(x) \ , \tag{124}$$

where $f(x)$ is a scaling function. By plotting $\chi(T, L)/L^{\gamma/\nu}$ versus the scaling variable x, one thus expects that the data for different T and L fall onto a master curve described by $f(x)$. This is a nice visual method for demonstrating the scaling properties.

For given L the maximum of $\chi(T, L)$ as a function of temperature happens at some $x_{\rm max}$. For the location $T_{\rm max}$ of the maximum this implies a FSS behaviour of the form

$$T_{\rm max} = T_c(1 - x_{\rm max}L^{-1/\nu} + \ldots) = T_c + cL^{-1/\nu} + \ldots \ . \tag{125}$$

This quantifies the shift of so-called pseudo-critical points which depends on the observables considered. Only in the thermodynamic limit $L \to \infty$ all quantities diverge at the same temperature T_c.

Further useful quantities in FSS analyses are the energetic fourth-order parameter

$$V(\beta) = 1 - \frac{\langle e^4 \rangle}{3\langle e^2 \rangle^2} , \tag{126}$$

the magnetic cumulants (Binder parameters)

$$U_2(\beta) = 1 - \frac{\langle \mu^2 \rangle}{3\langle |\mu| \rangle^2} , \tag{127}$$

$$U_4(\beta) = 1 - \frac{\langle \mu^4 \rangle}{3\langle \mu^2 \rangle^2} , \tag{128}$$

and their slopes

$$\begin{aligned} \frac{dU_2(\beta)}{d\beta} &= \frac{V}{3\langle |\mu| \rangle^2} \left[\langle \mu^2 \rangle \langle e \rangle - 2\frac{\langle \mu^2 \rangle \langle |\mu| e \rangle}{\langle |\mu| \rangle} + \langle \mu^2 e \rangle \right] \\ &= V(1 - U_2) \left[\langle e \rangle - 2\frac{\langle |\mu| e \rangle}{\langle |\mu| \rangle} + \frac{\langle \mu^2 e \rangle}{\langle \mu^2 \rangle} \right] , \end{aligned} \tag{129}$$

$$\frac{dU_4(\beta)}{d\beta} = V(1 - U_4) \left[\langle e \rangle - 2\frac{\langle \mu^2 e \rangle}{\langle \mu^2 \rangle} + \frac{\langle \mu^4 e \rangle}{\langle \mu^4 \rangle} \right] . \tag{130}$$

The Binder parameters scale according to

$$U_{2p} = f_{U_{2p}}(x)[1 + \ldots] , \tag{131}$$

i.e., for constant scaling variable x, U_{2p} takes approximately the same value for all lattice sizes, in particular $U_{2p}^* \equiv f_{U_{2p}}(0)$ at T_c. Applying the differentiation to this scaling representation, one picks up a factor of $L^{1/\nu}$ from the scaling function,

$$\frac{dU_{2p}}{d\beta} = (dx/d\beta) f'_{U_{2p}}[1 + \ldots] = L^{1/\nu} f_{U'_{2p}}(x)[1 + \ldots] . \tag{132}$$

As a function of temperature the Binder parameters for different L hence cross around (T_c, U_{2p}^*) with slopes $\propto L^{1/\nu}$, apart from corrections-to-scaling collected in $[1 + \ldots]$ explaining small systematic deviations. From a determination of this crossing point, one thus obtains a basically unbiased estimate of T_c, the critical exponent ν, and U_{2p}^*. Note that in contrast to the truly universal critical exponents, U_{2p}^* is only *weakly* universal. By this one means that the infinite-volume limit of such quantities does depend in particular on the boundary conditions and geometrical shape of the considered lattice, e.g., on the aspect ratio $r = L_y/L_x$.[150–157]

Further quantities with a useful FSS behaviour are the derivatives of the magnetization,

$$\frac{\mathrm{d}\langle|\mu|\rangle}{\mathrm{d}\beta} = V\left(\langle|\mu|e\rangle - \langle|\mu|\rangle\langle e\rangle\right) , \tag{133}$$

$$\frac{\mathrm{d}\ln\langle|\mu|\rangle}{\mathrm{d}\beta} = V\left(\frac{\langle|\mu|e\rangle}{\langle|\mu|\rangle} - \langle e\rangle\right) , \tag{134}$$

$$\frac{\mathrm{d}\ln\langle\mu^2\rangle}{\mathrm{d}\beta} = V\left(\frac{\langle\mu^2 e\rangle}{\langle\mu^2\rangle} - \langle e\rangle\right) . \tag{135}$$

These latter five quantities are good examples for expectation values depending on both e and μ. By applying the differentiation to the scaling form of $\langle|\mu|\rangle$, one reads off that

$$\frac{\mathrm{d}\langle|\mu|\rangle}{\mathrm{d}\beta} = L^{(1-\beta)/\nu} f_{\mu'}(x)[1+\ldots] , \tag{136}$$

$$\frac{\mathrm{d}\ln\langle|\mu|^p\rangle}{\mathrm{d}\beta} = L^{1/\nu} f_{\mathrm{d}\mu^p}(x)[1+\ldots] . \tag{137}$$

For first-order phase transitions similar considerations show[37,38,158–160] that there the delta function like singularities in the thermodynamic limit, originating from phase coexistence, are smeared out for finite systems as well.[161–165] They are replaced by narrow peaks whose height grows proportional to the volume $V = L^D$, analogously to (120) or (122), with a peak width decreasing as $1/V$ and a shift of the peak location from the infinite-volume transition temperature proportional to $1/V$, analogously to (125).[37,38,166–170]

6.3. *Organisation of the analysis*

To facilitate most flexibility in the analysis, it is advisable to store during data production the time series of measurements. Standard quantities are the energy and magnetization, but depending on the model at hand it may be useful to record also other observables. In this way the full dynamical information can be extracted still after the actual simulation runs and error estimation can be easily performed. For example it is no problem to experiment with the size and number of Jackknife bins. Since a reasonable choice depends on the *a priori* unknown autocorrelation time, it is quite cumbersome to do a reliable error analysis "on the flight" during the simulation. Furthermore, basing data reweighting on time-series data is more efficient since histograms, if needed or more convenient, can still be produced from this data but working in the reverse direction is obviously impossible.

For some models it is sufficient to perform for each lattice size a single long run at some coupling β_0 close to the critical point β_c. This is, however, not always the case and also depends on the observables of interest. In this more general case, one may use several simulation points β_i and combine the results by the multi-histogram reweighting method or may apply a recently developed finite-size adapted generalized ensemble method.[116,171] In both situations, one can compute the relevant quantities from the time series of the energies $e = E/V$ (if E happens to be integer valued, this should be stored of course) and $\mu = \sum_i \sigma_i/V$ by reweighting.

By using one of these techniques one first determines the temperature dependence of $C(\beta)$, $\chi(\beta)$, ..., in the neighborhood of the simulation point $\beta_0 \approx \beta_c$ (a reasonably "good" initial guess for β_0 is usually straightforward to obtain). Once the temperature dependence is known, one can determine the maxima, e.g., $C_{\max}(\beta_{\max_C}) \equiv \max_\beta C(\beta)$, by applying standard extremization routines: When reweighting is implemented as a subroutine, for instance $C(\beta)$ can be handled as a normal function with a continuously varying argument β, i.e., no interpolation or discretization error is involved when iterating towards the maximum. The locations of the maxima of C, χ, $\mathrm{d}U_2/\mathrm{d}\beta$, $\mathrm{d}U_4/\mathrm{d}\beta$, $\mathrm{d}\langle|\mu|\rangle/\mathrm{d}\beta$, $\mathrm{d}\ln\langle|\mu|\rangle/\mathrm{d}\beta$, and $\mathrm{d}\ln\langle\mu^2\rangle/\mathrm{d}\beta$ provide us with seven sequences of pseudo-transition points $\beta_{\max_i}(L)$ which all should scale according to $\beta_{\max_i}(L) = \beta_c + a_i L^{-1/\nu} + \dots$. In other words, the scaling variable $x = (\beta_{\max_i}(L) - \beta_c)L^{1/\nu} = a_i + \dots$ should be constant, if we neglect the small higher-order corrections indicated by $\dots$.

Notice that while the precise estimates of a_i do depend on the value of ν, the qualitative conclusion that $x \approx \mathrm{const}$ for each of the $\beta_{\max_i}(L)$ sequences does not require any *a priori* knowledge of ν or β_c. Using this information one thus has several possibilities to extract unbiased estimates of the critical exponents ν, α/ν, β/ν, and γ/ν from least-squares fits assuming the FSS behaviours (120), (121), (122), (132), (136), and (137).

Considering only the asymptotic behaviour, e.g., $\mathrm{d}\ln\langle|\mu|\rangle/\mathrm{d}\beta = aL^{1/\nu}$, and taking the logarithm, $\ln(\mathrm{d}\ln\langle|\mu|\rangle/\mathrm{d}\beta) = c + (1/\nu)\ln(L)$, one ends up with a linear two-parameter fit yielding estimates for the constant $c = \ln(a)$ and the exponent $1/\nu$. For small lattice sizes the asymptotic ansatz is, of course, not justified. Taking into account the (effective) correction term $[1 + bL^{-w}]$ would result in a *non-linear* four-parameter fit for a, b, $1/\nu$ and w. Even if we would fix w to some "theoretically expected" value (as is sometimes done), we would still be left with a *non-linear* fit which is usually much harder to control than a linear fit (where only a set of linear equations with a unique solution has to be solved, whereas a non-linear fit involves a numerical minimization of the χ^2-function, possessing possibly several local minima). The alternative method is to use the linear fit ansatz and to discard successively more and more small lattice sizes until the χ^2 per degree-of-freedom or the goodness-of-fit parameter[61] Q has reached an acceptable value and does not show any further trend. Of course, all this relies heavily on correct estimates of the statistical error bars on the original data for $\mathrm{d}\ln\langle|\mu|\rangle/\mathrm{d}\beta$.

Once ν is estimated one can use the scaling form $\beta_{\mathrm{max}_i}(L) = \beta_c + a_i L^{-1/\nu} + \ldots$ to extract β_c and a_i. As a useful check, one should repeat these fits at the error margins of ν, but usually this dependence turns out to be very weak. As a useful cross-check one can determine β_c also from the Binder parameter crossings, which is the most convenient and fastest method for a first rough estimate. As a rule of thumb, an accuracy of about $3-4$ digits for β_c can be obtained with this method without any elaborate infinite-volume extrapolations – the crossing points lie usually much closer to β_c than the various maxima locations. For high precision, however, it is quite cumbersome to control the necessary extrapolations and often more accurate estimates can be obtained by considering the scaling of the maxima locations. Also, error estimates of crossing points involve the data for two different lattice sizes which tends to be quite unhandy.

Next, similarly to ν, the ratios of critical exponents α/ν, β/ν, and γ/ν can be obtained from fits to (120), (121), (122), and (136). Again the maxima of these quantities or any of the FSS sequences β_{max_i} can be used. What concerns the fitting procedure the same remarks apply as for ν. The specific heat C usually plays a special role in that the exponent α is difficult to determine. The reason is that α is usually relatively small (3D Ising model: $\alpha \approx 0.1$), may be zero (logarithmic divergence as in the 2D Ising model) or even negative (as for instance in the 3D XY and Heisenberg models). In all these cases, the constant background contribution C_{reg} in (120) becomes important, which enforces a non-linear three-parameter fit with the just described problems. Also for the susceptibility χ, a regular background term cannot be excluded, but it is usually much less important since $\gamma \gg \alpha$. Therefore, in (121), (122), and (136), similar to the fits for ν, one may take the logarithm and deal with much more stable linear fits.

As a final step one may re-check the FSS behaviour of C, χ, $\mathrm{d}U_2/\mathrm{d}\beta, \ldots$ at the numerically determined estimate of β_c. These fits should be repeated also at $\beta_c \pm \Delta\beta_c$ in order to estimate by how much the uncertainty in β_c propagates into the thus determined exponent estimates. In (the pretty rare) cases where β_c is known exactly (e.g., through self-duality), this latter option is by far the most accurate one. This is the reason, why for such models numerically estimated critical exponents are usually quite precise.

When combining the various fit results for, e.g. β_c or ν, to a final average value, some care is necessary with the optimal weighted average and the final statistical error estimate, since the various fits for determining β_c or ν are of course correlated (since they all use the data from one and the same simulation). In principle this can be dealt with by applying a cross-correlation analysis.[172]

7. Applications

7.1. *Disordered ferromagnets*

Experiments on phase transitions in magnetic materials are usually subject to randomly distributed impurities. At continuous phase transitions, depending on the temperature resolution and the concentration of the impurities, the disorder may significantly influence measurements of critical exponents.[173] To emphasize this effect, in some experiments[174] non-magnetic impurities are introduced in a controlled way; see Fig. 11 for an example. Since the mobility of impurities is usually much smaller than the typical time scale of spin fluctuations, one may model the disorder effects in a completely "frozen", so-called "quenched" approximation. This limit is opposite to "annealed" disorder which refers to the case where the two relevant time scales are of the same order.

With the additional assumption that the quenched, randomly distributed impurities are completely uncorrelated, Harris[175] showed a long time ago under which conditions a *continuous* transition of an idealised pure material is modified by disorder coupling to the energy of the system. According to this so-called Harris criterion, the critical behaviour of the pure system around the transition temperature T_c is stable against quenched disorder when the critical exponent $\alpha_{\rm pure}$ of the specific heat, $C \propto |T-T_c|^{-\alpha_{\rm pure}}$, is negative. In renormalization-group language the perturbation is then "irrelevant" and the values of all critical exponents $\alpha, \beta, \gamma, \dots$ remain unchanged. On the other hand, when $\alpha_{\rm pure} > 0$, then quenched disorder should be "relevant" and the renormalization-group flow approaches a new disorder fixed point governed by altered critical exponents. An example is the three-dimensional (3D) Ising model universality class with $\alpha_{\rm pure} \approx 0.110 > 0$. The intermediate situation $\alpha_{\rm pure} = 0$ is a special, "marginal" case where no easy predictions can be made. A typical example for the latter situation is the two-dimensional (2D) Ising model where quenched disorder is known to generate logarithmic modifications.[176]

Figure 11 shows an experimental verification of the qualitative influence of disorder for a three-dimensional Ising-like system where the measured critical exponent $\gamma = 1.364(76)$ of the susceptibility $\chi \propto |T - T_c|^{-\gamma}$ is clearly different from that of the pure 3D Ising model, $\gamma_{\rm pure} = 1.2396(13)$. Theoretical results,

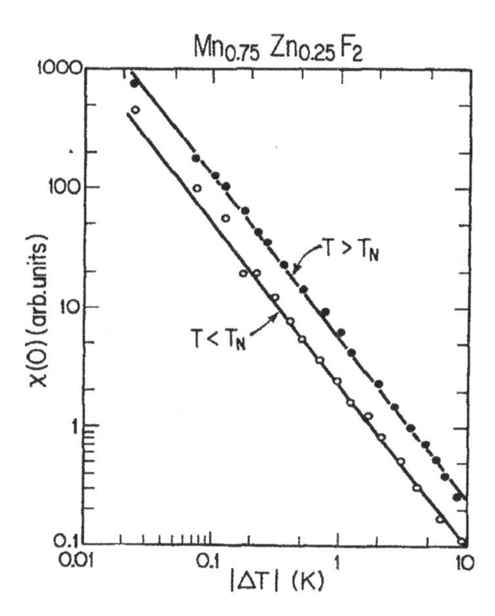

Fig. 11. Neutron scattering measurements of the susceptibility in $\mathrm{Mn}_{0.75}\mathrm{Zn}_{0.25}\mathrm{F}_2$ close to criticality, governed by the disorder fixed point of the Ising model, over the reduced temperature interval $4 \times 10^{-4} < |T/T_c - 1| < 2 \times 10^{-1}$. The solid lines show power-law fits with exponent $\gamma = 1.364(76)$ above and below T_c [after Mitchell *et al.* (Ref. 174)].

on the other hand, remained relatively scarce in 3D until recently. Most analytical renormalization group and computer simulation studies focused on the Ising model,[177,178] usually assuming *site* dilution when working numerically. This motivated us to consider the case of *bond* dilution[179–181] which enables one to test the expected universality with respect to the type of disorder distribution and, in addition, facilitates a quantitative comparison with recent high-temperature series expansions.[182–184]

The Hamiltonian (in a Potts model normalisation) is given as

$$-\beta\mathcal{H} = \sum_{\langle i,j \rangle} K_{ij}\delta_{\sigma_i,\sigma_j} \ , \tag{138}$$

where the spins take the values $\sigma_i = \pm 1$ and the sum goes over all nearest-neighbor pairs $\langle i, j\rangle$. The coupling strengths K_{ij} are drawn from the bimodal distribution

$$\wp[K_{ij}] = \prod_{\langle i,j \rangle} P(K_{ij}) = \prod_{\langle i,j \rangle} [p\delta(K_{ij} - K) + (1-p)\delta(K_{ij} - RK)] \ . \tag{139}$$

Besides bond dilution ($R = 0$), which we will consider here, this also includes random-bond ferromagnets ($0 < R < 1$) and the physically very different class of spin glasses ($R = -1$) as special cases. For the case of bond dilution, the couplings are thus allowed to take two different values $K_{ij} = K \equiv J\beta \equiv J/k_BT$ and 0 with probabilities p and $1-p$, respectively, with $c = 1-p$ being the concentration of missing bonds, which play the role of the non-magnetic impurities. The pure case thus corresponds to $p = 1$. Below the bond-percolation threshold[185] $p_c = 0.248\,812\,6(5)$ one does not expect any finite-temperature phase transition since without a percolating (infinite) cluster of spins long-range order cannot develop.

The model (138), (139) with $R = 0$ was studied by means of large-scale Monte Carlo simulations using the Swendsen-Wang (SW) cluster algorithm[39] (which in the strongly diluted case is better suited than the single-cluster Wolff variant). To arrive at final results in the quenched case, for each dilution, temperature and lattice size, the Monte Carlo estimates for $\langle Q_{\{J\}}\rangle$ of thermodynamic quantities $Q_{\{J\}}$ for a given random distribution $\{J\}$ of diluted bonds (realized as usual by averages over the time series of measurements) have to be averaged over many different disorder realisations,

$$Q \equiv [\langle Q_{\{J\}}\rangle]_{\rm av} = \frac{1}{\#\{J\}} \sum_{\{J\}} \langle Q_{\{J\}}\rangle \ , \tag{140}$$

where $\#\{J\}$ is the number of realisations considered. Denoting the empirically determined distribution of $\langle Q_{\{J\}}\rangle$ by $\mathcal{P}(\langle Q_{\{J\}}\rangle)$, this so-called quenched average can also be obtained from

$$Q = \int \mathcal{D}J_{ij}\wp(J_{ij})\langle Q_{\{J\}}\rangle = \int {\rm d}\langle Q_{\{J\}}\rangle \mathcal{P}(\langle Q_{\{J\}}\rangle)\langle Q_{\{J\}}\rangle \ , \tag{141}$$

where a discretized evaluation of the integrals for finite $\#\{J\}$ is implicitly implied. While conceptually straightforward, the quenched average in (140) is computationally very demanding since the number of realisations $\#\{J\}$ usually must be large, often of the order of a few thousands. In fact, if this number is chosen too small one may observe *typical* rather than average values[186] which may differ significantly when the distribution $\mathcal{P}(\langle Q_{\{J\}}\rangle)$ exhibits a long tail (which in general is hard to predict beforehand).

To get a rough overview of the phase diagram we first studied the dependence of the susceptibility peaks on the dilution, where the susceptibility $\chi = KV(\langle \mu^2\rangle - \langle|\mu|\rangle^2)$ with $\mu = (1/V)\sum_i \sigma_i$ is defined as usual. To this end we performed for $p = 0.95, 0.90, \ldots, 0.36$ and moderate system sizes SW cluster MC simulations with $N_{\rm MCS} = 2\,500$ MC sweeps (MCS) each. By performing quite elaborate analyses of autocorrelation times, this statistics was judged to be reasonable ($N_{\rm MCS} > 250\ \tau_e$). By applying single-histogram reweighting to the data for each of the 2 500 − 5 000 disorder realisation and then averaging the resulting $\chi(K)$ curves, we finally arrived at the data shown in Fig. 12.

From the locations of the maxima one obtains the phase diagram of the model in the $p - T$ plane shown in Fig. 13 which turned out to be in excellent agreement with a "single-bond effective-medium" (EM) approximation,[187]

$$K_c^{\rm EM}(p) = \ln\left[\frac{(1-p_c)e^{K_c(1)} - (1-p)}{p - p_c}\right] , \tag{142}$$

where $K_c(1) = J/k_BT_c(1) = 0.443\,308\,8(6)$ is the precisely known transition point of the pure 3D Ising model.[188] As an independent confirmation of (142), the phase diagram also coincides extremely well with recent results from high-temperature series expansions.[184]

The quality of the disorder averages can be judged as in Fig. 14 by computing running averages over the disorder realisations taken into account and looking at the distributions $\mathcal{P}(\chi_i)$. The plots show that the fluctuations in the running average disappear already after a few hundreds of realisations and that the dispersion of the χ_i values is moderate. The histogram also shows, however, that the distributions

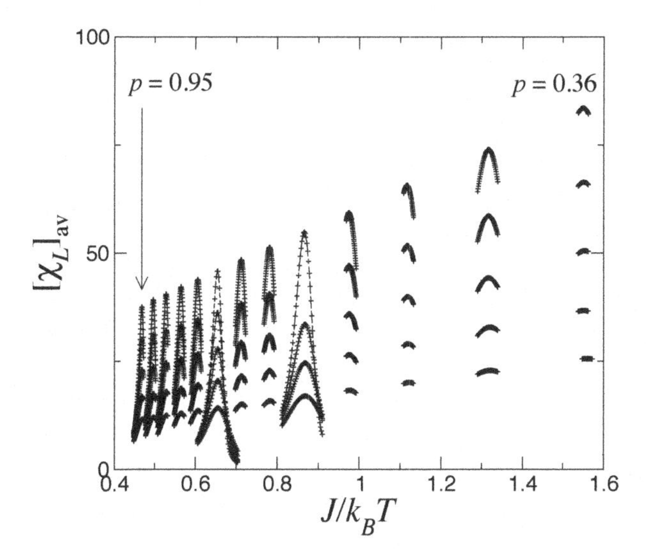

Fig. 12. The average magnetic susceptibility $[\chi_L]_{\rm av}$ of the 3D bond-diluted Ising model versus $K = J/k_BT$ for several concentrations p and $L = 8, 10, 12, 14, 16, 18$, and 20. For each value of p and each lattice size L, the curves are obtained by standard single-histogram reweighting of the simulation data at one value of K.

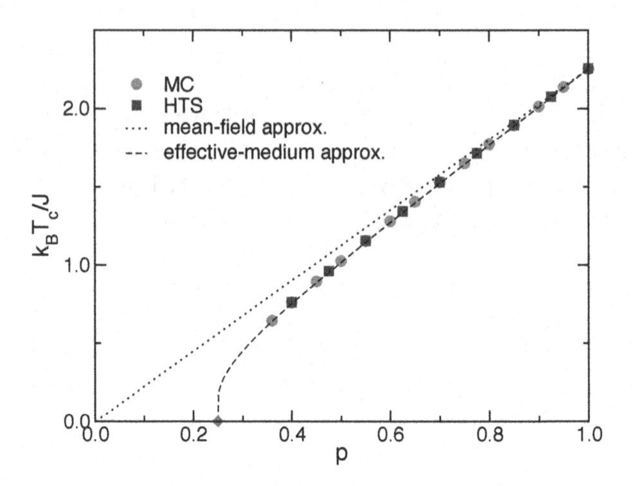

Fig. 13. Phase diagram of the bond-diluted Ising model on a three-dimensional simple cubic lattice in the dilution-temperature plane. The percolation point $p_c \approx 0.2488$ is marked by the diamond and $p = 1$ is the pure case without impurities. The results from the Monte Carlo (MC) simulations are compared with analyses of high-temperature series (HTS) expansions and with (properly normalized) mean-field and effective-medium approximations.

of physical observables typically do not become sharper with increasing system size at a finite-randomness disorder fixed point. Rather their relative widths stay constant, a phenomenon called non-self-averaging. More quantitatively, non-self-averaging can be checked by evaluating the normalized squared width $R_\chi(L) = V_\chi(L)/[\chi(L)]^2_{\rm av}$, where $V_\chi(L) = [\chi(L)^2]_{\rm av} - [\chi(L)]^2_{\rm av}$ is the variance of the susceptibility distribution. Figure 15 shows this ratio for three concentrations of the bond-diluted Ising model as a function of inverse lattice size. The fact that R_χ approaches a constant when L increases, as predicted by Aharony and Harris,[189] is the signature of a non-self-averaging system, in qualitative agreement with the results of Wiseman and Domany[190] for the site-diluted 3D Ising model.[i]

In order to study the critical behaviour in more detail, we concentrated on the three particular dilutions $p = 0.4$, 0.55, and 0.7. In a first set of simulations we focused on the FSS behaviour for lattice sizes up to $L = 96$. It is well known that *ratios* of critical exponents are almost equal for pure and disordered models, e.g., $\gamma/\nu = 1.966(6)$ (pure[191]) and $\gamma/\nu = 1.963(5)$ (disordered[192]). The only distinguishing quantity is the correlation length exponent ν which can be extracted, e.g., from the derivative of the magnetisation versus inverse temperature,

[i]Our estimate of R_χ is about an order of magnitude smaller since we worked with $\chi = KV(\langle\mu^2\rangle - \langle|\mu|\rangle^2)$ whereas in Ref. 190 the "high-temperature" expression $\chi' = KV\langle\mu^2\rangle$ was used.

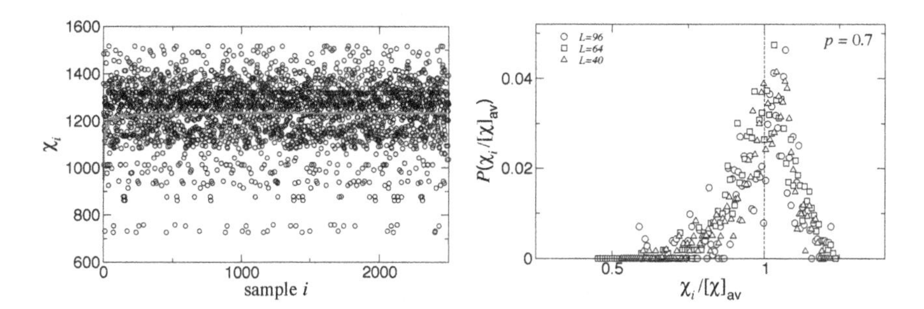

Fig. 14. *Left:* Susceptibility for the different disorder realisations of the three-dimensional bond-diluted Ising model for $L = 96$ and a concentration of magnetic bonds $p = 0.7$ at $K = 0.6535 \approx K_c(L)$. The running average over the samples is shown by the solid (red) line. *Right:* The resulting probability distribution of the susceptibility scaled by its quenched average $[\chi]_{\rm av}$, such that the results for the different lattice sizes $L = 40, 64$, and 96 collapse. The vertical dashed line indicates the average susceptibility $\chi_i/[\chi]_{\rm av} = 1$.

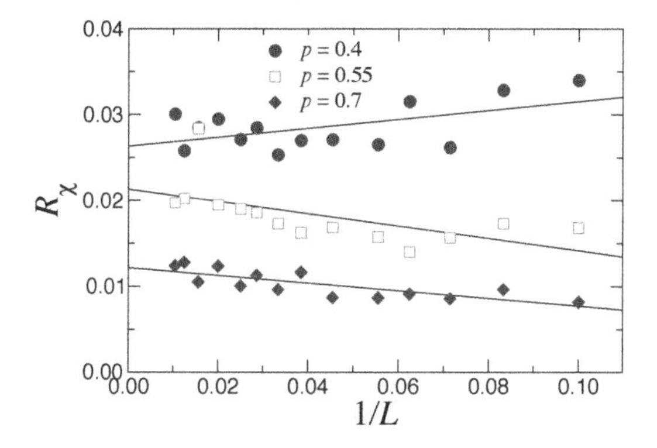

Fig. 15. Normalized squared width of the susceptibility distribution versus the inverse lattice size for the three concentrations $p = 0.4, 0.55$, and 0.7 at the effective critical coupling $K_c(L)$. The straight lines are linear fits used as guides to the eye.

$\mathrm{d}\ln[m]_{\rm av}/\mathrm{d}K \propto L^{1/\nu}$, at K_c or the locations of the susceptibility maxima. Using the latter unbiased option and performing least-square fits including data from $L_{\rm min}$ to $L_{\rm max} = 96$ we obtained the effective critical exponents shown in Fig. 16. For the dilution closest to the pure model ($p = 0.7$), the system is influenced by the pure fixed point with $1/\nu = 1.5863(33)$. On the other hand, when the bond concentration is small ($p = 0.4$), the vicinity of the percolation fixed point where $1/\nu \approx 1.12$ induces a decrease of $1/\nu$ below its expected disorder value. The dilution for which the cross-over effects are the least is around $p = 0.55$ which

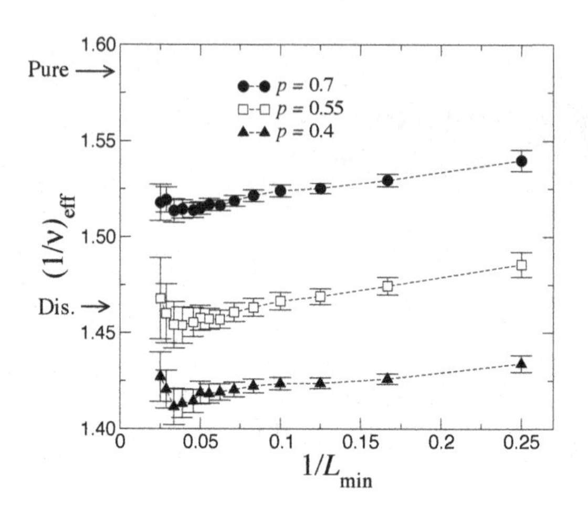

Fig. 16. Effective exponents $(1/\nu)_{\text{eff}}$ as obtained from fits to the behaviour of $\mathrm{d}\ln[m]_{\text{av}}/\mathrm{d}K \propto L^{1/\nu}$ as a function of $1/L_{\text{min}}$ for $p = 0.4$, 0.55, and 0.7. The upper limit of the fit range is $L_{\text{max}} = 96$.

suggests that the scaling corrections should be rather small for this specific dilution.

The main problem of the FSS study is the competition between different fixed points (pure, disorder, percolation) in combination with corrections-to-scaling terms $\propto L^{-\omega}$, which we found hard to control for bond dilution. In contrast to recent claims for the site-diluted model that $\omega \approx 0.4$, we were not able to extract a reliable estimate of ω from our data for bond dilution.

In a second set of simulations we examined the temperature scaling of the magnetisation and susceptibility for lattice sizes up to $L = 40$. This data allows direct estimates of the exponents β and γ whose relative deviation from the pure model is comparable to that of ν, e.g. $\gamma = 1.2396(13)$ (pure[191]) and $\gamma = 1.342(10)$ (disordered[192]). As a function of the reduced temperature $\tau = (K_c - K)$ ($\tau < 0$ in the low-temperature (LT) phase and $\tau > 0$ in the high-temperature (HT) phase) and the system size L, the susceptibility is expected to scale as

$$[\chi(\tau, L)]_{\text{av}} \sim |\tau|^{-\gamma} g_{\pm}(L^{1/\nu}|\tau|) \ , \tag{143}$$

where $g_{\pm}$ is a scaling function of the variable $x = L^{1/\nu}|\tau|$ and the subscript $\pm$ stands for the HT/LT phases. Assuming $[\chi(\tau)]_{\text{av}} \propto |\tau|^{-\gamma_{\text{eff}}}$ without any corrections-to-scaling terms, we can define a temperature dependent *effective* critical exponent $\gamma_{\text{eff}}(|\tau|) = -\mathrm{d}\ln[\chi]_{\text{av}}/\mathrm{d}\ln|\tau|$, which should converge towards the asymptotic critical exponent γ when $L \to \infty$ and $|\tau| \to 0$. Our results for $p = 0.7$

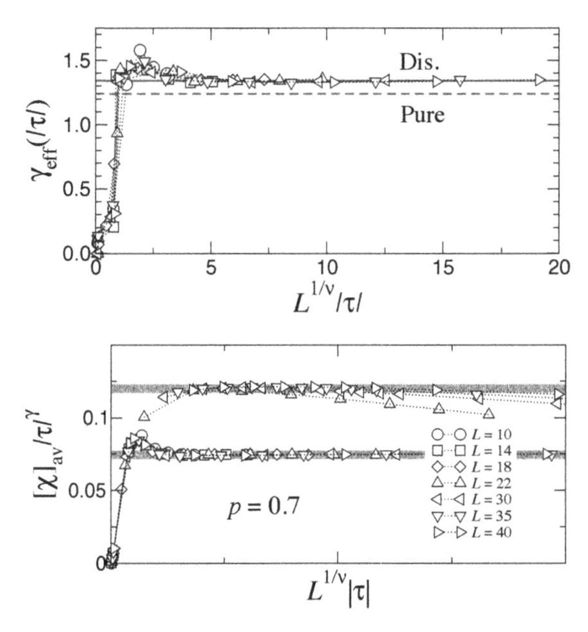

Fig. 17. *Top:* Variation of the temperature dependent effective critical exponent $\gamma_{\rm eff}(|\tau|) = -{\rm d}\ln[\chi]_{\rm av}/{\rm d}\ln|\tau|$ (in the low-temperature phase) as a function of the rescaled temperature $L^{1/\nu}|\tau|$ for the bond-diluted Ising model with $p = 0.7$ and several lattice sizes L. The horizontal solid and dashed lines indicate the site-diluted and pure values of γ, respectively. *Bottom:* The figure below shows the critical amplitudes $\Gamma_\pm$ above and below the critical temperature.

are shown in Fig. 17. For the greatest sizes, the effective exponent $\gamma_{\rm eff}(|\tau|)$ is stable around 1.34 when $|\tau|$ is not too small, i.e., when the finite-size effects are not too strong. The plot of $\gamma_{\rm eff}(|\tau|)$ vs. the rescaled variable $L^{1/\nu}|\tau|$ shows that the critical power-law behaviour holds in different temperature ranges for the different sizes studied. By analysing the temperature behaviour of the susceptibility, we also have directly extracted the power-law exponent γ using error weighted least-squares fits and choosing the temperature range that gives the smallest χ^2/d.o.f for several system sizes. The results are consistent with $\gamma \approx 1.34 - 1.36$, cf. Table 2.

From the previous expression of the susceptibility as a function of the reduced temperature and size, it is instructive to plot the scaling function $g_\pm(x)$. For finite size and $|\tau| \neq 0$, the scaling functions may be Taylor expanded in powers of the inverse scaling variable $x^{-1} = (L^{1/\nu}|\tau|)^{-1}$, $[\chi_\pm(\tau, L)]_{\rm av} = |\tau|^{-\gamma}[g_\pm(\infty) + x^{-1}g'_\pm(\infty) + O(x^{-2})]$, where the amplitude $g_\pm(\infty)$ is usually denoted by $\Gamma_\pm$.

Table 2. Critical exponents and critical amplitude ratio of the susceptibility as measured with different techniques.

Technique	γ	Γ_+/Γ_-	ω	Ref.
Neutron scattering	$1.44(6)$	2.2	0.5	193[a]
	$1.31(3)$	$2.8(2)$		194,195[b]
	$1.37(4)$	$2.40(2)$		174[c]
RG		2.2		196
	1.318		$0.39(4)$	197,198[d]
	$1.330(17)$		$0.25(10)$	199[e]
MC	$1.342(10)$		0.37	192[f]
	$1.34(1)$	$1.62(10)$		200[g]
	$1.342(7)$			201[h]
	$1.314(4)$	$1.67(15)$		202[i]
HTS	$1.305(5)$			184[j]

[a] $Fe_{1-x}Zn_xF_2$, $x = 0.4, 0.5$, $|\tau| \sim 10^{-2}$.
[b] $Fe_{0.46}Zn_{0.54}F_2$, $1.5 \times 10^{-3} \leq |\tau| \leq 10^{-1}$.
[c]*ic*5 $Mn_{0.75}Zn_{0.25}F_2$, $4 \times 10^{-4} \leq |\tau| \leq 2 \times 10^{-1}$.
[d] 4 loop approximation.
[e] 6 loop approximation, fixed dimension.
[f] site dilution, $p = 0.4$ to 0.8.
[g] bond dilution, $p = 0.7$. The correction to scaling is too small to be determined.
[h] site dilution, $p = 0.8$. The observed correction to scaling could be the next-to-leading one.
[i] site dilution, $p = 0.8$.
[j] bond dilution, $p = 0.6$ to 0.7.

Multiplying by $|\tau|^\gamma$ leads to

$$[\chi_\pm(\tau, L)]_{\rm av}|\tau|^\gamma = g_\pm(x) = \Gamma_\pm + O(x^{-1}) \ . \tag{144}$$

When $|\tau| \to 0$ but with L still larger than the correlation length ξ, one should recover the critical behaviour given by $g_\pm(x) = O(1)$. The critical amplitudes $\Gamma_\pm$ follow, as shown in the lower plot of Fig. 17. Some experimental and numerical estimates are compiled in Table 2.

To summarize, this application is a good example for how large-scale Monte Carlo simulations employing the cluster update algorithm can be used to investigate the influence of quenched bond dilution on the critical properties of the 3D Ising. It also illustrates how scaling and finite-size scaling analyses can be applied to a non-trival problem.

7.2. *Polymer statistics: Adsorption phenomena*

Polymers in dilute solutions are found at high temperatures typically in extended random coil conformations.[203–205] Lowering the temperature, entropy becomes less important and due to the monomer-monomer attraction globular conformations gain weight until the polymer collapses at the so-called θ-point in a cooperative rearrangement of the monomers.[203–205] The globular conformations are relatively compact with little internal structure. Hence, entropy does still play some role, and a further freezing transition towards low-degenerate crystalline energy dominated states is expected and indeed observed.[206,207] For sufficiently short-range interactions these two transitions may fall together,[208] but in general they are clearly distinct.

The presence of an attractive substrate adds a second energy scale to the system which introduces several new features. Apart from the adsorption transition,[209,210] it also induces several low-temperature structural phases by the competition between monomer-monomer and monomer-surface attraction whose details depend on the exact number of monomers. Theoretical predictions may guide future experiments on such small scales which appear feasible due to recent advances of experimental techniques. Among such sophisticated techniques at the nanometer scale are, e.g., atomic force microscopy (AFM), where it is possible to measure the contour length and the end-to-end distance of individual polymers[211] or to quantitatively investigate the peptide adhesion on semiconductor surfaces.[212] Another experimental tool with an extraordinary resolution in positioning and accuracy in force measurements are optical tweezers.[213,214]

With this motivation we recently performed a careful classification of thermodynamic phases and phase transitions for a range of surface attraction strengths and temperatures and compared the results for end-grafted polymers[215] with those of non-grafted polymers[216] that can move freely within a simulation box.[217] In these studies we employed a bead-stick model of a linear polymer with three energy contributions:

$$
\begin{aligned}
E = {} & 4\sum_{i=1}^{N-2}\sum_{j=i+2}^{N}\left(r_{ij}^{-12} - r_{ij}^{-6}\right) + \frac{1}{4}\sum_{i=1}^{N-2}\left(1 - \cos\vartheta_i\right) \\
& + \epsilon_s \sum_{i=1}^{N}\left(\frac{2}{15} z_i^{-9} - z_i^{-3}\right) \quad . \qquad (145)
\end{aligned}
$$

The first two terms are a standard 12-6 Lennard-Jones (LJ) potential and a weak bending energy describing the bulk behaviour. The distance between the monomers i and j is r_{ij} and $0 \le \vartheta_i \le \pi$ denotes the bending angle between the

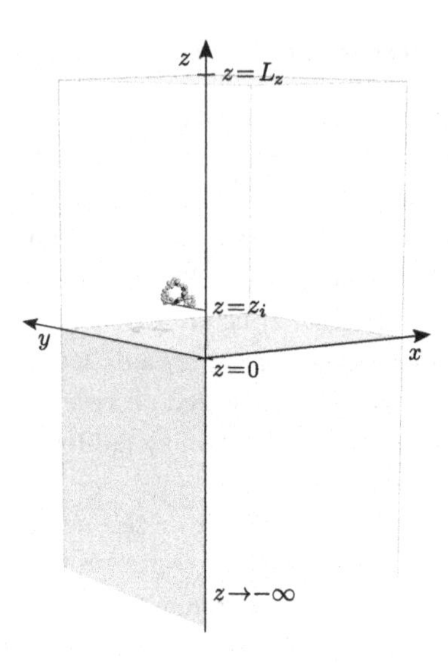

Fig. 18. Sketch of a single polymer subject to an attractive substrate at $z = 0$. The hard wall at $z = L_z$ prevents a non-grafted polymer from escaping.

ith, $(i + 1)$th, and $(i + 2)$th monomer. The third term is specific to an attractive substrate. This 9-3 LJ surface potential follows by integration over the continuous half-space $z < 0$ (cf. Fig. 18), where every space element interacts with each monomer by the usual 12-6 LJ expression.[218] The relative strength of the two LJ interactions is continuously varied by considering ϵ_s as a control parameter.

We employed parallel tempering simulations to a 40mer once grafted with one end to the substrate in the potential minimum and once freely moving in the space between the substrate and a hard wall a distance $L_z = 60$ away. There exist several attempts to optimise the choice of the simulation points β_i,[89,90] but usually one already gets a reasonable performance when observing the histograms and ensuring the acceptance probability to be around 50%, which approximately requires an equidistribution in β. We employed $64 - 72$ different replicas with 50 000 000 sweeps each, from which every 10th value was stored in a time series – the autocorrelation time in units of sweeps turned out to be of the order of thousands. Finally, all data are combined by the multi-histogram technique (using the variant of Ref. 219).

Apart from the internal energy and specific heat, a particular useful quantity for polymeric systems is the squared radius of gyration $R^2_{\rm gyr} = \sum_{i=1}^{N} (\vec{r}_i - \vec{r}_{\rm cm})^2$, with $\vec{r}_{\rm cm} = (x_{\rm cm}, y_{\rm cm}, z_{\rm cm}) = \sum_{i=1}^{N} \vec{r}_i/N$ being the center-of-mass of the polymer. In the presence of a symmetry breaking substrate, it is useful to also monitor the tensor components parallel and perpendicular to the substrate, $R_{\parallel}^2 = \sum_{i=1}^{N}[(x_i - x_{\rm cm})^2 + (y_i - y_{\rm cm})^2]$ and $R_{\perp}^2 = \sum_{i=1}^{N}(z_i - z_{\rm cm})^2$. As an indicator for adsorption one may take the distance of the center-of-mass of the polymer to the surface. Additionally, we also analyzed the mean number of monomers docked to the surface n_s where for the continuous substrate potential we defined a monomer i to be docked if $z_i < z_c \equiv 1.5$.

The main results are summarized in the phase diagram shown in Fig. 19. It is constructed using the profile of several canonical fluctuations as shown for the specific heat in Fig. 20. For the non-grafted polymer this plot clearly reveals the freezing and adsorption transitions. Freezing leads to a pronounced peak near $T = 0.25$ (we use units in which $k_B = 1$) almost independently of the surface attraction strengths. That this is indeed the freezing transition is confirmed by the very rigid crystalline structures found below this temperature. To differentiate between the different crystalline structures, the radius of gyration, its tensor components parallel and perpendicular to the substrate, and the number of surface contacts were analyzed. This revealed that the crystalline phases arrange in a different number of layers to minimize the energy. For high surface attraction strengths, a single layer is favored (AC1), and for decreasing ϵ_s the number of layers increases until for the 40mer a maximal number of 4 layers is reached (AC4), cf. the representative conformations depicted in the right panel of Fig. 19. The fewer layers are involved in a layering transition, the more pronounced is that transition. Raising the temperature above the freezing temperature, polymers form adsorbed and still rather compact conformations. This is the phase of adsorbed globular (AG) conformations that can be subdivided into droplet-like globules for surface interactions ϵ_s that are not strong enough to induce a single layer below the freezing transition and more pancake-like flat conformations (AG1) at temperatures above the AC1 phase. At higher temperatures, two scenarios can be distinguished. For small adsorption strength ϵ_s, a non-grafted polymer first desorbs from the surface [from AG to the desorbed globular (DG) bulk phase] and disentangles at even higher temperatures [from DG to the desorbed expanded bulk phase (DE)]. For larger ϵ_s, the polymer expands while it is still adsorbed to the surface (from AG/AG1 to AE) and desorbs at higher temperatures (from AE to DE). The collapse transition in the adsorbed phase takes place at a lower temperature compared to the desorbed phase because the deformation at the substrate leads to an effective reduction of the number of contacts.

Grafting the polymer to the substrate mainly influences the adsorption transition. Figure 20(b), e.g., reveals that it is strongly weakened for all ϵ_s. Due to grafting, the translational entropy for desorbed chains is strongly reduced. As a consequence adsorption of finite grafted polymers appears to be continuous, in

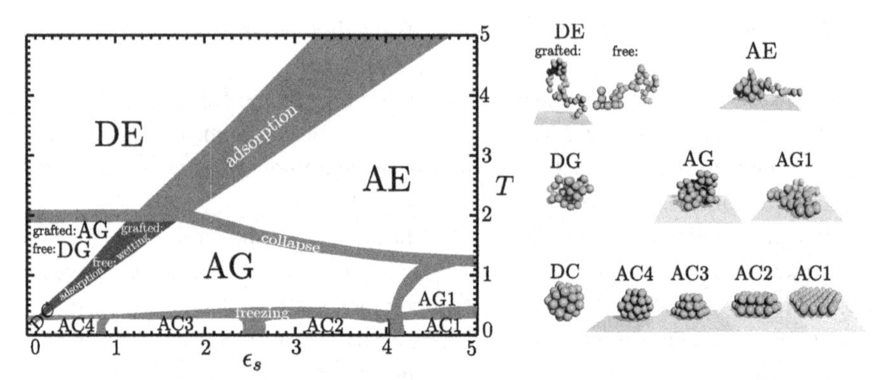

Fig. 19. The pseudo-phase diagram parametrized by adsorption strength ϵ_s and temperature T for a 40mer. The gray transition regions have a broadness that reflects the variation of the corresponding peaks of the fluctuations of canonical expectation values we investigated. Phases with an 'A/D' are adsorbed/desorbed. 'E', 'G' and 'C' denote phases with increasing order: expanded, globular and compact/crystalline. The right panel shows representative conformations of the individual phases.

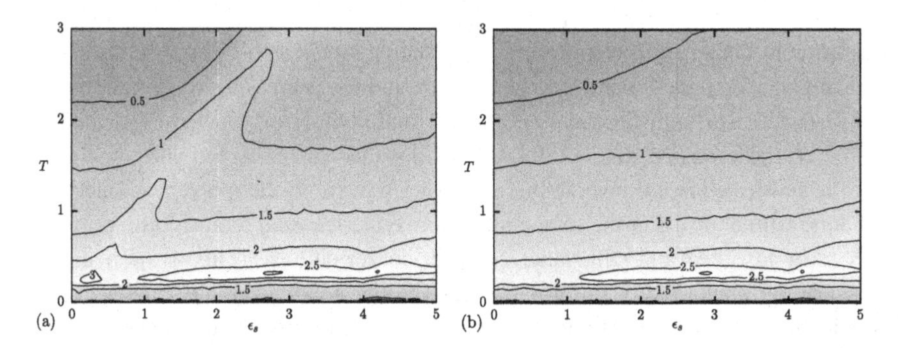

Fig. 20. Specific-heat profile, $c_V(\epsilon_s, T)$, for (a) the non-grafted and (b) the grafted polymer.

contrast to the non-grafted case where this behaviour becomes apparent for very long chains only. The reason is that *all* conformations of a grafted polymer are influenced by the substrate, because they cannot escape. Hence, the first-order-like conformational rearrangement of extended non-grafted polymers upon adsorption is not necessary and the adsorption is continuous.

The case of globular chains has to be discussed separately. While non-grafted globular chains adsorb continuously, for grafted globular chains it even is nontrivial to identify an adsorption transition. A globular chain attached to a substrate always has several surface contacts such that a "desorbed globule" stops to be a well-defined description here. For stronger surface attraction one might, however, identify the transition from attached globules that only have a few contacts

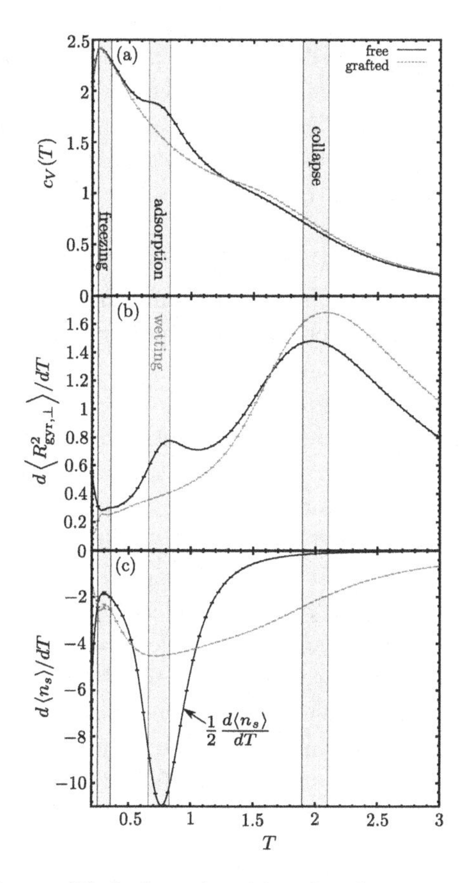

Fig. 21. (a) Specific heat $c_V(T)$, (b) fluctuation of the radius of gyration component perpendicular to the substrate $\mathrm{d}\left\langle R^2_{\mathrm{gyr},\perp}\right\rangle(T)/\mathrm{d}T$, and (c) fluctuation of the number of monomers in contact with the substrate $\mathrm{d}\left\langle n_s\right\rangle(T)/\mathrm{d}T$ for weak surface attraction, $\epsilon_s = 0.7$, where the adsorption occurs at a lower temperature than the collapse.

to docked conformations with the wetting transition. This roughly coincides with the position of the adsorption transition for the free chain between DG and AG in the phase diagram and is illustrated for $\epsilon_s = 0.7$ in Fig. 21. For a non-grafted polymer, at the adsorption transition a peak is visible in $c_V(T)$, $\mathrm{d}\left\langle R^2_{\mathrm{gyr},\perp}\right\rangle/\mathrm{d}T$ and $\mathrm{d}\left\langle n_s\right\rangle/\mathrm{d}T$. For the grafted polymer, on the other hand, the first two peaks disappear and with it the adsorption transition. Only a signal in the number of surface contacts is left. This change of surface contacts in an otherwise unchanged attached globule signals the wetting transition.

To summarize, this example was chosen to illustrate the application of extensive parallel tempering simulations to analyze and compare the whole phase diagram of a generic off-lattice model for grafted and non-grafted polymers as a function of temperature and surface interaction strength. The main differences between the two cases were found at and above the adsorption transition where the restriction of translational degrees of freedom due to grafting becomes important.

8. Concluding Remarks

The aim of this chapter is to give an elementary introduction into the basic principles underlying modern Markov chain Monte Carlo simulations and to illustrate their usefulness by two advanced applications to quenched, disordered spin systems and adsorption phenomena of polymers.

The simulation algorithms employing local update rules are very generally applicable but suffer from critical slowing down at second-order phase transitions. Non-local cluster update methods are much more efficient but more specialized. Some generalizations from Ising to Potts and O(n) symmetric spin models have been indicated. In principle also other models may be efficiently simulated by cluster updates, but there does not exist a general strategy for their construction. Reweighting techniques and generalized ensemble ideas such as simulated and parallel tempering, the multicanonical ensemble and Wang-Landau method can be adapted to almost any statistical physics problem where rare-event states hamper the dynamics. Well known examples are first-order phase transitions and spin glasses, but also some macromolecular systems fall into this class. The performance of the various algorithms can be judged by statistical error analysis which is completely general. Finally, also the outlined scaling and finite-size scaling analyses can be applied to virtually any model exhibiting critical phenomena as was exemplified for a disordered spin system.

Acknowledgements

I thank Yurij Holovatch for his kind invitation to present one of the Ising Lectures at the Institute for Condensed Matter Physics of the National Academy of Sciences of Ukraine, Lviv, Ukraine.

I gratefully acknowledge the contributions to the work reviewed here by my collaborators, in particular Michael Bachmann, Bertrand Berche, Pierre-Emmanuel Berche, Elmar Bittner, Christophe Chatelain, Monika Möddel, Thomas Neuhaus, Andreas Nußbaumer, Stefan Schnabel, and Martin Weigel, and thank Bernd Berg, Kurt Binder, David Landau, Yuko Okamoto, and Bob Swendsen for many useful discussions.

This work was partially supported by DFG Grant No. JA 483/24-3, DFG

Forschergruppe FOR877 under Grant No. Ja 483/29-1, DFG Sonderforschungsbereich/Transregio SFB/TRR 102 Project B04, Graduate School of Excellence GSC 185 "BuildMoNa", DFH-UFA German-French Graduate School under Grant No. CDFA-02-07, and the computer time Grant No. hlz17 of NIC, Forschungszentrum Jülich.

References

1. M. E. J. Newman and G. T. Barkema, *Monte Carlo Methods in Statistical Physics* (Clarendon Press, Oxford, 1999).
2. D. P. Landau and K. Binder, *Monte Carlo Simulations in Statistical Physics* (Cambridge University Press, Cambridge, 2000).
3. K. Binder and D. W. Heermann, *Monte Carlo Simulations in Statistical Physics: An Introduction*, 4th edition (Springer, Berlin, 2002).
4. B. A. Berg, *Markov Chain Monte Carlo Simulations and Their Statistical Analysis* (World Scientific, Singapore, 2004).
5. W. Lenz, Phys. Z. **21**, 613 (1920); E. Ising, Z. Phys. **31**, 253 (1925).
6. N. Metropolis and S. Ulam, J. Americ. Stat. Ass. **44**, 335 (1949).
7. J. M. Hammersley and D. C. Handscomb, *Monte Carlo Methods* (London, 1965).
8. A. A. Markov, Izv. Adad. Nauk SPB VI, 61 (1907).
9. N. G. van Kampen, *Stochastic Processes in Physics and Chemistry* (North-Holland, Amsterdam, 1981).
10. J. Honerkamp, *Statistical Physics – An Advanced Approach with Applications* (Springer, Berlin, 1998).
11. K. Binder, in *Monte Carlo Methods in Statistical Physics*, ed. K. Binder (Springer, Berlin, 1979).
12. D. W. Heermann, *Computer Simulation Methods in Theoretical Physics*, 2nd ed. (Springer, Berlin, 1990).
13. K. Binder (ed.), *The Monte Carlo Method in Condensed Matter Physics* (Springer, Berlin, 1992).
14. N. Metropolis, A. W. Rosenbluth, M. N. Rosenbluth, A. H. Teller, and E. Teller, J. Chem. Phys. **21**, 1087 (1953).
15. W. K. Hastings, Biometrika **57**, 97 (1970).
16. S. Schnabel, W. Janke, and M. Bachmann, J. Comp. Phys. **230**, 4454 (2011).
17. S. Kirkpatrick, C. D. Gelatt Jr., and M. P. Vecchi, Science **220**, 671 (1983).
18. D. Bouzida, S. Kumar, and R. H. Swendsen, Phys. Rev. A **45**, 8894 (1992).
19. M. A. Miller, L. M. Amon, W. P. Reinhardt, Chem. Phys. Lett. **331**, 278 (2000).
20. R. H. Swendsen, Physics Procedia **15**, 81 (2011).
21. G. O. Roberts, A. Gelman, and W. R. Gilks, Ann. Appl. Probab. **7**, 110 (1997).
22. M. Bédard, Stochastic Process. Appl. **118**, 2198 (2008).
23. W. Janke, *Pseudo random numbers: Generation and quality checks*, invited lecture notes, in Proceedings of the Euro Winter School *Quantum Simulations of Complex Many-Body Systems: From Theory to Algorithms*, eds. J. Grotendorst, D. Marx, and A. Muramatsu, John von Neumann Institute for Computing, Jülich, NIC Series, Vol. **10**, pp. 447–458 (2002), and references therein.

24. A. M. Ferrenberg, D. P. Landau, and Y. J. Wong, Phys. Rev. Lett. **69**, 3382 (1992).
25. R. J. Glauber, J. Math. Phys. **4**, 294 (1963).
26. W. Janke, in *Ageing and the Glass Transition*, ed. M. Henkel, M. Pleimling, and R. Sanctuary, Lect. Notes Phys. **716** (Springer, Berlin, Heidelberg, 2007), pp. 207–260.
27. H. E. Stanley, *Introduction to Phase Transitions and Critical Phenomena* (Oxford Press, Oxford, 1979).
28. J. J. Binney, N. J. Dowrick, A. J. Fisher, and M. E. J. Newman, *The Theory of Critical Phenomena* (Oxford University Press, Oxford, 1992).
29. D. A. Lavis and G. M. Bell, *Statistical Mechanics of Lattice Systems 2* (Springer, Berlin, 1999).
30. See the volumes of review articles edited by C. Domb and J. L. Lebowitz (eds.): *Phase Transitions and Critical Phenomena* (Academic Press, New York).
31. See, e.g., the photographs in Fig. 1.6 of Ref. 27.
32. J. Goodman and A. D. Sokal, Phys. Rev. Lett. **56**, 1015 (1986); Phys. Rev. D **40**, 2035 (1989).
33. R. G. Edwards, J. Goodman, and A. D. Sokal, Nucl. Phys. B **354**, 289 (1991).
34. A. D. Sokal, *Monte Carlo methods in statistical mechanics: Foundations and new algorithms*, Lecture Notes, Cours de Troisième Cycle de la Physique en Suisse Romande, Lausanne, Switzerland (1989).
35. A. D. Sokal, *Bosonic algorithms*, in *Quantum Fields on the Computer*, ed. M. Creutz (World Scientific, Singapore, 1992), p. 211.
36. W. Janke, *Nonlocal Monte Carlo algorithms for statistical physics applications*, Mathematics and Computers in Simulations **47**, 329–346 (1998).
37. W. Janke, *Recent developments in Monte Carlo simulations of first-order phase transitions*, in *Computer Simulations in Condensed Matter Physics VII*, eds. D. P. Landau, K. K. Mon, and H.-B. Schüttler (Springer, Berlin, 1994), p. 29.
38. W. Janke, *First-order phase transitions*, in *Computer Simulations of Surfaces and Interfaces*, NATO Science Series, II. Mathematics, Physics and Chemistry – Vol. **114**, eds. B. Dünweg, D. P. Landau, and A. I. Milchev (Kluwer, Dordrecht, 2003), pp. 111–135.
39. R. H. Swendsen and J.-S. Wang, Phys. Rev. Lett. **58**, 86 (1987).
40. P. W. Kasteleyn and C. M. Fortuin, J. Phys. Soc. Japan **26** (Suppl.), 11 (1969).
41. C. M. Fortuin and P. W. Kasteleyn, Physica **57**, 536 (1972).
42. C. M. Fortuin, Physica **58**, 393 (1972).
43. C. M. Fortuin, Physica **59**, 545 (1972).
44. D. Stauffer and A. Aharony, *Introduction to Percolation Theory*, 2nd ed. (Taylor and Francis, London, 1992).
45. J. Hoshen and R. Kopelman, Phys. Rev. B **14**, 3438 (1976).
46. P. L. Leath, Phys. Rev. B **14**, 5046 (1976).
47. M. E. J. Newman and R. M. Ziff, Phys. Rev. E **64**, 016706 (2001).
48. W. Janke and A. M. J. Schakel, Nucl. Phys. B **700**, 385 (2004); Comp. Phys. Comm. **169**, 222 (2005); Phys. Rev. E **71**, 036703 (2005); Phys. Rev. Lett. **95**, 135702 (2005); e-print `arXiv:cond-mat/0508734` and Braz. J. Phys. **36**, 708 (2006). For a review, see *Spacetime approach to phase transitions*, in *Order, Disorder and Critical-*

ity: Advanced Problems of Phase Transition Theory, Vol. 2, ed. Y. Holovatch (World Scientific, Singapore, 2007), pp. 123–180, and the extensive list of references to earlier work given therein.
49. U. Wolff, Phys. Rev. Lett. **62**, 361 (1989).
50. R. B. Potts, Proc. Camb. Phil. Soc. **48**, 106 (1952).
51. U. Wolff, Nucl. Phys. B **322**, 759 (1989).
52. M. Hasenbusch, Nucl. Phys. B **333**, 581 (1990).
53. U. Wolff, Nucl. Phys. B **334**, 581 (1990).
54. U. Wolff, Phys. Lett. A **228**, 379 (1989).
55. C. F. Baillie, Int. J. Mod. Phys. C **1**, 91 (1990).
56. M. Hasenbusch and S. Meyer, Phys. Lett. B **241**, 238 (1990).
57. R. H. Swendsen, J.-S. Wang, and A. M. Ferrenberg, in *The Monte Carlo Method in Condensed Matter Physics*, ed. K. Binder (Springer, Berlin, 1992).
58. W. Janke, Phys. Lett. A **148**, 306 (1990).
59. C. Holm and W. Janke, Phys. Rev. B **48**, 936 (1993).
60. X.-L. Li and A. D. Sokal, Phys. Rev. Lett. **63**, 827 (1989); *ibid.* **67**, 1482 (1991).
61. W. H. Press, S. A. Teukolsky, W. T. Vetterling, and B. P. Flannery, *Numerical Recipes in Fortran 77 – The Art of Scientific Computing*, 2nd edition (Cambridge University Press, Cambridge, 1999).
62. H. Müller-Krumbhaar and K. Binder, J. Stat. Phys. **8**, 1 (1973).
63. N. Madras and A. D. Sokal, J. Stat. Phys. **50**, 109 (1988).
64. T. W. Anderson, *The Statistical Analysis of Time Series* (Wiley, New York, 1971).
65. M. B. Priestley, *Spectral Analysis and Time Series*, 2 vols. (Academic, London, 1981), Chapters 5-7.
66. W. Janke, *Statistical analysis of simulations: Data correlations and error estimation*, invited lecture notes, in Proceedings of the Euro Winter School *Quantum Simulations of Complex Many-Body Systems: From Theory to Algorithms*, eds. J. Grotendorst, D. Marx, and A. Muramatsu, John von Neumann Institute for Computing, Jülich, NIC Series, Vol. **10**, pp. 423–445 (2002).
67. A. M. Ferrenberg, D. P. Landau, and K. Binder, J. Stat. Phys. **63**, 867 (1991).
68. W. Janke and T. Sauer, J. Stat. Phys. **78**, 759 (1995).
69. B. Efron, *The Jackknife, the Bootstrap and Other Resampling Plans* (Society for Industrial and Applied Mathematics [SIAM], Philadelphia, 1982).
70. R. G. Miller, Biometrika **61**, 1 (1974).
71. A. M. Ferrenberg and R. H. Swendsen, Phys. Rev. Lett. **61**, 2635 (1988).
72. A. M. Ferrenberg and R. H. Swendsen, Phys. Rev. Lett. **63**, 1658(E) (1989).
73. B. Kaufman, Phys. Rev. **76**, 1232 (1949).
74. A. E. Ferdinand and M.E. Fisher, Phys. Rev. **185**, 832 (1969).
75. P. D. Beale, Phys. Rev. Lett. **76**, 78 (1996).
76. N. Wilding, *Computer simulation of continuous phase transitions*, in *Computer Simulations of Surfaces and Interfaces*, NATO Science Series, II. Mathematics, Physics and Chemistry – Vol. **114**, eds. B. Dünweg, D. P. Landau, and A. I. Milchev (Kluwer, Dordrecht, 2003), pp. 161–171.
77. A. M. Ferrenberg and R. H. Swendsen, Phys. Rev. Lett. **63**, 1195 (1989).
78. S. Kumar, D. Bouzida, R. H. Swendsen, P. A. Kollman, and J. M. Rosenberg, J. Comp. Chem. **13**, 1011 (1992).

79. S. Kumar, J. M. Rosenberg, D. Bouzida, R. H. Swendsen, and P. A. Kollman, J. Comp. Chem. **16**, 1339 (1995).
80. T. Bereau and R. H. Swendsen, J. Comp. Phys. **228**, 6119 (2009).
81. E. Gallicchio, M. Andrec, A. K. Felts, and R.M. Levy, J. Phys. Chem. B **109**, 6722 (2005).
82. J. D. Chodera, W. C. Swope, J. W. Pitera, C. Seok, and K. A. Dill, J. Chem. Theory Comput. **3**, 26 (2007).
83. E. Marinari and G. Parisi, Europhys. Lett. **19**, 451 (1992).
84. A. P. Lyubartsev, A. A. Martsinovski, S. V. Shevkunov, and P. N. Vorontsov-Velyaminov, J. Chem. Phys. **96**, 1776 (1992).
85. R. H. Swendsen and J.-S. Wang, Phys. Rev. Lett. **57**, 2607 (1986).
86. J.-S. Wang and R. H. Swendsen, Prog. Theor. Phys. Suppl. **157**, 317 (2005).
87. C. J. Geyer, in *Computing Science and Statistics*, Proceedings of the 23rd Symposium on the Interface, ed. E. M. Keramidas (Interface Foundation, Fairfax, Virginia, 1991); pp. 156–163; C. J. Geyer and E. A. Thompson, J. Am. Stat. Assoc. **90**, 909 (1995).
88. K. Hukushima and K. Nemoto, J. Phys. Soc. Japan **65**, 1604 (1996).
89. H. G. Katzgraber, S. Trebst, D. A. Huse, and M. Troyer, J. Stat. Mech. P03018 (2006).
90. E. Bittner, A. Nußbaumer, and W. Janke, Phys. Rev. Lett. **101**, 130603 (2008).
91. J. Gross, W. Janke, and M. Bachmann, Comp. Phys. Comm. **182**, 1638 (2011); Physics Procedia **15**, 29 (2011).
92. B. A. Berg and T. Neuhaus, Phys. Lett. B **267**, 249 (1991).
93. B. A. Berg and T. Neuhaus, Phys. Rev. Lett. **68**, 9 (1992).
94. B. A. Berg, Fields Inst. Comm. **26**, 1 (2000).
95. B. A. Berg, Comp. Phys. Comm. **147**, 52 (2002).
96. W. Janke, Physica A **254**, 164 (1998).
97. W. Janke, *Histograms and all that*, invited lectures, in *Computer Simulations of Surfaces and Interfaces*, NATO Science Series, II. Mathematics, Physics and Chemistry – Vol. **114**, eds. B. Dünweg, D. P. Landau, and A. I. Milchev (Kluwer, Dordrecht, 2003), pp. 137–157.
98. W. Janke, Int. J. Mod. Phys. C **3**, 1137 (1992).
99. K. Binder, in *Phase Transitions and Critical Phenomena*, Vol. 5b, eds. C. Domb and M. S. Green (Academic Press, New York, 1976), p. 1.
100. G. M. Torrie and J. P. Valleau, Chem. Phys. Lett. **28**, 578 (1974); J. Comp. Phys. **23**, 187 (1977) 187; J. Chem. Phys. **66**, 1402 (1977).
101. B. A. Berg and W. Janke, Phys. Rev. Lett. **80**, 4771 (1998).
102. B. A. Berg, A. Billoire, and W. Janke, Phys. Rev. B **61**, 12143 (2000); Phys. Rev. E **65**, 045102 (RC) (2002); Physica A **321**, 49 (2003).
103. E. Bittner and W. Janke, Europhys. Lett. **74**, 195 (2006).
104. W. Janke (ed.), *Rugged Free Energy Landscapes: Common Computational Approaches to Spin Glasses, Structural Glasses and Biological Macromolecules*, Lecture Notes in Physics **736** (Springer, Berlin, 2008).
105. W. Janke, B. A. Berg, and M. Katoot, Nucl. Phys. B **382**, 649 (1992).
106. A. Nußbaumer, E. Bittner, and W. Janke, Europhys. Lett. **78**, 16004 (2007).
107. E. Bittner, A. Nußbaumer, and W. Janke, Nucl. Phys. B **820**, 694 (2009).
108. B. A. Berg, J. Stat. Phys **82**, 323 (1996).

109. B. A. Berg and W. Janke, unpublished notes (1996).
110. B. A. Berg, Comp. Phys. Comm. **153**, 397 (2003).
111. J. Goodman and A. D. Sokal, Phys. Rev. Lett. **56**, 1015 (1986); Phys. Rev. D **40**, 2035 (1989).
112. W. Janke and T. Sauer, Phys. Rev. E **49**, 3475 (1994).
113. W. Janke and S. Kappler, Nucl. Phys. B (Proc. Suppl.) **42**, 876 (1995).
114. W. Janke and S. Kappler, Phys. Rev. Lett. **74**, 212 (1995).
115. M. S. Carroll, W. Janke, and S. Kappler, J. Stat. Phys. **90**, 1277 (1998).
116. B. A. Berg and W. Janke, Phys. Rev. Lett. **98**, 040602 (2007).
117. T. Neuhaus and J. S. Hager, J. Stat. Phys. **113** 47 (2003).
118. K. Binder and M. H. Kalos, J. Stat. Phys. **22**, 363 (1980).
119. H. Furukawa and K. Binder, Phys. Rev. A **26**, 556 (1982).
120. M. Biskup, L. Chayes, and R. Kotecký, Europhys. Lett. **60**, 21 (2002); Comm. Math. Phys. **242**, 137 (2003); J. Stat. Phys. **116**, 175 (2003).
121. K. Binder, Physica A **319**, 99 (2003).
122. A. Nußbaumer, E. Bittner, T. Neuhaus, and W. Janke, Europhys. Lett. **75**, 716 (2006).
123. A. Nußbaumer, E. Bittner, and W. Janke, Phys. Rev. E **77**, 041109 (2008).
124. K. Leung and R. K. P. Zia, J. Phys. A **23**, 4593 (1990).
125. F. Wang and D. P. Landau, Phys. Rev. Lett. **86**, 2050 (2001); Phys. Rev. E **64**, 056101 (2001).
126. R. B. Griffiths, Phys. Rev. Lett. **24**, 1479 (1970).
127. G. S. Rushbrooke, J. Chem. Phys. **39**, 842 (1963).
128. R. B. Griffiths, Phys. Rev. Lett. **14**, 623 (1965).
129. B. D. Josephson, Proc. Phys. Soc. **92**, 269 (1967); *ibid.* 276 (1967).
130. M. E. Fisher, Phys. Rev. **180**, 594 (1969).
131. L. P. Widom, J. Chem. Phys. **43**, 3892 (1965); *ibid.* 3898 (1965).
132. L. P. Kadanoff, Physics **2**, 263 (1966).
133. K. G. Wilson and J. Kogut, Phys. Rep. C **12**, 75 (1974).
134. L. Onsager, Phys. Rev. **65**, 117 (1944).
135. B. M. McCoy and T. T. Wu, *The Two-Dimensional Ising Model* (Harvard University Press, Cambridge, 1973).
136. R. J. Baxter, *Exactly Solved Models in Statistical Mechanics* (Academic Press, New York, 1982).
137. L. Onsager, Nuovo Cimento (Suppl.) **6**, 261 (1949); see also the historical remarks in Refs. 135,136.
138. C. N. Yang, Phys. Rev. **85**, 808 (1952).
139. C. H. Chang, Phys. Rev. **88**, 1422 (1952).
140. B. Nickel, J. Phys. A: Math. Gen. **32**, 3889 (1999); J. Phys. A: Math. Gen. **33**, 1693 (2000); W. P. Orrick, B. G. Nickel, A. J. Guttmann, and J. H. H. Perk, Phys. Rev. Lett. **86**, 4120 (2001); J. Stat. Phys. **102**, 795 (2001).
141. S. Boukraa, A. J. Guttmann, S. Hassani, I. Jensen, J.-M. Maillard, B. Nickel, and N. Zenine, J. Phys. A: Math. Theor. **41**, 455202 (2008) [the coefficients of low- and high-T susceptibility series expansions up to order 2000 are given on `http://www.ms.unimelb.edu.au/~iwan/ising/Ising_ser.html`].

142. Y. Chan, A. J. Guttmann, B. G. Nickel, and J. H. H. Perk, J. Stat. Phys. **145**, 549 (2011) [e-print `arXiv:1012.5272v3` (cond-mat.stat-mech) with additional informations].
143. R. Kenna, D. A. Johnston, and W. Janke, Phys. Rev. Lett. **96**, 115701 (2006); *ibid.* **97**, 155702 (2006) [Publisher's Note: *ibid.* **97**, 169901(E) (2006)].
144. F. Y. Wu, Rev. Mod. Phys. **54**, 235 (1982).
145. F. Y. Wu, Rev. Mod. Phys. **55**, 315(E) (1983).
146. M. Weigel and W. Janke, Phys. Rev. B **62**, 6343 (2000).
147. M. E. Barber, in *Phase Transitions and Critical Phenomena*, Vol. 8, eds. C. Domb and J. L. Lebowitz (Academic Press, New York, 1983), p. 146.
148. V. Privman (ed.), *Finite-Size Scaling and Numerical Simulations of Statistical Systems* (World Scientific, Singapore, 1990).
149. K. Binder, in *Computational Methods in Field Theory*, Schladming Lecture Notes, eds. H. Gausterer and C. B. Lang (Springer, Berlin, 1992), p. 59.
150. G. Kamieniarz and H. W. J. Blöte, J. Phys. A **26**, 201 (1993).
151. J. Salas and A. D. Sokal, J. Stat. Phys. **98**, 551 (2000).
152. X. S. Chen and V. Dohm, Phys. Rev. E **70**, 056136 (2004).
153. V. Dohm, J. Phys. A **39**, L259 (2006).
154. W. Selke and L. N. Shchur, J. Phys. A **38**, L739 (2005).
155. M. Schulte and C. Drope, Int. J. Mod. Phys. C **16**, 1217 (2005).
156. M. A. Sumour, D. Stauffer, M. M. Shabat, and A. H. El-Astal, Physica A **368**, 96 (2006).
157. W. Selke, Eur. Phys. J. B **51**, 223 (2006); J. Stat. Mech. P04008 (2007).
158. J. D. Gunton, M. S. Miguel, and P. S. Sahni, in *Phase Transitions and Critical Phenomena*, Vol. 8, eds. C. Domb and J. L. Lebowitz (Academic Press, New York, 1983).
159. K. Binder, Rep. Prog. Phys. **50**, 783 (1987).
160. H. J. Herrmann, W. Janke, and F. Karsch (eds.): *Dynamics of First Order Phase Transitions* (World Scientific, Singapore, 1992).
161. M. E. Fisher and A. N. Berker, Phys. Rev. B **26**, 2507 (1982).
162. V. Privman, M. E. Fisher, J. Stat. Phys. **33**, 385 (1983).
163. K. Binder and D. P. Landau, Phys. Rev. B **30**, 1477 (1984).
164. M. S. S. Challa, D. P. Landau, and K. Binder, Phys. Rev. B **34**, 1841 (1986).
165. V. Privman and J. Rudnik, J. Stat. Phys. **60**, 551 (1990).
166. C. Borgs and R. Kotecky, J. Stat. Phys. **61**, 79 (1990).
167. J. Lee and J. M. Kosterlitz, Phys. Rev. Lett. **65**, 137 (1990).
168. C. Borgs, R. Kotecky, and S. Miracle-Solé, J. Stat. Phys. **62**, 529 (1991).
169. C. Borgs and W. Janke, Phys. Rev. Lett. **68**, 1738 (1992).
170. W. Janke, Phys. Rev. B **47**, 14757 (1993).
171. E. Bittner and W. Janke, Phys. Rev. E **84**, 036701 (2011).
172. M. Weigel and W. Janke, Phys. Rev. Lett. **102**, 100601 (2009); Phys. Rev. E **81**, 066701 (2010).
173. J. Cardy, *Scaling and Renormalization in Statistical Physics* (Cambridge University Press, Cambridge, 1996), Chap. 8.
174. P. W. Mitchell, R. A. Cowley, H. Yoshizawa, P. Böni, Y. J. Uemura, and R. J. Birgeneau, Phys. Rev. B **34**, 4719 (1986).
175. A. B. Harris, J. Phys. C **7**, 1671 (1974).

176. For a review, see B. Berche and C. Chatelain, *Phase transitions in two-dimensional random Potts models*, in *Order, Disorder and Criticality: Advanced Problems of Phase Transition Theory*, Vol. 1, ed. Y. Holovatch (World Scientific, Singapore, 2004), pp. 147–199.
177. W. Selke, L. N. Shchur, and A. L. Talapov, in *Annual Reviews of Computational Physics I*, ed. D. Stauffer (World Scientific, Singapore, 1994), pp. 17–54.
178. For a recent overview, see R. Folk, Y. Holovatch, and T. Yavors'kii, Physics Uspiekhi **173**, 175 (2003) [e-print `arXiv:cond-mat/0106468`].
179. C. Chatelain, B. Berche, W. Janke, and P.-E. Berche, Phys. Rev. E **64**, 036120 (2001).
180. P.-E. Berche, C. Chatelain, B. Berche, and W. Janke, Comp. Phys. Comm. **147**, 427 (2002); Eur. Phys. J. B **38**, 463 (2004).
181. B. Berche, P.-E. Berche, C. Chatelain, and W. Janke, Condens. Matter Phys. **8**, 47 (2005).
182. M. Hellmund and W. Janke, Comp. Phys. Comm. **147**, 435 (2002); Nucl. Phys. B (Proc. Suppl.) **106&107**, 923 (2002).
183. M. Hellmund and W. Janke, Phys. Rev. E **67**, 026118 (2003).
184. M. Hellmund and W. Janke, Phys. Rev. B **74**, 144201 (2006).
185. C. D. Lorenz and R. M. Ziff, Phys. Rev. E **57**, 230 (1998).
186. B. Derrida, Phys. Rep. **103**, 29 (1984); A. Aharony and A. B. Harris, Phys. Rev. Lett. **77**, 3700 (1996); S. Wiseman and E. Domany, Phys. Rev. Lett. **81**, 22 (1998).
187. L . Turban, Phys. Lett. A **75**, 307 (1980); J. Phys. C **13**, L13 (1980).
188. A. L. Talapov and H. W. J. Blöte, J. Phys. A **29**, 5727 (1996).
189. A. Aharony and A. B. Harris, Phys. Rev. Lett. **77**, 3700 (1996).
190. S. Wiseman and E. Domany, Phys. Rev. Lett. **81**, 22 (1998); Phys. Rev. E **58**, 2938 (1998).
191. R. Guida and J. Zinn-Justin, J. Phys. A **31**, 8103 (1998).
192. H. G. Ballesteros, L. A. Fernández, V. Martín-Mayor, A. Muñoz Sudupe, G. Parisi, and J. J. Ruiz-Lorenzo, Phys. Rev. B **58**, 2740 (1998).
193. R. J. Birgeneau, R. A. Cowley, G. Shirane, and H. Yoshizawa, J. Stat. Phys. **34**, 817 (1984).
194. D. P. Belanger, A. R. King, and V. Jaccarino, Phys. Rev. B **34**, 452 (1986).
195. D. P. Belanger, Braz. J. Phys. **30**, 682 (2000).
196. S. A. Newlove, J. Phys. C: Solid State Phys. **16**, L423 (1983).
197. R. Folk, Yu. Holovatch, and T. Yavors'kii, J. Phys. Stud. **2**, 213 (1998).
198. R. Folk, Y. Holovatch, and T. Yavors'kii, Phys. Rev. B **61**, 15114 (2000).
199. A. Pelissetto and E. Vicari, Phys. Rev. B **62**, 6393 (2000).
200. P.-E. Berche, C. Chatelain, B. Berche, and W. Janke, Eur. Phys. J. B **38**, 463 (2004).
201. P. Calabrese, V. Martín-Mayor, A. Pelissetto, and E. Vicari, Phys. Rev. E **68**, 036136 (2003).
202. D. Ivaneyko, J. Ilnytskyi, B. Berche, and Yu. Holovatch, Condens. Matter Phys. **8**, 149 (2005).
203. P.-G. de Gennes, *Scaling Concepts in Polymer Physics* (Cornell University Press, Ithaca and London, 1979).
204. J. des Cloizeaux and G. Jannink, *Polymers in Solution* (Clarendon Press, Oxford, 1990).
205. A. Y. Grosberg and A. R. Khokhlov, *Statistical Physics of Macromolecules* (American Institute of Physics, New York, 1994).

206. W. Paul, T. Strauch, F. Rampf, and K. Binder, Phys. Rev. E **75**, 060801(R) (2007).
207. T. Vogel, M. Bachmann, and W. Janke, Phys. Rev. E **76**, 061803 (2007).
208. M. P. Taylor, W. Paul, and K. Binder, Phys. Rev. E **79**, 050801 (2009).
209. E. Eisenriegler, K. Kremer, and K. Binder, J. Chem. Phys. **77**, 6296 (1982).
210. E. Eisenriegler, *Polymers near Surfaces: Conformation Properties and Relation to Critical Phenomena* (World Scientific, Singapore, 1993).
211. F. Kuhner, M. Erdmann, and H. E. Gaub, Phys. Rev. Lett. **97**, 218301 (2006).
212. M. Bachmann, K. Goede, A. Beck-Sickinger, M. Grundmann, A. Irbäck, and W. Janke, Angew. Chem. Int. Ed. **122**, 9721 (2010).
213. D. E. Smith, S. J. Tans, S. B. Smith, S. Grimes, D. L. Anderson, and C. Bustamante, Nature **413**, 748 (2001).
214. K. Kegler, M. Salomo, and F. Kremer, Phys. Rev. Lett. **98**, 058304 (2007).
215. M. Möddel, W. Janke, and M. Bachmann, Macromolecules **44**, 9013 (2011).
216. M. Möddel, M. Bachmann, and W. Janke, J. Phys. Chem. B **113**, 3314 (2009).
217. M. Möddel, W. Janke, and M. Bachmann, Phys. Chem. Chem. Phys. **12**, 11548 (2010).
218. W. A. Steele, Surface Sci. **36**, 317 (1973).
219. M. K. Fenwick, J. Chem. Phys. **129**, 125106 (2008).

Chapter 4

Ising Model on Connected Complex Networks

Krzysztof Suchecki
IFISC, Instituto de Física Interdisciplinar y Sistemas Complejos (CSIC-UIB), Campus Universitat Illes Balears, E-07122 Palma de Mallorca, Spain
ksucheck@ifisc.uib-csic.es

Janusz A. Hołyst
Faculty of Physics, Center of Excellence for Complex Systems Research, Warsaw University of Technology, Koszykowa 75, PL–00-662 Warsaw, Poland
jholyst@if.pw.edu.pl

Ising dynamics for a system of two weakly connected scale-free networks is analytically investigated using a properly tailored mean field approach. Since order parameters in both networks can be different there are three states of possible spin configurations that correspond to parallel ordered, antiparallel ordered and disordered phases. Transition temperatures between these states are calculated. There is a first-order (discontinuous) phase transition between a phase when both networks possess opposite order parameters and a phase when both networks are parallel ordered. At higher temperature a continuous transition to a paramagnetic phase takes place. The temperature of the first-order phase transition diminishes with the increasing inter-network links density and it becomes zero when the density reaches a critical value. Analytical results based on mean-field approximation are backed up in part with numerical Monte-Carlo simulations.

Contents

1. Introduction

The Ising model is a simple approach created to describe the basic behavior of magnetic materials, i.e. the emergence of ferromagnetic ordering.[1] Because of its very simple formulation, it has been considered outside its originally intended use, as an elementary model for processes where a macroscopic spontaneous order appears due to microscopic interactions between constituting particles or agents.

It has been also applied in the social science context, where e.g. spins represent some binary opinions and spin interactions correspond to social contacts.[2] With the rise of popularity of complex networks,[3–5] the Ising model has been naturally considered in such a topology, where there is neither regular lattice nor a specified dimensionality. The model in such topologies has been shown to behave somewhat different than in the regular lattices, with critical temperatures depending on a network size[6,7] or on topological parameters such as an exponent γ of scale-free node degree distributions.[8–10] Many different aspects have been studied, from antiferromagnetic interactions[11,12] and spin-glasses[13,14] to a directed structure of a network[15] or complex modular, hierarchical, fractal topologies.[16] Besides a large interest in properties of the Ising model and Ising-like models from the statistical physics point of view, they are also important for the opinion formation modeling.[2,17–19] The Ising model exhibits what can be described as a majority rule dynamics, where spins align themselves in the same way as their neighbors they interact with. This feature can often be found in social systems, where a given person changes his/her opinion to fit it to his/her neighbors' majority. Since it is common that social networks possess a modular structure of weakly coupled clusters[20] it is interesting to study the Ising model for such a topology.[21–23] The issue of dynamical models on connected networks is especially challenging, warranting investigation also for non-Ising models.[24] In fact, recent years brought a large increase of interest in critical properties of so-called interdependent networks (for a review see[25]) that can model catastrophic behavior of coupled technological systems.[26]

The chapter presents some aspects of the Ising model dynamics on a pair of connected networks[27,28] and shows some of the issues encountered while studying dynamical systems in complex network environments.

2. Ising Model

The Ising model describes magnetic materials as a lattice of spins with mutual interactions. Because exchange interactions, the strongest interactions between quantum spins, decay strongly with the inter-spin distance, the Ising model usually is limited to local interactions between the closest neighbors. The next simplification is the single direction that describes the spin component. A possible external magnetic field acts along this direction and spins can only place themselves parallel to this direction, either "up" or "down", reducing the spin variables to single bits of information.

The spins are usually localized in sites of regular lattices with interactions happening only between closest neighbors. This can be thought as if spins are vertices, while interactions are edges, forming a graph, e.g. a regular lattice graph. Each vertex contains a spin, so referring to spin or vertex are equivalent. Each spin can be assigned a label i to make them distinguishable. If so, each spin i has a value s_i that can be either -1 or $+1$. The Hamiltonian of the Ising model can thus be written as

$$H = -\sum_{j>i} J_{ij} s_i s_j - h \sum_i s_i \tag{1}$$

where s_i,s_j are spin values, h is an external magnetic field and J_{ij} is the exchange integral between the two spins - an effective energy of mutual interactions.

In a typical Ising model it is assumed that $J_{ij} = J\epsilon_{ij}$, where J is a constant exchange integral between neighbors and $\epsilon_{ij} = 1$ for neighboring i and j and is equal 0 otherwise. This means that all neighboring spin pairs interact in the same way. In a more general case values of J_{ij} may depend on the spin pair ij, that may lead to the complex patterns of equilibrium spin configurations corresponding to so-called spin-glass states.[13]

The simple Ising model, with a positive constant J, has two phases: a low temperature ferromagnetic phase where majority of spins are aligned in one direction (either -1 or $+1$) and a high temperature paramagnetic phase where spins have essentially random values. If the exchange integral J is negative, the ordered state is antiferromagnetic, i.e. nearest neighboring spins possess opposite directions or (if triangles are present in the lattice) a spin-glass state. The actual state of the system will depend on thermal fluctuations of the spin values and therefore on temperature T. When the number of sites in a regular lattice is very large the critical temperature T_c that separates ferromagnetic or antiferromagnetic states from a paramagnetic one is nearly independent from the system size. Further in the chapter we only consider ferromagnetic interactions with $J \geq 0$.

3. Ising Model on Complex Networks

When the network possesses a complex topology (see Appendix A for a short introduction into complex networks), critical properties of the Ising model are different as compared to Ising models at regular lattices.

One of the staple complex network models is the Barabasi-Albert (BA) scale-free network.[29] It is one of early models that possesses a scale-free degree distribution of vertices. It is an evolving network model, what means that it starts from initial conditions and is later on changing according to a rule. In the case of the BA network, the rule is relatively simple: each step t a new vertex is added to the network and m new edges are created, connecting the new vertex with already existing vertices. The exact vertex to connect to is chosen at random, using the *preferential attachment* rule. This means the edges connect to a vertex i with a probability proportional to the present vertex degree $P(i,t) \sim k_i(t)$. Starting from m fully connected vertices and using this rule one obtains a network that possesses a scale-free degree distribution with the exponent $\gamma = 3$.

$$P(k) = \frac{2m^2}{k^3}. \tag{2}$$

It is worth to note that the BA model is an almost uncorrelated random network, i.e. degrees of nearest-neighbor nodes are nearly independent one from another. Since correlations in BA model are small and intangible thus it allows us to use it as a proxy for an uncorrelated random network for numerical simulation purposes. In our analytical investigations we consider a generic uncorrelated scale-free network with a scaling exponent $\gamma = 3$. The reason for using the BA model in numerical simulations is because such a network is always connected — there are no vertices that are disconnected from other vertices in the network. This assures that the subnetworks we consider to be internally interacting are in fact doing so.

Although for large regular lattices the temperature of phase transitions in the Ising model is independent from the system size, the critical temperature for transition between ferromagnetic and paramagnetic states for Ising models at scale-free networks can depend on number of network nodes.[6] It is worth to note that the temperature considered here is not truly a critical temperature in a strict sense, since the system is not in the thermodynamic limit. In the case of BA scale-free network, the "critical" temperature turns out to be logarithmically dependent on the number of spins N

$$T_c \sim \ln N. \tag{3}$$

This size effect is not observed in regular graphs or random exponential networks (where vertex degrees are all the same or similar). Analytical investigations show that the effect is due to highly heterogeneous vertex degrees. Looking at the Ising model Hamiltonian (1) we can write the energy tied to each single spin i as

$$E_i(s_i) = -\sum_j J_{ij} s_i s_j - h s_i \,. \tag{4}$$

Since $E_i(-s_i) = -E_i(s_i)$ thus the average spin value in the canonical ensemble is

$$\langle s_i \rangle = \frac{\exp(-\beta E_i(1)) - \exp(\beta E_i(1))}{\exp(-\beta E_i(1) + \exp(\beta E_i(1))} = \tanh\left(\beta \sum_j \left(J_{ij} \langle s_j \rangle\right) + \beta h\right) \tag{5}$$

where $\beta = 1/(k_{\mathcal{B}}T)$ and the mean of s_j is only over canonical ensemble, not over different j. From now on, we shall use rescaled temperature and put $k_{\mathcal{B}} = 1$, to simplify notation and avoid with confusion k_B introduced later on. We assume a constant exchange integral $J_{ij} = J\epsilon_{ij}$ and use a mean field approximation, so that instead of considering a specific vertices i or j, we consider an "average vertex". To do that instead of a specific $\epsilon_{ij} \in \{0, 1\}$, we use an average $\tilde{\epsilon}_{ij}$ that is equal to probability of connection existing between vertices i and j. In uncorrelated random networks, this depends on vertex degrees

$$\tilde{\epsilon}_{ij} = \frac{k_i k_j}{E} \tag{6}$$

where E is double total number of edges (and equivalent to number of directed edges if each bidirectional edge is considered as two directed ones). We can finally write

$$\langle s_i \rangle = \tanh\left(\beta J \left(\sum_j \frac{k_i k_j}{E} \langle s_j \rangle\right) + \beta h\right). \tag{7}$$

Note that when using the mean field approach for regular lattices, in equation analogous to the above, the $\sum_j \tilde{\epsilon}_{ij} \langle s_j \rangle$ simply becomes $k \langle s \rangle$ where $\langle s \rangle$ is average spin. In regular random graphs, where all vertices have same degree k, there is $k_i k_j = k^2$ and the sum can be replaced by the factor N, what reduces the whole expression to $k \langle s \rangle$. In both cases, the equation takes the form of

$$\langle s_i \rangle = \tanh\left(\beta J k \langle s \rangle + \beta h\right) \tag{8}$$

where one can consider any vertex, not specifically i and obtain self-consistent equation for average spin

$$\langle s \rangle = \tanh\left(\beta J k \langle s \rangle + \beta h\right). \tag{9}$$

In non-regular graphs, with degree heterogeneity, the calculations are more complicated. The important step now is to introduce a *weighted spin*

$$S = \sum_i k_i s_i / E. \tag{10}$$

Using this definition, we can insert S into Eq. (7) and get

$$\langle s_i \rangle = \tanh\left(\beta J k_i S + \beta h\right). \tag{11}$$

To obtain self-consistent equation for the weighted spin S, we can just multiply above by k_i/E and sum over i, what results in

$$S = \sum_i \left(\frac{k_i}{E} \tanh\left(\beta J k_i S + \beta h\right) \right). \tag{12}$$

This equation is a self-consistent equation for the weighted spin in a mean-field approximation. The weighted spin S is a generalization of average spin for heterogeneous systems. Several other models keep their properties in heterogeneous uncorrelated networks when a weighted spin is considered instead of regular average spin, for example the voter model.[30] This happens only when the impact of the spin depends on its connectivity. In cases where it does not, such as for a voter model with *edge update*, a regular spin is still the relevant variable.

We are interested in the ferromagnetic-paramagnetic transition and in its critical temperature T_c in the absence of the external magnetic field ($h = 0$). This transition is a second order phase transition and occurs when the order parameter is approaching zero. Therefore we can assume $|S| \ll 1$ what will make us possible to linearize the Eq. (12).

Let us remind that the non-linearized hyperbolic tangent equation

$$x = \tanh\left(ax\right) \tag{13}$$

has one solution $x = 0$ for $a \leq 1$ and two additional non-zero solutions for $a > 1$ (Fig.1). Our case is much more complicated, since it contains a sum over different tangents with additional factor k_i. However, using the fact that S is close to zero, we can use the linear approximation for hyperbolic tangent and perform this sum easily.

We get

$$S = \sum_i \left(\frac{k_i}{E} \beta J k_i S \right) \tag{14}$$

which is further simplified by performing the sum over the only variables dependent on i — k_i, and substituting $E = \langle k \rangle N$, obtaining

$$S = \beta J \frac{\langle k^2 \rangle}{\langle k \rangle} S. \tag{15}$$

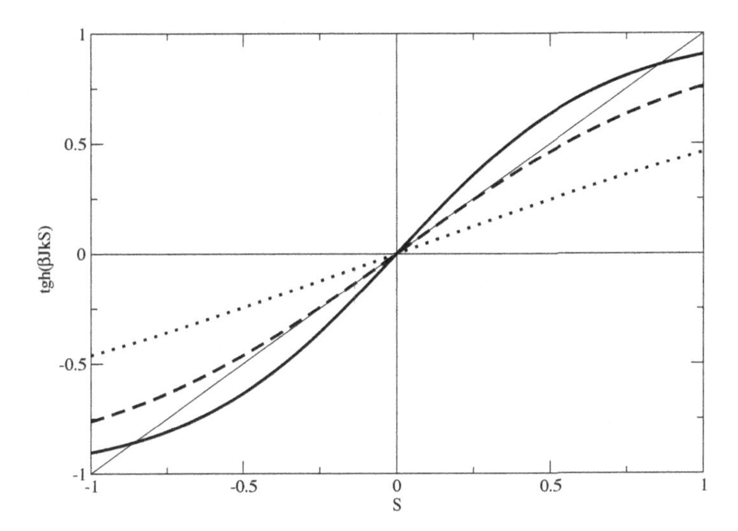

Fig. 1. Function $\tanh(ax)$ against x for three different values of a. For $a < 1$ (dotted line) the equation $x = \tanh(ax)$ has only one solution $x = 0$. For $a > 1$ (continuous line) the equation has three solutions – one unstable $x = 0$ and two non-zero stable solutions. The case $a = 1$ (broken line) is special border case. The parameter a is equal to slope of hyperbolic tangent near $x = 0$, what means the slope of the line in its linear approximation.

The original equation (12) will have non-zero solutions for

$$\beta J \langle k^2 \rangle / \langle k \rangle > 1. \tag{16}$$

The critical temperature T_c is therefore given by the equation

$$T_c = J \frac{\langle k^2 \rangle}{\langle k \rangle} \tag{17}$$

The dependence of the critical temperature on the system size is hidden in the above equation in the expression $\langle k^2 \rangle / \langle k \rangle$. In the thermodynamic limit of infinite system size, the second moment of the degree distribution for BA network is infinite

$$\langle k^2 \rangle = \int_m^{+\infty} k^2 P(k) dk = \int_m^{+\infty} 2m^2 k^2 / k^3 dk = 2m^2 \ln(+\infty) \tag{18}$$

where m is the minimum degree of vertices in the network (and it is a parameter for BA network generation). Obviously, real networks are always finite and have therefore finite second moment of degree distribution. A finite BA network will have the largest hub with degree k_{max}. The value can be found using a relation

$$\int_{k_{max}}^{+\infty} P(k) dk = \frac{1}{N} \tag{19}$$

what means that the expected total number of vertices with degree of k_{max} or higher is one. It is impossible to predict the actual degree of the largest hub in a single network representation and it can vary significantly from k_{max}, but an analysis based on this estimate gives qualitatively correct results. From the Equation (19) we find that

$$k_{max} = m\sqrt{N}. \tag{20}$$

Using that value to calculate the degree distribution moments yields the following

$$\frac{\langle k^2 \rangle}{\langle k \rangle} = \frac{m^2 \ln N}{2m(1 - 1/\sqrt{N})} \approx \frac{m}{2} \ln N \tag{21}$$

which can be put into Equation (17) to obtain the sought dependence of critical temperature on the system size in a logarithmic way

$$T_c = J\frac{m}{2} \ln N. \tag{22}$$

This result is obviously only relevant for finite systems. In the thermodynamical limit, the BA scale-free network is always ordered for any finite temperature T and no phase transition exists. It is worth to note however, that the dependence of critical temperature on the system size is logarithmic. This means that even for very large number of vertices, comparable to number of particles in macroscopic amount of substance ($\sim 10^{23}$) the temperature is not exceedingly high. Additionally, if we consider the Ising model on complex network structure in context of social applications, then the largest conceivable system size is of the order of number of people in the world ($\sim 10^{10}$). Because of that, considering finite size effects is as important as understanding the thermodynamic limit behavior.

4. Ising Model on Connected Networks

4.1. *Analytic approach*

Now we consider a pair of connected networks A and B (Figure 2). The parameters describing both networks can be split into four groups - two describe internal properties of each network, and two describe network-network interactions. We introduce the following notation: s_{Ai} and s_{Bi} are spins in networks A and B, J_{AA} and J_{BB} are coupling constants between spins in networks A and B respectively, $J_{AB} = J_{BA}$ are the coupling constants between spins in different networks, k_{AAi} and k_{BBi} are intra-network vertex degrees, k_{ABi} and k_{BAi} are inter-network vertex degrees, E_{AA} and E_{BB} are twice the total numbers of all intra-network edges in A and B, $E_{AB} = E_{BA}$ is the number of edges between the networks.

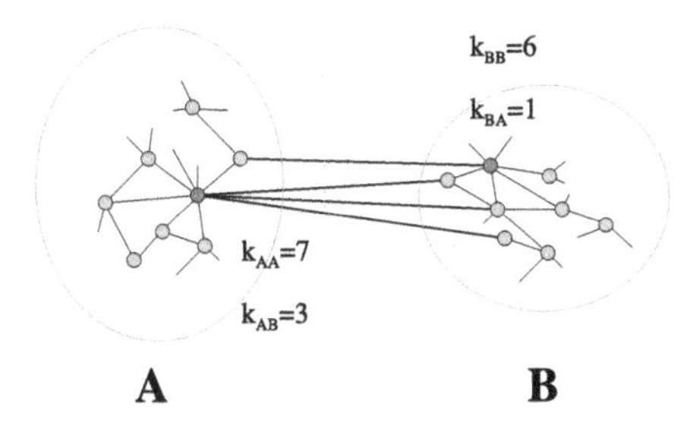

Fig. 2. Two connected Barabasi-Albert networks. A few vertices from each network are shown. The intra-network degrees k_{AA} and k_{BB} as well as inter-network degrees k_{AB} and k_{BB} for two sample vertices are presented.

Now we extend the Equation (7) by introducing the influence of the second network. Similar to the single network case, we use the mean-field approximation, where actual exchange integrals J_{AAij} are substituted by constant value J_{AA} multiplied by expected edge number (6). All other specific integrals are treated in the same way, using appropriate vertex degrees — internal or external. This way we can obtain equations for average spins in both networks:

$$\langle s_{Ai} \rangle = \tanh \left(\beta J_{AA} k_{AAi} \sum_j \frac{k_{AAj} \langle s_{Aj} \rangle}{E_{AA}} + \right. \\ \left. + \beta J_{BA} k_{ABi} \sum_j \frac{k_{BAj} \langle s_{Bj} \rangle}{E_{BA}} \right) , \tag{23}$$

$$\langle s_{Bi} \rangle = \tanh \left(\beta J_{BB} k_{BBi} \sum_j \frac{k_{BBj} \langle s_{Bj} \rangle}{E_{BB}} + \right. \\ \left. + \beta J_{AB} k_{BAi} \sum_j \frac{k_{ABj} \langle s_{Aj} \rangle}{E_{AB}} \right) . \tag{24}$$

Similar to the case of single complex network, we introduce weighted spin, but in case of connected networks we need *four* weighted spins, because each vertex i has two different degrees - internal (k_{AAi},k_{BBi}) and external (k_{ABi}, k_{BAi}):

$$S_{AA} = \sum_i k_{AAi} \langle s_{Ai} \rangle / E_{AA} , \tag{25}$$

$$S_{BB} = \sum_i k_{BBi} \langle s_{Bi} \rangle / E_{BB} , \tag{26}$$

$$S_{AB} = \sum_i k_{ABi} \langle s_{Ai} \rangle / E_{AB}\,, \tag{27}$$

$$S_{BA} = \sum_i k_{BAi} \langle s_{Bi} \rangle / E_{BA}\,. \tag{28}$$

The weighted spins S_{AB} and S_{BA} have non-obvious meaning. They represent the weighted spin of network A and B respectively, but from the perspective of the other network. They show the effective spin of a network as seen by the other. It is easy to imagine some vertices may be exceedingly well connected to the other network, while some may be not connected at all, thus making the weighted spin observable by a second network different than its internal weighted spin.

After introducing weighted spins, we proceed as in the case of single network, multiplying equations by degree and performing a sum. However, we have two degrees per vertex so we end up with four equations, each for different weighted spin. Equation (23) is multiplied by k_{AAi} and k_{ABi} and Equation (24) is multiplied by k_{BBi} and k_{BAi}. After multiplication and performing sum, we obtain the following four equations

$$S_{AA} = \sum_i \left(\frac{k_{AAi}}{E_{AA}} \tanh\left(\beta J_{AA} k_{AAi} S_{AA} + \beta J_{BA} k_{ABi} S_{BA}\right) \right), \tag{29}$$

$$S_{BB} = \sum_i \left(\frac{k_{BBi}}{E_{BB}} \tanh\left(\beta J_{BB} k_{BBi} S_{BB} + \beta J_{AB} k_{BAi} S_{AB}\right) \right), \tag{30}$$

$$S_{AB} = \sum_i \left(\frac{k_{ABi}}{E_{AB}} \tanh\left(\beta J_{AA} k_{AAi} S_{AA} + \beta J_{BA} k_{ABi} S_{BA}\right) \right), \tag{31}$$

$$S_{BA} = \sum_i \left(\frac{k_{BAi}}{E_{BA}} \tanh\left(\beta J_{BB} k_{BBi} S_{BB} + \beta J_{AB} k_{BAi} S_{AB}\right) \right). \tag{32}$$

Following the same methodology, we now assume that the phase transition occurs when the spins and therefore weighted spins are near-zero. By assuming near-zero weighted spins, we can approximate hyperbolic tangent with linear function and simplify the equations to

$$S_{AA} = \beta J_{AA} \sum_i \frac{k_{AAi}^2}{E_{AA}} S_{AA} + \beta J_{BA} \sum_i \frac{k_{ABi} k_{AAi}}{E_{AA}} S_{BA} \tag{33}$$

$$S_{BB} = \beta J_{BB} \sum_i \frac{k_{BBi}^2}{E_{BB}} S_{BB} + \beta J_{AB} \sum_i \frac{k_{BAi} k_{BBi}}{E_{BB}} S_{AB} \tag{34}$$

$$S_{AB} = \beta J_{AA} \sum_i \frac{k_{AAi} k_{ABi}}{E_{AB}} S_{AA} + \beta J_{BA} \sum_i \frac{k_{ABi}^2}{E_{AB}} S_{BA} \tag{35}$$

$$S_{BA} = \beta J_{BB} \sum_i \frac{k_{BBi} k_{BAi}}{E_{BA}} S_{BB} + \beta J_{AB} \sum_i \frac{k_{BAi}^2}{E_{BA}} S_{AB}. \tag{36}$$

If we assume that

$$k_{ABi} = p_A k_{AAi} , \tag{37}$$

$$k_{BAi} = p_B k_{BBi} , \tag{38}$$

what means that the number of edges outside the network is proportional to the number of edges within the network, we can further simplify our four equations. If the intra-network edges and inter-network edges have the same nature, it is natural to assume that vertices with high degree will also have many inter-network connections. Since all edges work in the same way, as interactions for the same model, it is also logical they have the same nature.

The probabilities p_A and p_B are fixed numbers, although the values are not independent and are connected with the number of edges between networks in following way: $p_A E_{AA} = E_{AB} = E_{BA} = p_B E_{BB}$. Using this assumption, we do not need to consider the cross-network weighted spins S_{AB} and S_{BA} as they are proportional to S_{AA} and S_{BB}. We can simplify our notation

$$S_{AA} \equiv S_A , \tag{39}$$

$$S_{BB} \equiv S_B . \tag{40}$$

Using this notation our first two equations get the form

$$S_A = \beta J_{AA} S_A \sum_i \frac{k_{AAi}^2}{E_{AA}} + \beta J_{BA} S_B \sum_i \frac{k_{AAi}^2 p_A}{E_{AA}} , \tag{41}$$

$$S_B = \beta J_{BB} S_B \sum_i \frac{k_{BBi}^2}{E_{BB}} + \beta J_{AB} S_A \sum_i \frac{k_{BBi}^2 p_B}{E_{BB}} . \tag{42}$$

This equation array can be written as a single matrix equation.

$$\mathbf{S} = \beta \hat{\Lambda} \mathbf{S}, \tag{43}$$

where $\mathbf{S}$ is a vector $\begin{bmatrix} S_A \\ S_B \end{bmatrix}$ describing the state of the system and $\hat{\Lambda}$ is a matrix describing effective interaction strengths between spins belonging to the same or to different networks

$$\hat{\Lambda} = \begin{bmatrix} \Lambda_{AA} & \Lambda_{BA} \\ \Lambda_{AB} & \Lambda_{BB} \end{bmatrix} = \begin{bmatrix} J_{AA} \frac{\langle k_{AA}^2 \rangle}{\langle k_{AA} \rangle} & p_B J_{BA} \frac{\langle k_{AA}^2 \rangle}{\langle k_{AA} \rangle} \\ p_A J_{AB} \frac{\langle k_{BB}^2 \rangle}{\langle k_{BB} \rangle} & J_{BB} \frac{\langle k_{BB}^2 \rangle}{\langle k_{BB} \rangle} \end{bmatrix} . \tag{44}$$

In the case of a single network A, solutions other than $S_A = 0$ can exist only if $\beta J_{AA} \frac{\langle k_{AA}^2 \rangle}{\langle k_{AA} \rangle} > 1$. In the case of two connected networks, this condition corresponds to an eigenvalue of Equation (43) greater than 1. The eigenvalues are

$$\lambda_{\pm} = \frac{\beta}{2} \left(\Lambda_{AA} + \Lambda_{BB} \pm \sqrt{(\Lambda_{AA} - \Lambda_{BB})^2 + 4\Lambda_{BA}\Lambda_{AB}} \right). \tag{45}$$

Comparing these eigenvalues with 1, we get the following critical temperatures

$$T_{c\pm} = \frac{\Lambda_{AA} + \Lambda_{BB} \pm \sqrt{(\Lambda_{AA} - \Lambda_{BB})^2 + 4\Lambda_{BA}\Lambda_{AB}}}{2}. \tag{46}$$

To shorten the above result, we introduce the following notation

$$\mathcal{A} = (\Lambda_{AA} + \Lambda_{BB})/2\,, \tag{47}$$

$$\mathcal{D} = (\Lambda_{AA} - \Lambda_{BB})/2\,, \tag{48}$$

$$\mathcal{C} = \sqrt{\Lambda_{BA}\Lambda_{AB}}\,. \tag{49}$$

The value $\mathcal{A}$ ("average") describes an average interactions between spins inside both networks, $\mathcal{D}$ ("difference") is the difference between those interaction strengths, $\mathcal{C}$ ("coupling") describes a strength of inter-network interactions. Using this notation the critical temperatures can be written shortly as

$$T_{c\pm} = \mathcal{A} \pm \sqrt{\mathcal{D}^2 + \mathcal{C}^2}\,. \tag{50}$$

Let us now consider eigenvectors associated with $\lambda_{\pm}$. They are proportional to the magnetization of both networks that appears below a given temperature and disappears above it. The unnormalized eigenvectors are

$$\mathbf{S}_{\pm} = \begin{bmatrix} 1 \\ \frac{-\mathcal{D} \pm \sqrt{\mathcal{D}^2 + \mathcal{C}^2}}{\Lambda_{BA}} \end{bmatrix}. \tag{51}$$

The eigenvectors' components tell us the relation between weighted spins in both networks in pure states corresponding to the eigenvectors. A real state of the system will be a linear combination of pure states. Given a pure state can exist below its temperature $T_{c\pm}$ and vanishes above. The eigenvector $\mathbf{S}_+$ has the same signs of the components and corresponds to ferromagnetic order with parallel weighted spins, while the eigenvector $\mathbf{S}_-$ has opposite signs of its components and corresponds to ferromagnetic order with antiparallel weighted spins of both networks. We can therefore distinguish three stable states: *antiparallel ferromagnetic* where $\mathbf{S}_-$ is dominant, *parallel ferromagnetic* where $\mathbf{S}_+$ is dominant and *paramagnetic* where the system is disordered. The antiparallel state is stable below T_{c-}, parallel below $T_{c+} > T_{c-}$ and paramagnetic above T_{c+}. Note that below T_{c-} both parallel and antiparallel states are stable. The fully ordered state $\mathbf{S}_+$ is the ground state, while antiparallel $\mathbf{S}_-$ can be treated as a metastable state. We can say the system is bistable below T_{c-} and monostable above T_{c+}. As explained further below, the temperature T_{c-} is not the real temperature where bistable-monostable transition occurs. In fact the real behavior of the system corresponds to a first-order (discontinuous) phase transition.

In the limit of vanishing inter-network connections ($\mathcal{C} \equiv \Lambda_{BA}\Lambda_{AB} = 0$) the eigenvalues of the matrix (44) are simply its diagonal elements Λ_{AA}, Λ_{BB} and associated normalized eigenvectors are correspondingly $\mathbf{S} = \begin{bmatrix} 1 \\ 0 \end{bmatrix}$, and $\mathbf{S} = \begin{bmatrix} 0 \\ 1 \end{bmatrix}$. This means that in this limit two stable states of the system correspond to the *individual* ordering of every networks, and there is no relation between their order parameters. It shows our approach gives correct results in this specific case.

It is worth to note that the diagonal matrix elements Λ_{AA} and Λ_{BB} are critical temperatures for single networks

$$\Lambda_{AA} = T_{cA}\,, \tag{52}$$

$$\Lambda_{BB} = T_{cB}\,, \tag{53}$$

while

$$\Lambda_{AB}\Lambda_{BA} = p_A p_B T_{cA} T_{cB} \tag{54}$$

thanks to the assumptions represented by Equations (37), (38). We can use this to write the critical temperatures as

$$T_{c\pm} = \frac{T_{cA} + T_{cB} \pm \sqrt{(T_{cA} - T_{cB})^2 + 4p_A p_B T_{cA} T_{cB}}}{2}. \tag{55}$$

This analytic result holds true for any random uncorrelated network, where the probability of an edge existing between any two vertices i and j is proportional to the product $k_i k_j$, and where an inter-network degree of a given node is proportional to its intra-network degree.

The analytic calculations presented so far have considered paramagnetic-ferromagnetic transition, where $S \approx 0$, so the linear approximation of hyperbolic tangent could be used. This assumption is valid for the ferromagnetic-paramagnetic phase transition at T_{c+}, above this temperature the only stable pure state $\mathbf{S}_+$ becomes unstable. However, our second temperature T_{c-} describes the place where only one of the two stable pure states $\mathbf{S}_+$ and $\mathbf{S}_-$ becomes unstable. The temperature T_{c-} would be correct, if the weighted spin values and thus linear approximation were correct. But since the pure state $\mathbf{S}_+$ is stable both below and above this temperature, it means that the linear approximation is incorrect in this case and the value of T_{c-} may be wrong. Further on, we show that the transition from parallel to antiparallel occurs at a temperature $T_{c1} < T_{c-}$. The antiparallel ordering solution indeed exists up to temperature T_{c-}, but it becomes unstable at T_{c1}. Figure 3 shows how the spin of the system depends on the temperature, both in real case and in case where antiparallel solution is artificially stabilized up till T_{c-}.

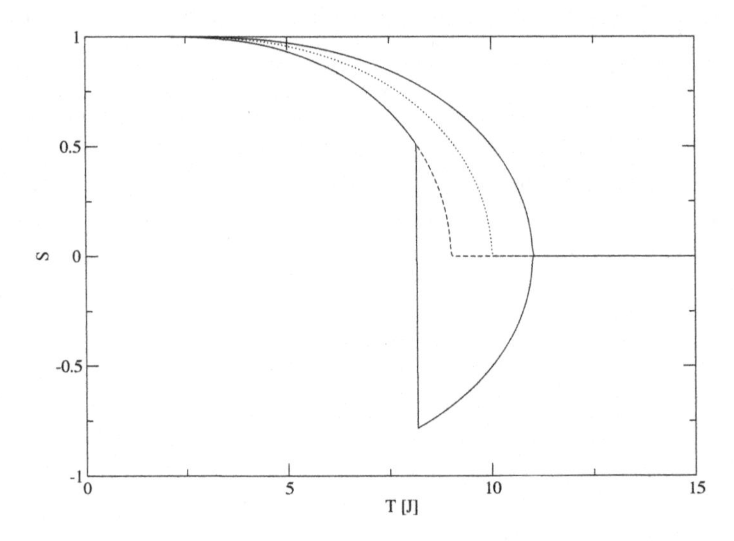

Fig. 3. Plots of $S_A(T)$ for $N_A = N_B = 5000$ and $k = \text{const.} = 10$, created using map iterations (77), (78). The dotted line is for $p = 0$ (unconnected networks) and the rest are for $p = 0.1$. The dashed line is for the graph with artificially forced $S_A = -S_B$ (in such case the weighted spin S is forced to go to zero continuously) while the solid lines are without such forcing, for parallel and antiparallel initial ordering. The discontinuous behavior of the weighted spin S is evident during the bistable-monostable transition and reflects the presence of the first-order phase transition in this system.

Let us go back to the equations (29) – (32), where the assumption about the small weighted spin values was not applied. Using equations (37), (38) and the notations (39), (40) we obtain the following equations for the weighted average spins

$$S_A = \sum_i \frac{k_{AAi}}{E_{AA}} \tanh\left(\beta J_{AA} k_{AAi} S_A + \beta J_{BA} p_A k_{AAi} S_B\right) , \tag{56}$$

$$S_B = \sum_i \frac{k_{BBi}}{E_{BB}} \tanh\left(\beta J_{BB} k_{BBi} S_B + \beta J_{BA} p_B k_{BBi} S_A\right) . \tag{57}$$

The equations can not be solved analytically, because of the sum of the hyperbolic tangents. Let us consider a case where networks have the same size

$N_A = N_B$, the same edge density and constant degrees $k_{BBi} = k_B = k_{AAi} = k_A = \text{const.}$ i.e. they represent regular random graphs. These assumptions are a serious simplification, but we later show that the behavior of the non-simplified system is similar. Please note that these conditions induce $p_A = p_B = p$.

Using the fact that the degrees are constant we can omit the sum and can write the simplified equations

$$S_A = \tanh\left(\beta J_{AA} k_A S_A + \beta J_{BA} p k_A S_B\right) , \tag{58}$$

$$S_B = \tanh\left(\beta J_{BB} k_A S_B + \beta J_{BA} p k_A S_A\right) . \tag{59}$$

The right hand side of the equation (58) is hyperbolic tangent, shifted by the value $H = J_{BA} k_A p_A S_B$ along the x axis.

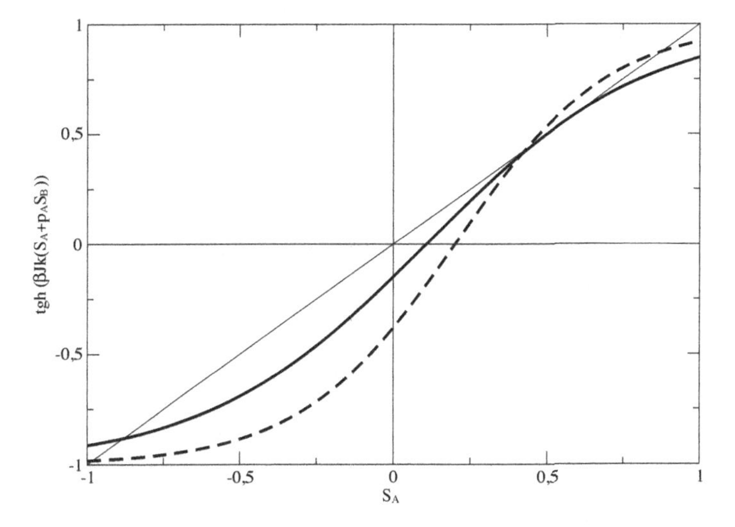

Fig. 4. Hyperbolic tangent plot. Dashed line is Eq. 60 for $T < T_c$ when it has three solutions, two of them stable. The solid line is for $T = T_c$, when one of the two stable solutions merges with the unstable one and the resulting solution is not stable. Thin lines are axes and $y = x$ line. The upper right intersection of the plot with the $y = x$ line is the point S_A^*.

Eqs. (58) and (59) can be further simplified if we assume that $J_{AA} = J_{BB} = J_{BA} = J$ and introduce a renormalized temperature $\tilde{T} = T/(J_{AA} k_A)$. Then we have

$$S_A = \tanh\left((S_A + pS_B)/\tilde{T}\right) , \tag{60}$$

$$S_B = \tanh\left((S_B + pS_A)/\tilde{T}\right) . \tag{61}$$

When the temperature $\tilde{T}$ is low the equation 60 has three solutions (dashed line at Fig.4), just as normal hyperbolic tangent has (Eq. 13) except with a certain shift induced by the presence of the term pS_B. Since both equations have same form, for low enough temperature, both have a stable nonzero solutions S_A^*, S_B^*, whose values depend on each other. Since our system is symmetric, with both network being exactly the same and started from antiparallel initial conditions, for low temperatures the actual spin values are $S_B^* = -S_A^*$. If we naively use this symmetry and forget that our system consists of two networks then we receive from Eq. 61 a standard mean field equation for a one variable

$$S_B^* = \tanh\left(S_B^*(1-p)/\tilde{T}\right). \tag{62}$$

The stabilty analysis[31] of Eq. 62 leads to two nonzero symmetrical solutions $S_{B1}^* < 0$ and $S_{B2}^* = -S_{B1}^*$ that exist only in the temperature region $\tilde{T} < 1 - p$. This simplification is responsible for a *continous* decrease of the order parameter S_B^* when the temperature $\tilde{T}$ reaches a critical value $\tilde{T}_c = 1 - p$. This is exactly the temperature $\tilde{T}_{c-}$, derived from linear approximation earlier (Eq. 55), only rescaled. Our full system consists of two coupled networks thus the stability analysis should be performed for a system of two coupled Eqs. 60, 60. Let us treat the solution S_{B1}^* of Eq. 62 as *a constant term*, insert it into Eq. 60 and consider what happens when the temperature increases. The hyperbolic tangent becomes gradually flatter. This decreases the value of S_A^*. Now, it is important to note that the second network, which is actually responsible for the shift, undergoes similar changes, causing $|S_{B1}^*|(\tilde{T})$ to decrease accordingly. Thus, the value of the shift decreases as the temperature increases. However, both values do not approach 0 continuously. At certain temperature $\tilde{T}_{c1}$, the hyperbolic tangent becomes tangential to the $y = x$ line (solid line in Fig.4). At this point, the opposite solution S_A^* and $S_B^* = -S_A^*$ become unstable and fluctuations of actual S_A or S_B around these values cause the system to leave the antiparallel state and order in parallel, destroying relation $S_B^* = -S_A^*$. This change is discontinuous and corresponds to the first-order phase transition in our system. As we shall show below the critical temperature $\tilde{T}_{c1}$ of this transition is lower than a critical temperature $\tilde{T}_c = \tilde{T}_{c-}$ received from Equation 62.

At T_{c1}, the tangent is tangential to the $y = x$ line. We can write the conditions for $\beta_{c1} = 1/T_{c1}$ and $S_{Ac} = S_A|_{T=Tc1}$

$$\frac{\tanh(\beta_{c1} J k_A S_{Ac} + \beta_{c1} J k_A p_A S_B)}{S_{Ac}} = 1\,, \tag{63}$$

$$\frac{\partial \tanh(\beta_{c1} J k_A S_{Ac} + \beta_{c1} J k_A p_A S_B)}{\partial S_A} = 1\,. \tag{64}$$

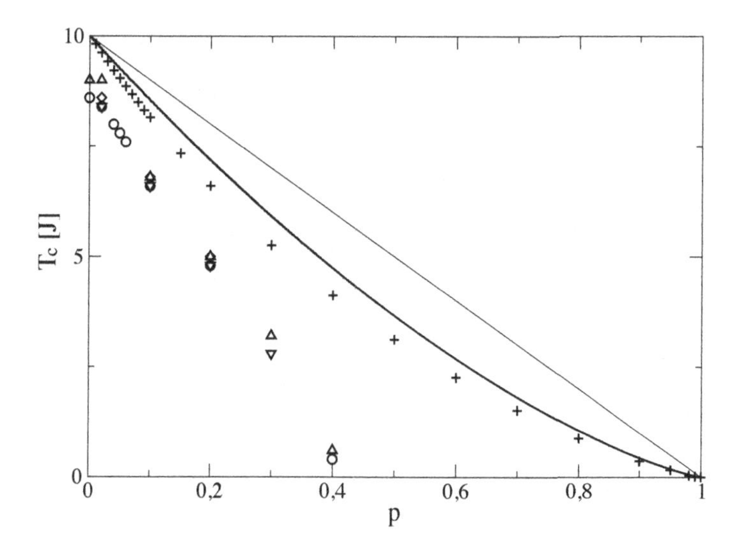

Fig. 5. Dependence of temperature T_{c1} on the parameter p for two constant degree networks $k = 10$. The thin straight line is $T_{c1}(p)$ calculated by using linear approximation (55). The solid curved line is analytical prediction (68), the plus symbols are map iterations, while circles, triangles and diamonds are results of numerical Monte-Carlo simulations. Networks of size $N_A = N_B = 5000$ were used. Circles are for $\tau = 100$, triangles up are for $\tau = 30$, triangles down are for $\tau = 200$. Diamonds are for $\tau = 100$ but for networks of size $N = N_A = N_B = 50000$ and they merge with triangles up. Increasing the density of inter–network connections decreases the temperature T_{c1}. While analytical mean-field approach and map iterations suggest that this temperature tends to zero only for $p = 1$ the Monte-Carlo simulations show that T_{c1} vanishes above some critical inter–network links density $p \approx 0.4$. It means that the bistable phase can exist only for weakly coupled networks.

For $S_B = -S_A$ we can calculate β_{c1} and S_{Ac} and obtain

$$S_{Ac} = \frac{\ln\left(\sqrt{\beta_{c1}Jk_A} + \sqrt{\beta_{c1}Jk_A - 1}\right)}{\beta_{c1}Jk_A(1 - p_A)}, \tag{65}$$

$$\beta_{c1} = \frac{\ln\frac{1+S_{Ac}}{1-S_{Ac}}}{2Jk_A(1 - p_A)S_{Ac}}. \tag{66}$$

This set of equations determines the point (T_{c1}, S_{Ac}) for the transition between the bistable phase where the antiparallel state is stable and the monostable phase where this state is not stable.

If we multiply (65) by β_{c1} and (66) by S_{Ac} we get $\beta_{c1}S_{Ac}$ in both and can compare the right sides, obtaining a relation

$$S_{Ac} = \frac{\beta_{c1}Jk_A + \sqrt{\beta_{c1}Jk_A(\beta_{c1}Jk_A - 1)} - 1}{\beta_{c1}Jk_A + \sqrt{\beta_{c1}Jk_A(\beta_{c1}Jk_A - 1)}}. \tag{67}$$

Comparing this with Eq. (65) we obtain a single implicit equation for β_{c1} and p_A, that can be simplified to get

$$p_A = 1 - \frac{1 + \sqrt{1 - 1/(\beta_{c1} J k_A)}}{\beta_{c1} J k_A + \sqrt{\beta_{c1} J k_A (\beta_{c1} J k_A - 1)} - 1} \times \ln\left(\sqrt{\beta_{c1} J k_A} + \sqrt{\beta_{c1} J k_A - 1}\right). \tag{68}$$

Drawing $p_A(\beta_{c1})$ and changing axes yields a dependence of T_{c1} on parameter p_A (see Figure 5).

We can also approximate the behavior of the solution for small p_A. Our conditions (63)–(64) can be written

$$\tanh(\beta_{c1} J k_A (1 - p_A) S_{Ac}) = S_{Ac}\,, \tag{69}$$

$$\cosh^2(\beta_{c1} J k_A (1 - p_A) S_{Ac}) = \beta_{c1} J k_A\,. \tag{70}$$

If we multiply the equations' sides, we obtain a single equation for the product $X = \beta_{c1} J k_A S_{Ac}$

$$\sinh(2(1 - p_A)X) = 2X\,. \tag{71}$$

We know that for very small p_A there is $T_{c1} \approx T_c^+$ thus the value of S_{Ac} around this temperature is very small. It follows X and the whole argument of hyperbolic sinus is also small and we can expand it around zero

$$2(1 - p_A)X + (2(1 - p_A)X)^3/6 \approx 2X\,. \tag{72}$$

As result we get as an approximate value of X

$$X \approx \sqrt{\frac{3}{2} p_A}\,. \tag{73}$$

Putting (73) into (70) we obtain the following

$$\cosh^2\left((1 - p_A)\sqrt{(3/2)p_A}\right) = \beta_{c1} J k_A\,. \tag{74}$$

Since the argument of $\cosh^2$ is very small thanks to small p_A value, we can expand $\cosh^2 x \approx 1 + x^2$ as well as $1/(1 + x) \approx 1 - x$ and finally obtain

$$T_{c1} \approx J k_A (1 - (3/2)p_A)\,. \tag{75}$$

This sets the expected dependence of T_{c1} for very small p_A and therefore for very weakly interacting networks. Let us compare this result for the first-order phase transition temperature T_{c1} with the estimation of the temperature T_{c-} for the second order phase transition given by (46). The last result can be written for identical networks as:

$$T_{c-} = J k_A (1 - p_A)\,. \tag{76}$$

It follows that the first order transition takes place for lower temperatures than the second order one.

So far, we have concentrated on a case of constant vertex degree k and two networks of same size. Without such simplifications, the equations are very hard to solve analytically. We have studied more complex cases using map iterations and Monte-Carlo simulations.

4.2. *Map iterations*

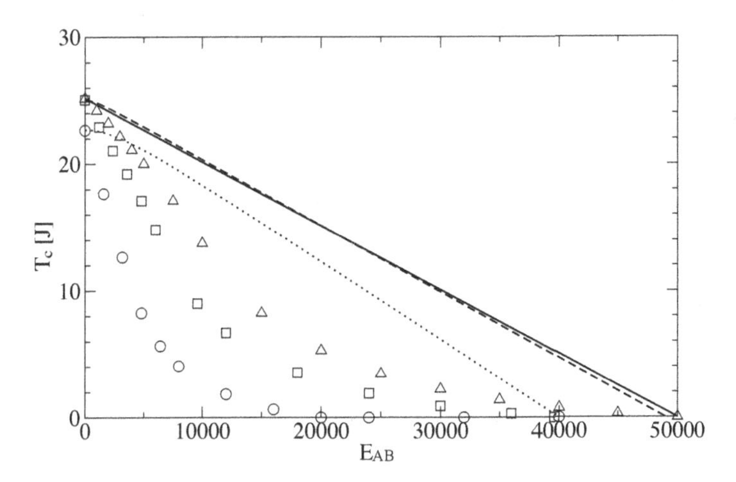

Fig. 6. Dependence of temperature T_{c1} on the number of inter-network connections E_{AB} for two Barabasi-Albert networks. Lines are analytic predictions using linear approximation (55), while symbols are temperatures obtained from map iterations (77), (78). Solid line and triangles correspond to $N_A = N_B = 5000$, dashed line and squares correspond to $N_A = 6000$, $N_B = 4000$, while dotted line and circles correspond to $N_A = 8000$, $N_B = 2000$.

Since the problem of the exact temperature of the transition could not be solved fully analytically, we have used numerical methods. We consider a two-dimensional map to solve the issue. A map is a dynamic system, where a time-dependent variable changes in discrete steps, according to a defined equation. An example is the logistic map with a variable $x(t)$, defined by the equation $x(t+1) = rx(t)(1 - x(t))$, where $0 < r \leq 4$ is a constant parameter (see for example[32]). The variable $x(t)$ given by the logistic map may behave in several ways depending on the selected value of the parameter r. It can converge to a stable fixed point or to a periodic orbit but it can also behave chaotically. For our purposes, the important point is that if the map has a fixed point, the variable will converge to its value, provided the initial conditions are in the attractor basin of that fixed point. In our case, we consider a two-dimensional map, meaning a map with two time-dependent variables and defined by two equations. We base the map on Equations (56), (57). The stable points in the map are solutions of these equations, and their attractor basins span the whole variable space. This allows us to find these solutions by iterating the map many times, until we reach a limit of numerical accuracy. Our map is defined by the following equations

$$S_A(t+1) = \sum_{k_A} P(k_A)\frac{k_A}{E_A}\tanh\left(\beta J_{AA}k_A S_A(t) + \beta J_{BA}p_A k_A S_B(t)\right), \tag{77}$$

$$S_B(t+1) = \sum_{k_B} P(k_B)\frac{k_B}{E_B}\tanh\left(\beta J_{BB}k_B S_B(t) + \beta J_{AB}p_B k_B S_A(t)\right), \tag{78}$$

where the $S_A(t)$, $S_B(t)$ are time-dependent variables and the rest are constant parameters, including given degree distributions $P(k_A)$ and $P(k_B)$. With our definition of weighted spin $S = 1/E\sum_i k_i s_i$, where s_i are spin values of vertices i, k_i are degrees and E is twice the number of edges in network, it can have values from range $[-1, 1]$. We assume $J_{AA} = J_{BB} = J_{AB} = J_{BA} = J$ and express all temperatures in units of coupling constant J over Boltzmann constant k_B, so we can omit these constants in the equations and have $\beta = 1/T$.

We investigate the dependence of a stable point spin on the temperature $S(T)$ assuming the antiparallel initial condition $S_A(t=0) = 1$, $S_B(t=0) = -1$. Since the system is fully symmetric, below T_c we have $S(T) = S_A(T) = -S_B(T)$. At T_{c1}, the antiparallel state becomes unstable point and the system jumps to the parallel ordering, what corresponds to the first-order phase transition. We note that in our map iterations the system always jumped to parallel ordering with the negative spin values and never positive. This consistency is obvious, considering the deterministic nature of the map iterations. The fact that it ordered with negative spins does not have any meaning and comes from the technical aspects of the map iteration implementation – number representation and function calculation accuracy on the machine we ran the map iterations on. Since we do not rely on map iterations to find the spin values, only the critical parameter value, it is of no importance.

By observing $S(T)$ we can find the temperature T_{c1} of the first-order phase transition, where the antiparallel state disappears and a jump between positive and negative spin values occurs (see Fig.3). We define our solution as $S_A(T) = (S_A(t_{max}))_T$, $S_B(T) = (S_B(t_{max}))_T$, where the time $t_{max} = 1000$ is the number of iterations of the map that have been performed before we assume it reached the stable point. The assumed t_{max} provides us with enough accuracy in finding the value of T_{c1}.

We investigated various T ranges, usually around the temperature T_{c1}, with the temperature step $\Delta T = 0.2$. Our networks were of size $N_A = N_B = 5000$ and usually possessed a power law degree distribution $P(k_A) = P(k_B)$ taken from a Barabasi-Albert network growth simulation or constant degree $k =$ const. for testing of the analytical equations. Since the networks are same thus $p_A = p_B = p$.

We have investigated the dependence of the temperature T_{c1} for two networks with constant $k = 10$. The results are in Figure 5. As can be seen, the map iterations do not agree with analytical equations exactly. This is probably due to the limited accuracy of numerical calculations, that near such critical point can play crucial role. The numerical noise can tip the system over the edge into the parallel state. The purpose of introducing the iterated maps was to calculate the behavior of the system in the more complex case, where vertices do not have a constant degree. We have performed the iterations of the map for the scale-free distribution of degrees. The distributions used were generated by Barabasi-Albert network creation algorithm, same as used in Monte-Carlo simulations.

Looking at Figure 6 it is evident, that the linear approximation (Eq.55) leads to incorrect results. For small p, the discontinuous transition temperature is linearly dependent on the parameter p, but with different factor than predicted by *linear approximation*. For higher inter-network connection number, the dependence is no longer linear.

4.3. *Monte-Carlo simulations*

Previous sections tackled the issue of critical temperatures of the Ising model in connected network analytically. In this section, we present results of numerical Monte-Carlo simulations to support analytical findings.

We have used asynchronous Glauber Ising dynamics on two connected networks. In most cases we use two Barabasi-Albert networks with the same number of vertices $N_A = N_B = 5000$ and twice the number of edges $E_{AA} = E_{BB} = 50000$ ($\langle k_{AA} \rangle = \langle k_{BB} \rangle = 10$). The initial conditions can be either parallel, where all spins are set to the same value $s_{Ai} = s_{Bj} = +1$, or antiparallel, where $s_{Ai} = +1$, $s_{Bj} = -1$. After setting the initial conditions we start the simulation and allow the system to relax for τ time steps and then average weighted spin or other investigated value over next τ time steps. This procedure is repeated multiple times, usually 100, and the final result is an average from all the runs. Each time, the network is re-created, thus being valid for network model in statistical sense, rather than for particular realization. One time step equals $N_A + N_B$ random single vertex updates, what means on average one update per vertex per time step. The relaxation and measurement time τ used was either $\tau = 20$ time steps for investigating T_{c+} or $\tau = 100$ time steps for investigating T_{c1}. The time τ has been increased for investigation of T_{c1} since allowing sufficient time for the system to flip between antiparallel and parallel states is essential. It should be not too high though, since the system should flip only when it's in unstable point or very close to it, not from random fluctuation when still far away from the unstable point. The number of inter-network edges E_{AB} is variable, but we follow assumption of inter-network degree k_{ABi} being proportional to intra-network degree k_{AAi} (equations (37), (38)), by choosing both endpoints for inter-network edges preferentially (with the selection probability proportional to intra-network degree).

The first part of our numeric investigation concerns temperatures T_{c+} and T_{c-}. Since we know that T_{c-} is not the point where the transition antiparallel-parallel occurs, we can not find this point through Monte-Carlo simulations. However, we can investigate how much the temperature T_{c-} differs from the numerically obtained transition temperature T_{c1}. For now, we focus on T_{c+} and assume parallel initial condition, since they practically exclude possibility of antiparallel order emerging, especially within the time $\tau = 20$. Finding the exact value of critical temperature T_{c+} from numerical simulations is not straightforward. If one observes the dependence of the weighted magnetization on rising temperature and tries to fit the magnetization decay to a linear or to an exponential function the results strongly depend on relaxation time τ. To overcome this problem we observed the temperature dependence of the system *susceptibility* χ. In fact, by comparison to standard models of magnetic systems, one can expect that the initial susceptibility diverges at $T = T_{c+}$. In our finite system we are looking simply for the maximum of χ. To estimate χ we are using two methods. First we compare average weighted spin $S = (S_A + S_B)/2$ for a small external field $h = 0.05J$ and the value of S_0 with no external field. It follows $\chi = (S_h - S_0)/h$. Because such results are strongly fluctuating as a function of system history and temperature (Figure 7), we calculate running average over 30 temperature points and find the maximum of χ by fitting a parabolic curve. The top of the parabola corresponds to the position of the critical temperature T_{c+}. We found that these values are independent on the relaxation time τ used in our numerical experiment. The second method of finding the critical temperature T_{c+} is observation of the time average $\langle S^2 \rangle - \langle S \rangle^2$, where we average over one relaxation period τ. The magnitude of the fluctuations is proportional to the susceptibility $\chi \sim \langle S^2 \rangle - \langle S \rangle^2$ according to the fluctuation-dissipation theorem.[33] Similarly to the previous method, we calculate running average over 10 points and find the maximum. We find those values much more stable than the ones calculated using the first method (see Figure 9).

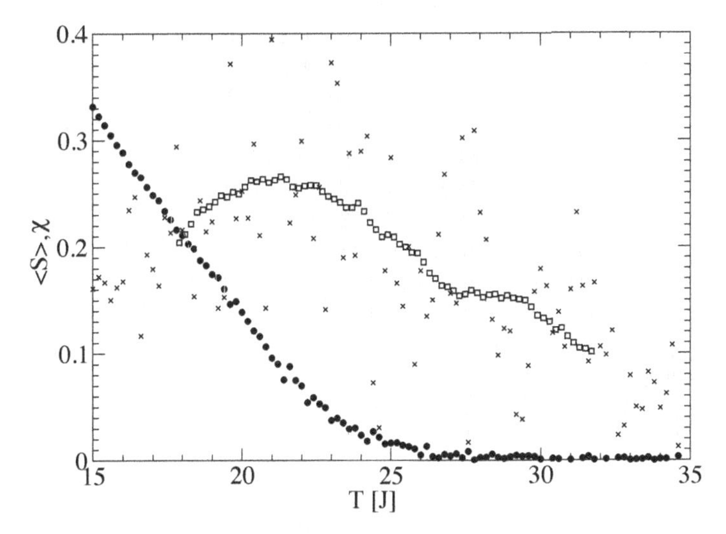

Fig. 7. The dependence of the average weighted spin $S = S_A + S_B$ and its susceptibility for small external field $h = 0.05J$ in the case of two Barabasi-Albert networks with $E_{AB} = 1000$ connections between them. An initial condition for each temperature is a parallely ordered system. The full symbols depict S, the X symbols correspond to susceptibility $\chi = (S_{0.05} - S_0)/h$ (the lines are just to guide eye), the empty symbols are 30-point running average of the susceptibility. The parabolic fit was used to find the susceptibility maximum.

Now let us concentrate on finding the numerical results for the temperature T_{c1} corresponding to the fist-order phase transition. This is the temperature where the system switches from the antiparallel to the parallel order. We set the initial conditions as antiparallel and observe at what temperature the system will switch. We also increase the relaxation and averaging time τ to $\tau = 100$, to allow enough time for system to switch and reach equilibrium afterwards. Figure 8 shows the example results when considering a regular graph. We measure $|S| = |(S_A + S_B)/2|$ and $|S_A|$ because the transition we look for can be spotted on plots of these values. Since we start from antiparallel ordering, $|S|$ is close to zero below T_{c1}, as both networks have same sizes and weighted spin S values, only of opposite sign, so the total is close to zero. It is not exactly zero because of fluctuations. Since we measure the absolute value, those fluctuations do not cancel each other, but add up, resulting in non-zero total value of weighted spin. At T_{c1} a transition from bistable to monostable phase occurs and the antiparallel state is no longer stable, thus the system immediately switches to the parallel state that is still stable. This can be spotted as a jump of the $|S|$, since both networks' weighted spins do not cancel each other anymore. We have to use the absolute value, since different simulations order either with positive or negative spin with equal probability, so if we didn't use average of absolute value, we would not be able to see the transition point. As the temperature grows higher, the networks order in parallel with lesser and lesser value of spin and finally at temperature T_{c+} they become paramagnetic.

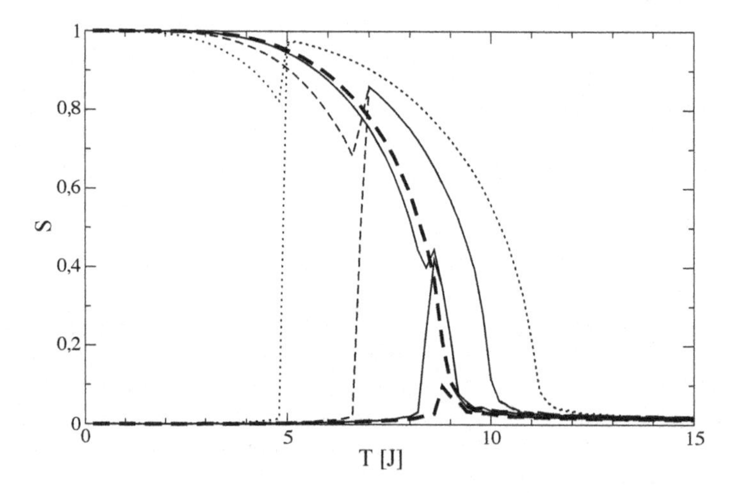

Fig. 8. Dependence of $|S| = |(S_A + S_B)/2|$ (lower, near-zero lines) and $|S_A|$ (upper lines) on the temperature for Monte-Carlo simluations. The thick dashed lines are for $p = 0$, solid lines are for $p = 0.02$, thin dashed for $p = 0.1$ and dotted for $p = 0.2$. The simulations start from antiparallel ordering $S_A = -S_B = 1$, while $N_A = N_B = 5000$ and $\langle k_A \rangle = \langle k_B \rangle = 10$. In all cases $\tau = 100$.

In the early numerical studies we have assumed that the maximum of the value $|S| = |(S_A + S_B)/2|$ occurs at the temperature T_{c1} and values found in this way can be found in Figure 9. In the later studies, we have used slightly different definition, described further.

We have investigated the dependence of temperatures T_{c+} and T_{c1} on the number of inter-network edges, using methods of determining T_{c+} and T_{c1} explained above. The results are shown on Figure 9.

It is worth to note that the analytical expectations for temperatures T_{c+} and T_{c1} are based on equation (21), not on the actual value of $\frac{\langle k^2 \rangle}{\langle k \rangle}$. This is because the mean field methods give higher critical temperature values for the Ising model in Barabasi-Albert network than it really is. The $\langle k^2 \rangle$ value for finite Barabasi-Albert network is significantly higher than expected because of the largest hub usually being higher degree than it is given by (19) due to fluctuations during

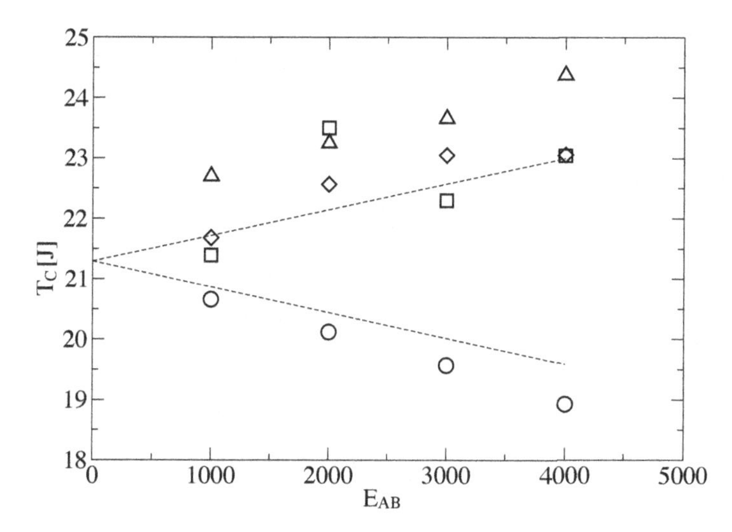

Fig. 9. The dependence of temperatures $T_{c\pm}$ on number of edges between networks E_{AB}. The dashed lines are analytic predictions (46). The symbols are numeric results. Circles correspond to T_{c-}. Squares and diamonds both correspond to T_{c+} and are calculated from susceptibility $\chi = (S_h - S_0)/h$ (two different measurements). Triangles correspond to T_{c+} and are calculated from susceptibility $\chi \sim \langle S^2 \rangle - \langle S \rangle^2$. Note that large differences between squares and diamonds show that calculating χ from a difference between S_h and S_0 is inherently inaccurate.

the network growth. As mentioned before, we did not investigate the topological fluctuations and therefore used a value that is based on the statistical topology of the whole network, rather than being dominated by single hub.

We can see that the numerically determined values do not agree perfectly with analytical expectations. However, we can conclude that the dependence of the temperatures on the inter-network connections is the same as expected for T_{c+}. The temperature T_{c1} declines with increasing E_{BA} faster than expectations for T_{c-}, but remembering that for small E_{AB} and constant-degree networks the analytical expectations for T_{c1} were to decline as $T_{c1} \sim 1 - (3/2)p_A$, the actual results seem to follow that expectations, even though the degree distribution is different.

The transition at T_{c+} is a typical ferromagnetic-paramagnetic transition, also found in single networks and regular lattices. The transition at T_{c1} is a transition between bistable region where both antiparallel and parallel ferromagnetic ordering is possible and a monostable region where only parallel ferromagnetic ordering is possible. We have investigated the transition temperature T_{c1} further, for larger range of E_{AB}.

First, we have investigated how well simulations compare to analytic results and map iterations. The results (Fig.5) indicate, that while temperatures T_{c1} are different than predicted analytically, the error is not large. The fact, that the temperatures drop to zero at around $p \approx 0.4$, not at $p = 1$ shows that mean-field method does not describe the dynamics of the system accurately. In the full system, some vertices in one network may be more connected to the second network than to their own. In our analysis we considered them part of network A or B a priori, without considering actual edges. We did not investigate topological fluctuations and their influence on the transition temperatures.

Our main results concern the case of the Barabasi-Albert networks. The weighted spin against temperature for several different interconnection densities is shown in Figure 10. The figure clearly shows that a discontinuous transition takes place. Moreover we can see that the maximum of S is not necessarily a best indicator of the transition. Since we are working with much larger values of $E_{AB} \sim p$ than before, we have found that a better way to determine temperature T_{c1} is not to find a maximum of $|S| = |(S_A + S_B)/2|$, but a minimum of $|S_A|$ for simulations starting from antiparallel ordering. The value of $|S_A|$ decreases as the temperature increases, reflecting weakening of ferromagnetic order within network due to temperature. At temperature T_{c1} there is a sudden change. The antiparallel ordered networks start to order in parallel. This increases the value of $|S_A|$ since the resulting parallel ferromagnetic ordering is stronger than antiparallel, an obvious fact since in antiparallel state the networks weaken each other's order, while reinforcing it in parallel state. Thus the temperature point where $|S_A|$ rapidly increases corresponds to the temperature where switching from antiparallel to parallel state starts to occur, which is the sought temperature T_{c1}.

We have investigated the dependence of T_{c1} on the number of inter-network edges E_{AB}. We used antiparallel initial conditions and $\tau = 100$. We used the method of finding T_{c1} through minimum of $|S_A|$. The results are shown in Figure 11. As seen on the plot, the numerically found temperature T_{c1} is much lower than predicted by either analytical calculations or map iterations. This is because the analytical results as well as map iterations were based on numerical expectations of $\langle k^2 \rangle / \langle k \rangle$ and as explained before, this gives higher temperature values than it should. Another reason for this discrepancy could be the commonly observed fact that the mean-field approximation overestimates values of critical temperatures since it neglects critical fluctuations. We re-scale our results by a constant factor, so that the value for unconnected networks is the same allowing us to concentrate on the effect of the network interactions and disregard issues concerning single network inaccuracies. After that we obtain a relatively good agreement with map iterations for a small number of internetwork edges E_{AB}. When $E_{AB} \approx 15000$, the results from map iterations and simulations start to differ strongly. This is

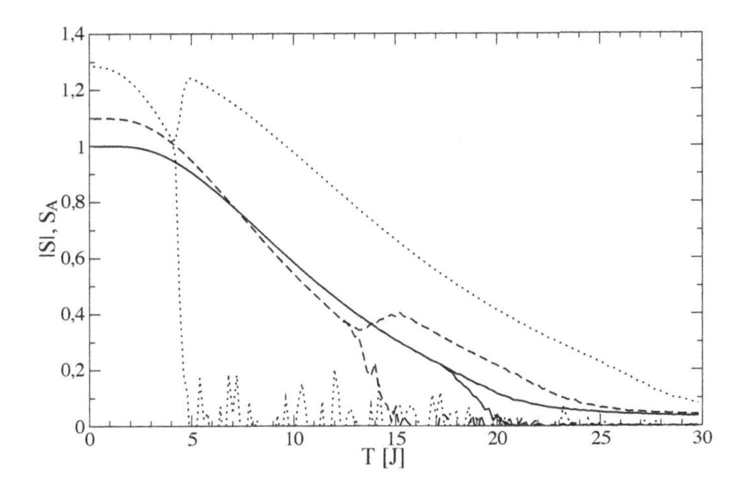

Fig. 10. Dependence of weighted spin absolute value $|S| = |(S_A + S_B)/2|$ and of weighted spin of single network S_A on temperature T. Both lines are same for $T < T_c$ while above T_c the upper lines are $|S|$ and lower (near-zero) are S_A. The results are for $N_A = N_B = 5000$ and $\langle k_A \rangle = \langle k_B \rangle = 10$. The solid lines are for $E_{AB} = 0$, dashed for $E_{AB} = 5000$ and dotted for $E_{AB} = 15000$. The weighted spin values above 1 result from increasing $\langle k \rangle$ due to interconnections while the normalization uses static $\langle k \rangle$ values taken from unconnected networks.

the result of the limited system size - the delicate antiparallel ordering is quickly destroyed by fluctuations and strong influence of the other network, and system reverts to parallel ordering that has lower energy. However, for lower interconnection densities, where interaction strengths are not so large, the simulations agree with map iterations, showing the existence of discontinuous transition and that the temperatures are correctly approximated by analytical approach.

Please note that T_{c1} approaches to zero around $p \approx 0.4$ not only in Barabasi-Albert networks (Fig.11) but also in random graphs (Fig.5). The first–order phase transition between the bistable and monostable phases is thus possible only for weakly coupled networks.

5. Conclusions

We show that in the system of two connected scale-free networks the ferromagnetic Ising model has more states than simple ordered and disordered state. In low temperatures, the system is in a bistable phase, with possible parallel and antiparallel ordered states possible. In the parallel ordered state, both parts of the system are ordered ferromagnetically in the same way. In the antiparallel ordered state the system possesses ferromagnetic order inside the particular networks, but

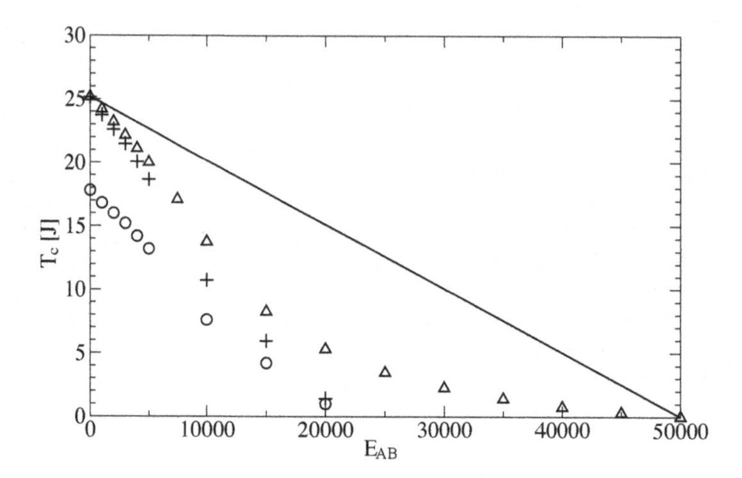

Fig. 11. Dependence of T_{c1} on the inter-network edge number $E_{AB} \sim p$. The line is analytic prediction of second order phase transition model, triangles are map iterations that display first order phase transition, while circles are data obtained from Monte-Carlo simulations. The plus symbols are same as circles, but re-scaled to have same value at $E_{AB} = 0$ as map iterations.

both networks possess opposite magnetizations. Above certain temperature T_{c1}, the antiparallel state ceases to be stable and antiparalel ordered system will transit discontinuously into a parallel ordered state as result of the first-order transition. If the system is in a monostable phase then above a temperature $T_{c+} > T_{c1}$ the parallel state also ceases to be stable and the system enters continuously a disordered paramagnetic phase. Both transition temperatures depend on network parameters — sizes and densities as well as interaction strength between the networks. Increasing the density of inter–network connections increases the temperature T_{c+} but decreases the temperature T_{c1} that becomes zero above some critical inter–network links density. It means that the bistable phase and the resulting first order phase transition can exist only for weakly coupled networks. We show that this behavior can be explained analytically with some help of numerical methods, although the temperature values can not be accurately predicted by the approach we have used.

Let us compare our results with selected studies of other authors. In[34] an analysis of transitions from a metastable to a stable state for two connected Ising networks with Erdös-Rényi topology in the presence of external magnetic was presented. Metastable initial simulations conditions corresponded to spin ordering opposite to an external magnetic field and in the course of time the system relaxed to a stable configuration (majority of spins along the external field) as a result of nucleation processes. It was observed that when the networks were weakly connected the system switched from the metastable to the stable state in two steps, each corresponding to switching of one network. This is analogous to temporary appearance of antiparallel ordering in the system. When the density of inter-networks links was over some critical value the nucleation process took place in a single step, i.e. no intermediate phase was observed. This corresponds to the situations where inter-network connections are sparse enough to be in bistable phase for given temperature, and to the situation where they are dense enough to be in monostable phase.

In[35] a modular system consisting of many small sub-networks (modules) was considered. It was observed that for weakly interacting modules a local ordered phase exists for a range of temperatures below T_{c+}, where networks are ordered internally, but disordered relative to each other. It is worth to note that in that study, the local ordered phase exists *above* a certain temperature T_c^g. This may seem contradictory to existence of an antiparallel order below T_{c1} observed in our simulations, but it is worth to note that the paper[35] investigates only most stable state and the simulated system consisted of 16 clusters of 32 spins while in our case there were only 2 networks, but with at least 5000 spins each. The antiparallel order observed in our system below T_{c1} is not a stable state but a specific metastable state, that loses its stability above T_{c1}.

Recently there is also a large interest in *percolative* transitions in interconnected networks since the first–order phase transition was observed in the model proposed in.[26] For a recent review of this subject see.[25]

In short, we show that the Ising model in two connected complex networks possesses three distinct states: parallel order, antiparallel order and disorder, and we calculate the temperatures at which transitions between them occur. The first–order phase transitions can take place in such a system if the connections density between both networks is below some critical value.

Acknowledgments

Janusz A. Hołyst was supported by FP7 FET Open project Dynamically Changing Complex Networks – DynaNets, EU Grant Agreement Number 233847, and a corresponding matching fund from Polish Ministry of Science and Higher Education, Grant Number 1029/7.PR UE/2009/7.

Appendix A. Introduction to Complex Networks

Complex networks is a relatively new topic of study, in practice started with papers of Watts-Strogatz[36] and Barabasi-Albert.[29] The topic encompasses studies of wide range of networks, from physical networks like power grid to very abstract ones like protein interaction networks. All the diverse networks are understood and studied using the approach of Graph Theory. The following introduction briefly explains main concepts related to complex networks, focusing on those used in this chapter. For a more comprehensive overview, refer to.[3–5]

A *graph* is a set of *vertices* that are pairwise connected by *edges*, an example being a set of computers connected by network cables. In complex networks research, vertices are sometimes also called *nodes* and edges *links*. Vertices and edges could represent things from an unlimited range of entities and relations between them. Vertices can represent physical objects or places, or a certain configuration of a protein. Edges can be physical wire connections, WWW hyperlinks or possible transitions between protein configurations. Each study has its own definitions of vertices and edges, and what they represent can change how they are referred to, such as vertices being called spins when investigating Ising Model (as in this chapter), agents for agent-based models on networks, etc. and edges being called relations when considering social networks, connections when investigating the Internet, etc. The vertices connected to a given vertex by edges are called its *neighbors*.

Edges can be either *undirected* (where both connected vertices are treated equally) or *directed* (where the interaction the edge is representing is not symmetric).

Both vertices and edges can have their own intrinsic properties, for example a computer having certain limited computational ability or a network cable having limited transfer speed. The properties usually investigated can be expressed as numbers, often called *weights*. Edge weights usually define the strength of a relation between the vertices they connect, with weight equal 0 being the same as if the edge did not exist, although it is not always the case in a particular study (in this chapter, if we consider exchange integral J a weight, it is true). If edges are not weighted then usually, a pair of directed edges connecting two vertices in both directions is equivalent to an undirected edge.

One can study a wide variety of vertex and link properties derived from the overall network topology. One of the simplest is *vertex degree* k_i, which is simply a number of edges that connect to the given vertex i. Note that in the case of a directed network (containing directed edges), two measures can be introduced, one counting outgoing edges and one ingoing. In case of weighed network (where edges have varying weights), when the zero edge weight is equivalent to an edge absence, the measure of vertex degree may be not important or informative. Sometimes the measure of sum of the incident edge weights can be used instead.

The complex networks are usually considered in relation to a *random network* model by Erdös-Rényi (E-R).[37] The E-R network consists of any number N of vertices, with edges existing between any pair i, j at random with a fixed probability p. This model is often considered a null model for complex networks – most measurements done for complex networks are compared to those of random graph. Measurement values are often designated high if they are significantly higher for investigated system than for random network of same size and number of connections. Similarly, a network can be said to contain some feature (such as presence of very high degree vertices or network discontinuity) if it is present there and absent in random network. If it is present in random network, it is treated as natural, and its absence would be worthy of notice instead (such as absence of small-world behavior).

The thing that makes complex network complex is a non-trivial topology of the edges between the vertices. If a graph is not a random graph and is not a regular, repetitive lattice, it could be considered complex. Many measures to characterize complex networks were invented. One of the most basic ones are *a degree distribution* and *an average path length.*

The degree distribution is simply a distribution of degrees of vertices in the whole network. Because a degree is a discrete number, the distribution is a discrete distribution, but in many cases they are approximated by continuous distributions in analysis, especially when degrees span a large range and therefore its discrete nature isn't that important. A regular square lattice degree distribution would be a delta distribution, with only a fixed degree k_0 vertices present. The E-R network (a random graph) has a binomial degree distribution, usually close to Poissonian distribution.

The *average path length* is an average of shortest path lengths between all pairs of vertices in the network. The shortest path length between two vertices i and j is the minimum number of edges one would have to cross to "travel" from vertex i to vertex j. It is also sometimes called *chemical distance.* In regular lattices, average path length scales the same as linear lattice size, what means as a square root of network size (number of vertices) for 2D lattice, as a cube root of network size for 3D lattice and so on. In the case of random graph (E-R), the average path length scales as logarithm of the network size, displaying so called *small-world* behavior. In reverse, this means that the number of vertices that can be reached increases exponentially with the distance.

Most complex networks display at least one of the following characteristics: scale-free degree distribution or small-world behavior. The scale-free degree distribution is a broad, highly skewed distribution following power-law, usually noted as $P(k) \sim k^{-\gamma}$. The γ is a scaling exponent that depends on the particular network model and parameters. It is important to note that if $\gamma < 3$, then the second moment of the distribution is infinite. Obviously, for any finite network, the actual second moment is finite, but can be very susceptible to random factors in the network creation. Moreover, the scale-free degree distributions of actual networks usually exhibit this behavior only above certain degree, while lower degrees either do not exist at all, or have different shaped distribution. Since scale-free distribution is treated as sign of existence of high-degree hubs, the low-degree parts of the distribution are often disregarded or not investigated. A network can be said to have scale-free degree distribution even if only the higher degrees display such behavior.

A few network models warrant special mention, as the examples of networks featuring the basic characteristics considered a sign of complexity. Watts-Strogatz model[36] is a network that is a hybrid between a regular lattice and a random graph. Basically, starting from a regular lattice, every existing edge has a fixed chance p to be rewired, connecting to a random vertex anywhere in the network. The resulting network features narrow degree distribution with an average equal to original lattice degree and a small-world behavior, even for low values of p, where the *clustering* of the original lattice is retained. The clustering can be defined as the number of triangles (paths of length 3) in the network relative to the possible number of triangles. Clustering is considered high, if it's significantly higher than that of a random network of same size and average degree.

The Barabasi-Albert model[29] (BA) is a second network considered as a basic one. It is an evolving network system, starting from initial conditions equivalent to a small completely conencted seed cluster and later on growing according to a rule. In the case of the BA network, the rule is relatively simple: each step t a new vertex is added to the network and m new edges are created, connecting the new vertex with already existing vertices. The exact vertex to connect to is chosen at random, using the *preferential attachment* rule. This means the edges connect to a vertex i with a probability proportional to the present vertex degree $P(i,t) \sim k_i(t)$. Starting from m fully connected vertices and using this rule one obtains a network that possesses a scale-free degree distribution $P(k) = 2m^2/k^3$ and exhibits the small-world behavior.

In general, a model that exhibits certain feature, such as the scale-free degree distribution or the small-world behavior, can be thought as belonging to a special graph category. We can therefore define *a random network* as any network where existence of edges is random (and therefore can be described statistically instead of looking at specific cases), *a scale-free network* as any network that has scale-free degree distribution, *an exponential network* as any network where its degree distribution has a short exponential tail (as opposed to fat, power-law tails of scale-free network degree distributions), *small-world network* as any where the average path length is logarithmically or slower diverging with the network size and so on. In general almost all complex network can be considered random networks. Barabasi-Albert network is a scale-free, small-world network, while Watts-Strogatz model is an exponential, small-world network with high clustering.

References

1. R. M. White, Quantum Theory of Magnetism, 2nd Ed. Springer-Verlag, 1983.
2. M. Lewenstein, A. Nowak and B. Latane, Statistical mechanics of social impact, *Phys. Rev. A.* **45** (2), 763–776, (1992).
3. R. Albert and A. L. Barabasi, Statistical mechanics of complex networks, *Reviews of Modern Physics* **74**, 47-97, (2002).
4. S. N. Dorogovtsev and J. F. F. Mendes, *Evolution of Networks: From biological networks to the Internet and WWW.* (Oxford University Press, 2003), ISBN 0-19-851590-1
5. G. Caldarelli, *Scale-Free Networks.* (Oxford University Press, 2007), ISBN 0-19-921151-7
6. A. Aleksiejuk, J. A. Hołyst and D. Stauffer, Ferromagnetic phase transition in Barabasi-Albert networks, *Physica A.* **310**, 260–266, (2002).
7. G. Bianconi, Mean field solution of the Ising model on a Barabasi-Albert network, *Phys. Lett. A.* **303** (2-3), 166–168, (2002).
8. S. N. Dorogovtsev, A. V. Goltsev and J. F. F. Mendes, Ising model on networks with an arbitrary distribution of connections, *Phys. Rev. E.* **66**, 016104, (2002).
9. A. V. Goltsev, S. N. Dorogovtsev, J. F. F. Mendes, Critical phenomena in networks, *Phys. Rev. E.* **67**, 026123, (2003).
10. C. P. Herrero, Ising model in scale-free networks: A Monte Carlo simulation, *Phys. Rev. E.* **69**, 067109, (2004).
11. B. Tadić, K. Malarz and K. Kułakowski, Magnetization reversal in spin patterns with complex geometry, *Phys. Rev. Lett.* **94**, 137204, (2005).
12. C. P. Herrero, Antiferromagnetic Ising model in small-world networks, *Phys. Rev. E.* **77**, 041102, (2008).
13. D. H. Kim, G. J. Rodgers, B. Kahng and D. Kim, Spin-glass phase transition on scale-free networks, *Phys. Rev. E.* **71**, 056115, (2005).
14. T. Jörg, H. G. Katzgraber and F. Krząkała, Behavior of Ising spin glasses in a magnetic field, *Phys. Rev. Lett.* **100**, 197202, (2008).
15. M. A. Sumour and M. M. Shabat, Monte Carlo simulation of Ising model on directed Barabasi-Albert network, *Int. J. Mod. Phys. C.* **16** (4), 585-589 (2005).
16. M. Hinczewski, Griffiths singularities and algebraic order in the exact solution of an Ising model on a fractal modular network, *Phys. Rev. E.* **75**, 061104, (2007).
17. J. A. Hołyst, K. Kacperski and F. Schweitzer, Phase transitions in social impact models of opinion formation, *Physica A.* **285** (1), 199–210, (2000).
18. J. A. Hołyst, K. Kacperski and F. Schweitzer, Social Impact Models of Opinion Dy-

namics. In *Annual Review of Comput. Phys.* **9**, 253–273 (2001). ISBN:978-981-02-4537-5.
19. S. Galam, From Galam-Mauger law to a powerful mean field scheme, *J. Applied Physics*. **87**, 7040–7042, (2000).
20. M. Girvan and M. E. J. Newman, Community structure in social and biological networks, *Proc. Natl. Acad. Sci. USA*. **99** (12), 7821–7826, (2002).
21. B. Karrer, E. Levina and M. E. J. Newman, Robustness of community structure in networks, *Rhys. Rev. E*. **77**, 046119, (2008).
22. L. Danon, A. Arenas and A. Diaz-Guilera, Impact of community structure on information transfer, *Phys. Rev. E*. **77**, 036103, (2008).
23. L. Luthi, E. Pestelacci and M. Tomassini, Cooperation and community structure in social networks, *Physica A*. **387** (4), 955-966, (2008).
24. R. Lambiotte, M. Ausloos and J. A. Hołyst, Majority model on a network with communities, *Phys. Rev. E*. **75**, 030101(R), (2007).
25. J. Gao, S. V. Buldyrev, H. E. Stanley and S. Havlin, Networks formed from interdependent networks, *Nature Physics* **8**, 40-48 (2012).
26. S. V. Buldyrev, R. Parshani, G. Paul, H. E. Stanley and S. Havlin, Catastrophic cascade of failures in interdependent networks, *Nature* **464**, 1025-1028 (2010).
27. K. Suchecki and J. A. Hołyst, Ising model on two connected Barabasi-Albert networks, *Phys. Rev. E*. **74**, 011122, (2006).
28. K. Suchecki and J. A. Hołyst, Bistable-monostable transition in the Ising model on two connected complex networks, *Phys. Rev. E*. **80**, 031110, (2009).
29. A. L. Barabasi and R. Albert, Emergence of scaling in random networks, *Science*. **286**, 509–512, (1999).
30. K. Suchecki, V. M. Equíluz, M. San Miguel, Conservation laws for the voter model in complex networks, *Europhysics Letters*. **69**, 228–234, (2005).
31. H. E. Stanley, *Introduction to Phase Transitions and Critical Phenomena*. (Oxford University Press, Oxford 1971)
32. H. G. Schuster, *Deterministic chaos*. (VCH Verlagsgeselschaft, Weinheim, 1988).
33. D. Chowdhury and D. Stauffer, *Principles of Equilibrium Statistical Mechanics*. (Wiley-VCH, Berlin, 2000).
34. H. Chen and Z. Hou, Optimal modularity for nucleation in a network-organized Ising model, *Phys. Rev. E*. **83**, 046124, (2011).
35. S. Dasgupta, R. K. Pan, S. Sinha, Phase of Ising spins on modular networks analogous to social polarization, *Phys. Rev. E*. **80**, 025101(R), (2009).
36. D. J. Watts and S. H. Strogatz, Collective dynamics of 'small-world' networks, *Nature*. **393**, 409–410, (1998).
37. P. Erdös and A. Rényi, On the evolution of random graphs, *Publications of the Mathematical Institute of the Hungarian Academy of Sciences* **5**, 17–61, (1960).

Chapter 5

Minority Game: An "Ising Model" of Econophysics

František Slanina

Institute of Physics, Academy of Sciences of the Czech Republic,
CZ-182 21 Prague, Czech Republic
slanina@fzu.cz

Within the discipline of Econophysics, the model of agents' behaviour called Minority Game plays an eminent role, as it is easily formulated and can be exactly solved. After a short motivation, we expose the exact solution in detail and finish the chapter by numerous variants of Minority Game, adapted for modeling situations in Economics and Society in general.

Contents

1. Rules of the Game

1.1. *Econophysics*

Statistical physics is a fairly universal tool. Nowadays, its interdisciplinary applications can hardly be a surprise. There are at least three good reasons for applying physics outside physics, notably in economics and other social sciences, which is the direction we pursue in this article. First, practically, we may find that the mathematical description of a known physical problem overlaps with the mathematical description of an unrelated problem in economics. Then, it is natural to translate all known results from one field to the other, thus saving a lot of work. Second, methodologically, both physical systems dealt with in statistical physics, and social systems studied by economists, consist of large number of elements. The relevant phenomena result from the collective behaviour of "elementary particles". Therefore, it is natural to expect that the approaches will share many common features in the two fields, even though the details may differ and the interpretation may require skills that are special to either physics or economics. Third, from a more philosophical point of view, if we believe in essential unity of knowledge, whether based on atomistic postulates of not, anything which is known about a corner of the Universe must somehow be related to what is known about another, however separate corner.

Anyhow, whatever level of thinking we prefer, we may find that a dialogue between physicists and social scientists, in particular economists, is fruitful. One of the trees which bring such fruits is the discipline of econophysics, which, as the name suggests, is a merger of physics and economics. It is quite old now and several good books describe what it is about.[1–7] In this article, we shall describe one successful model which emerged within econophysics. It is called the Minority Game (MG). For specialised books on MG see.[8–10]

1.2. *Inductive thinking*

A good deal of abstraction is needed to believe that humans are guided by totally rational reasoning. Nothing like that can be found in reality. Yet the idea is so deeply rooted in the orthodox economic theory that it is assumed at least as a benchmark or a starting point to develop a correct theory.

The limits imposed to our rational capabilities stem from information deficit. If we knew what the governor of a central bank decided just an hour ago, we would be better off that those of our colleagues who should wait until tomorrow morning for official communication.

Indeed, more transparency in the business environment would likely be beneficiary. The analysts, including empirical econophysicists, would certainly be happy if they knew all details about the myriads of transactions, from huge to tiny, which are now covered by the business secrecy.

But even if information were here, it may not be enough to imply a reliable statement on, e. g., the probability that a business will fail next month. Of course, the human actions are also influenced by random noise, weather, car accidents, falling rocks and the like. In short, the sources of irrational aspects in our decisions are multiple.

On the other hand, there is a more deep hindrance to our rational aspiration. Everyone must agree that all economic decisions are made for future, while all the information they are based upon, resides in the past. Inferring the future events from the past ones is an inductive rather than deductive procedure. Inductive thinking is a much more difficult endeavour than mere deduction with limited information. That is why induction requires specific approaches and specially designed models.

Looking more closely how inductive thinking works in real situation we soon notice that the decisions are based on pattern recognition. Humans possess certain ensembles of "images" stored in their memories, associating certain outcomes with preceding situations. The images are not collected systematically nor they are rationally classified. We cannot be much wrong if we consider them random. The images, or patterns, are then used to make decisions, by matching them with the reality and choosing among those, which are compatible with the current situation. This implies that we can attribute a quality to the images, highlighting those who in the past suggested a beneficial decision. The important point in the inductive thinking is that we constantly assess and re-assess the mental images we carry in our heads and this way we continuously react to changing rules of the world around us. Even if the teacher at a high school taught us a virtually random collection of facts, as soon as we master the art of selecting among them what is most likely useful now, and even more importantly, if we are capable of constant updating of the usefulness of what we know, we cannot fail totally in our adaptation for life.

1.2.1. *Bar El Farol*

W. Brian Arthur deserves merit for bringing the inductive thinking into economy in terms of a simple model.[11] It was inspired by the *El Farol* bar in Santa Fe, every Thursday night attracting fans of live Irish music. Suppose there are 100 people considering to go to the bar but there are only 60 seats available. It is annoying to go to the bar if it is overcrowded and it is also pity to stay home while vacant chairs are longing for us there.

But how can the individuals coordinate so that the bar attendance is close to its capacity, if they cannot communicate to each other before they decide about their night programme? We would expect that the number of visitors will be completely random and the source of entertainment will be used ineffectively. But if the people are given the record of number of visitors in the past weeks, they may infer inductively how many they are about to be there next time. If the prediction is below 60, the decision is to go, if it exceeds 60, it is better to stay.

Arthur's pioneering idea was that we can model the situation supposing each person has certain fixed, and relatively small, number of predictors. Based on the past sequence of attendances, say ...,51,36,82,45,66,49, the predictors compute the next attendance. Predictors can process the information of the past attendances in many ways. For example they can say that next time the attendance will be a) the same as the last one, or b) rounded average of the last three attendances, or c) the same as p weeks ago (detecting cycles of period p), and the reader will surely invent many more rules. If the predictor expects the attendance below 60, it suggests the agent to go to the bar, otherwise it suggests to stay home.

An important point is that the agents are given several such predictors and in each step attribute plus or minus points to the predictors, depending on their success in suggesting the right action. The agent then decides to use the predictor which has maximum points. Such mechanism emulates the bounded intelligence of the music fans. They will never know for sure what to do, but they will adapt on the conditions of the world. Let us stress again that the predictors are random and do not contain anything that could be a priori useful to guess the correct answer.

In a computer simulation, Arthur showed that the agents soon self-organise so that the attendance fluctuates around the optimal value 60. The equilibrium is not found by a predefined algorithm, and the imprecision in the inference is not due to an external noise. The correct answer emerges spontaneously from the collective action of the agents.

Moreover, the agents themselves provide both signal to be deciphered and noise obscuring it. Thus, they constitute a complete self-sustained ecology, much the same as humans themselves provide the environment and determine conditions of life for humankind. The micro-cosmos of El Farol is a distillate of the whole human society.

1.3. *The algorithm*

Arthur's analysis of the El Farol bar problem was a lucid demonstration of how the agents find the equilibrium by induction. When we turn to investigation of the fluctuations around the stationary state, things become much more complicated. The complexity stems from the intrinsic *frustration* (see the Appendix). Indeed, it is impossible to find a general optimum strategy for the El Farol bar visitor, as this would imply that all people would do the same action and the attendance would be either zero or full. Surely that is not close to optimum resource utilisation. The idea of representative agent is of no use here.

Instead, we need to know how the agents adapt to each other, keeping in mind that their strategies are better or worse only conditioned to other agents' action.

This way we can investigate the crucial question of how close the system can approach the optimum, or how large are the fluctuations around the equilibrium.

1.3.1. *Minority wins*

To this end, the rules will be further simplified.[12,13] To keep the optimum state trivial we assume that the agents can take one of two possibilities. Like the bar visitors could go out or stay home, here we let the agents go one step up or down. To keep in touch with reality we can imagine the step up to be an order to buy a unit of a commodity, and conversely the step down to be an equally sized sell order.

Having N agents, denote $a_i(t) \in \{-1, +1\}$ the action of the i-th agent at time t. The aggregate movement of the whole ensemble of agents

$$A(t) = \sum_{i=1}^{N} a_i(t) \tag{1}$$

will be called attendance, in analogy with the El Farol bar model, although now it is rather related to the shift in a commodity price. If the attendance is positive the price rises and those who decided to sell, i. e. agents who acted as $a_i(t) = -1$ can feel rewarded, as they get better price than one step before. On the contrary, buyers, $a_i(t) = +1$ suffer a loss, as they spent more money than they would if they had bought one step earlier. Conversely, negative attendance rewards buyers and punishes sellers. Therefore, it is always beneficiary to go against the trend and stay in the minority. This is why the model was named Minority Game. To avoid the ambiguity in telling which group makes the minority, we always expect the number of agents N is odd. Obviously, the optimal situation is reached when the size of the minority group is as large as possible, or, equivalently, when the attendance is closest possible to zero.

The minority reward rule is expressed formally as the prescription for the update of agents' wealth $W_i(t)$ as follows

$$W_i(t) - W_i(t-1) = -a_i(t) \operatorname{sign} A(t) \ . \tag{2}$$

Various variants of the rule can be implemented, replacing $\operatorname{sign} A(t)$ by $G(A(t))$, where $G(x)$ is a anti-symmetric non-decreasing function, $G(x) = -G(-x)$. It can be interpreted as impact of the demand-offer disequilibrium on the commodity price. Very natural choice is the linear price impact $G(x) = x$ and we shall use it later in analytical solution of the Minority Game. As we shall see, the precise form of $G(x)$ is of little importance and one may choose one or another depending on the particular question asked.

The reward rule in Minority Game is very transparent and simple to implement because of its immediacy. The players collect their points at the same moment (or, to be more precise, infinitesimally short time after) they make their actions. This is perfectly appropriate for bar attendance, but if we want to interpret the Minority Game as a trading model, a complication arises. We should note that the gain expressed by (2) is potential, rather than actual. Indeed, in order to know how much the seller really earned we needed to know at which price she acquired the commodity at a more or less distant past. Similarly, the buyer will know the outcome only after she sells the commodity back in the future. To compute the financial effect of an action at time t it is necessary to know the price movements within longer time span, not only the immediate change suggested by the attendance $A(t)$ just at the same step as the action was taken. In the Minority Game, such multitemporal nature of stock-market activity is drastically simplified.

1.3.2. *Multiple strategies*

So far we dealt with the gain the agents receive or loss they must suffer for their actions. Obviously, they want to act so that they get rewarded, but how to choose what to do? To this end, each agent is given a set of S strategies $S = 2$ will be enough) predicting the right action in the next game, based on the outcomes in the last M steps. The parameter M measures the length of the agent's memory. As the game goes on, the agents learn which of the S strategies is worth using. We shall explain soon how they manage that, but now let us see how the strategies are made up.

To reduce the information content to manageable extent, we shall store not the full sequence of attendances $A(t')$, but only the series of profitable choices

$$\chi(t') = -\text{sign}\, A(t') \tag{3}$$

for $t' = t-1, t-2, \ldots, t-M$. A strategy is a prescription which predicts the next right action from the binary string of M past right actions

$$\mu(t) = [\chi(t-1), \ldots, \chi(t-M)] \tag{4}$$

kept in the agents' memory and therefore it is a map $\{-1,+1\}^M \to \{-1,+1\}$. Distinguishing the S strategies of the agent i by index s (especially for $S = 2$ we can choose $s \in \{-1,+1\}$) we denote $a^{\mu}_{s,i}$ the action suggested by s-th strategy of agent i provided the sequence of past outcomes is μ.

The number of possible combinations of outcomes in M steps is $P = 2^M$ and the number of all possible strategies is a very large number 2^{2^M} already for moderate memory length M. This means that the collection of agents is indeed very heterogeneous and to find two agents equipped by the same strategies is practically impossible. On the other hand, the strategies, albeit different, may be significantly correlated. A pair of strategies can, for example, differ only for a few cases of the past outcomes μ, which may even never occur in the course of the game. If we imagine each strategy as a P-component vector, there may be at maximum P mutually orthogonal, i. e. uncorrelated strategies. The effective size of the strategy space P should be then compared with the number of agents N among whom the strategies are distributed. So, we can anticipate that the properties of the Minority Game will depend on the scaling parameter

$$\alpha = \frac{P}{N} = \frac{2^M}{N} \tag{5}$$

when both the number of agents and the memory length go to infinity.

1.3.3. *Dynamics of scores*

Well, we know that the strategies take responsibility for the agents' behaviour and every agent has to choose somehow which of her S strategies is the best at a particular moment. Now we must decide how to measure the quality of the strategy. A very simple method is comparing the outcomes of the game with the predictions of all strategies. Every strategy of each agent records of its score $S_{s,i}$ and receives $+1$ or -1 points depending on whether it suggested right or wrong action. The strategy with maximum points is then actually used by the agent. Important thing is that also the strategies which are not currently used are continuously tested, so that the player learns which of her patterns of behaviour is better suited for current circumstances.

We can formalise this rule for $S = 2$ by introducing the difference $q_i(t) = (S_{+,i} - S_{-,i})/2$ in scores of the two strategies of the agent i. (The factor $1/2$ is used for future convenience). First, we introduce some notation which will be useful throughout the rest of this chapter

$$\begin{aligned} \omega_i^\mu &= \frac{1}{2}\left(a_{+,i}^\mu + a_{-,i}^\mu\right) , \\ \xi_i^\mu &= \frac{1}{2}\left(a_{+,i}^\mu - a_{-,i}^\mu\right) , \\ \Omega^\mu &= \sum_{i=1}^{N} \omega_i^\mu . \end{aligned} \tag{6}$$

Given the scores' difference, the agents choose among their strategies according to

$$s_i(t) = \operatorname{sign} q_i(t) \tag{7}$$

and their actions are

$$a_i(t) = a^{\mu(t)}_{s_i(t)} = \omega_i^{\mu(t)} + s_i(t)\xi_i^{\mu(t)} . \tag{8}$$

The attendance follows immediately from (8)

$$A(t) = \Omega^{\mu(t)} + \sum_{i=1}^{N} \xi_i^{\mu(t)} s_i(t) . \tag{9}$$

The score differences are updated in each step according to

$$q_i(t+1) - q_i(t) = -\xi_i^{\mu(t)} \operatorname{sign} A(t) \tag{10}$$

which can be written, using (7), and (9), in a compact form

$$q_i(t+1) - q_i(t) = -\xi_i^{\mu(t)} \operatorname{sign}\left(\Omega^{\mu(t)} + \sum_{j=1}^{N} \xi_j^{\mu(t)} \operatorname{sign} q_j(t)\right) . \tag{11}$$

The equation (11), together with the prescription (3) and (4) for the memories $\mu(t)$, fully describes the Minority Game.

To summarise, Minority Game consists in coupled dynamical processes $q_i(t)$ and $\mu(t)$ defined above. Note that the actual wealth of the agents collected according to (2) does not enter the dynamics. It is rather a secondary by-product, while the virtual points attributed to the strategies play the primary role.

1.3.4. *Initial conditions*

The last piece completing the picture of the Minority Game is the question of initial conditions. The question is less trivial than one might think. Indeed, we shall soon see that the most intriguing feature of the basic Minority game is the presence of a phase transition from an efficient but non-ergodic regime to an inefficient ergodic phase. The ergodicity breaking implies dependence on initial condition which persists for infinitely long time.

The canonic choice of initial conditions is randomly drawn $\mu(0)$ with uniform probability distribution and all strategies having the same (e. g. zero) score. It means $q_i(0) = 0$ for all agents i. All the results of numerical simulations in the next section will use these initial conditions. We shall shortly comment on the influence of non-zero initial values for q_i in due course.

Given the assignment of strategies $a^{\mu}_{s,i}$ and initial conditions, the dynamics of the Minority Game is fully deterministic. In actual simulations we must average the results over a very small subset of all possible strategy assignments and choices of initial memory string $\mu(0)$. This introduces a casual element into otherwise non-random dynamics. We shall see later, that adding some stochasticity into the

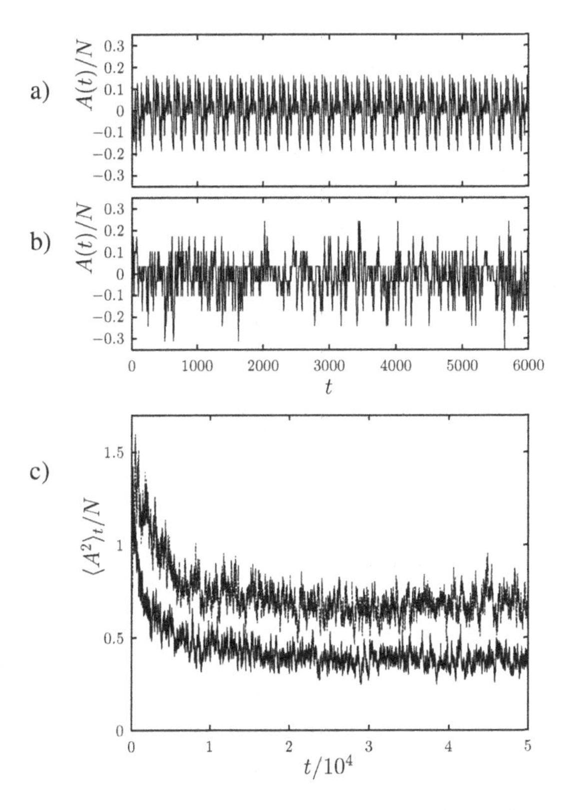

Fig. 1. Time series for the attendance (panels a) and b)) and local time average of the square attendance (panel c)). In the upper left panel, the memory is $M = 5$ and the number of agents is $N = 1001$, while in the lower left panel $M = 5$ and $N = 29$. In the right panel, the memory is $M = 10$ and the number of agents $N = 1001$ (solid line) and $N = 4001$ (dashed line). The parameter of the time average is $\lambda = 0.01$.

MG rules, especially to the strategy choice (7) makes the game "softer" and more amenable to analytic investigation.

1.4. *Phase transition*

It is indeed very easy and straightforward to embody the algorithm of the canonic Minority Game, as expressed by the equations (11), (3), and (4), into a computer code and observe the results. MG owes much of its appeal to the simplicity of the code and easiness with which a newcomer can touch and feel the behaviour of the agents.

1.4.1. *Adaptation*

The first thing to observe is the evolution of attendance $A(t)$. Already on a very qualitative level we can distinguish two types of behaviour, as seen in Fig. 1. For fixed memory length M, the time series becomes periodic, after an initial transient, if the number of agents is large enough. On the contrary, if N is small, the attendance follows a rather chaotic course.

The most important quantitative measure in the Minority game is the volatility

$$\sigma^2 = \lim_{T\to\infty} \frac{1}{T} \sum_{t=1}^{T} A^2(t) . \tag{12}$$

It measures how effectively the agents utilise the available resources. Indeed, if exactly half of the population is in the minority, the average gain of all agents is maximum (zero) and the volatility reaches its minimum value, which is zero. In the simulations, we always observe positive volatility, and it is easy to check that σ^2 grows when the average gain decreases.

The principal question is, whether the agents are able to adapt on-line on the behaviour of other agents so that the volatility is kept as small as possible. To this end, we introduce a time-local version of the volatility, computed during the simulations as $\langle A^2\rangle_t = \lambda A^2(t) + (1-\lambda)\langle A^2\rangle_{t-1}$. In the example shown in Fig. 1 we can clearly see how the initially large volatility decreases, until it saturates at certain level, which depends on the parameters N and M. The suppression of the volatility is a clear evidence of self-organisation due to individual learning of the agents.

1.4.2. *Minimum of volatility*

The greatest surprise of MG comes when we plot the dependence of the stationary volatility on the memory length M. We can see a pronounced minimum, indicating, that there is certain optimum size of the memory, beyond which the agent do not become more "intelligent". On the contrary, too much information leads to confusion.[14,15]

Interestingly, if we plot the reduced volatility σ^2/N against the scaling variable $\alpha = 2^M/N$, all results fall onto a single curve, as shown in Fig. 2. The minimum sharpens as we increase the system size and eventually approaches a singularity at the critical value $\alpha_c \simeq 0.34$. This marks a dynamical phase transition of rather unusual character. There is no way how to define a free energy in this model, so the character of the phase transition cannot be extracted from the singularities of the free energy. Nevertheless, the simulations suggest and as we shall see soon, the calculations confirm, that there is a singularity in the quantity

σ^2. This indicates the presence of a phase transition. To learn more about it, we must first determine the corresponding order parameter.

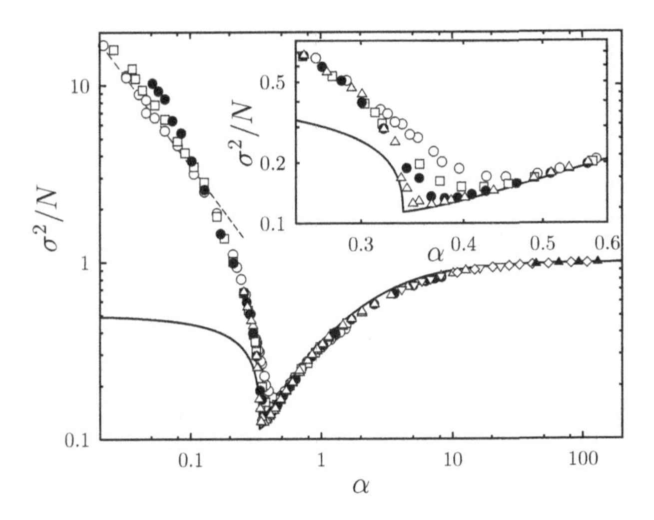

Fig. 2. Dependence of the volatility in the Minority Game on the scaling parameter $\alpha = 2^M/N$, for several combinations of memory M and number of agents N. In the inset, detail of the same data close to the transition point, with $M = 5$ (○), 6 (□), 8 (●), 10 (△). The full line is the result of analytic calculations using the replica method, Eqs. (45), (48), and (49). The dashed line is the dependence $\sim \alpha^{-1}$.

1.4.3. *Efficiency and order parameters*

Economics is largely concerned in market efficiency. The market is considered efficient, if there is no information left in the price signal. This means that you cannot use some freely accessible information and make profit of it. In our study, we completely disregard the natural question if real markets are efficient or not, or to which extent they can be efficient at all, or if the rhetoric about market efficiency is pure and empty propaganda. We shall consider efficiency a technical concept, denoting the amount of statistically discernible information (or neg-entropy) contained in a time sequence of numbers. These numbers may be prices, or they may be attendances on a round of Minority Game.

In this spirit, we can investigate the efficiency of the agents in extracting and destroying the information stored in the sequence of attendance. More precisely, we shall ask what is the average sign of the attendance, with fixed memory pattern

μ

$$\langle \mathrm{sign} A|\mu\rangle = \lim_{T\to\infty} \frac{\sum_{t=1}^{T} \delta_{\mu\,\mu(t)} \mathrm{sign} A(t)}{\sum_{t=1}^{T} \delta_{\mu\,\mu(t)}} \,. \tag{13}$$

If for some pattern μ the average is significantly different from zero, it is possible to predict which will be the most probable winning side. If the non-zero average persists, it means that the agents are not able to use this information and the system as a whole is not efficient.[15] The global measure of the efficiency is the average over all 2^M memory patterns

$$\theta^2 = \frac{1}{2^M} \sum_{\mu=0}^{2^M-1} \langle \mathrm{sign} A|\mu\rangle^2 \,. \tag{14}$$

It may be appropriately called predictability, as it measures the amount of remaining information in the time series.[16]

The data shown in Fig. 3 demonstrate that in the crowded phase, $\alpha < \alpha_c$, the system is efficient, as all information form the signal is eliminated, while for $\alpha > \alpha_c$ the system is inefficient. We can see it qualitatively, observing rather large deviations from 0 in $\langle \mathrm{sign} A|\mu\rangle$, when the number of agents is relatively small, and also quantitatively, in the dependence of θ^2 on the scaling parameter α. In the crowded and efficient phase the perdictability is virtually zero, while for $\alpha > \alpha_c$ it continuously grows. Therefore, θ^2 can be considered as an order parameter.

1.4.4. *Frozen agents*

It often happens in dynamical phase transitions that the order parameter is not unique. It is the case also here. Let us look at the way an agent uses her two strategies. She can alter them more or less regularly, so that both strategies are used equally often. But it may also happen that one of the strategies is chosen more often than the other. After long enough time the preference of one strategy should be clearly visible in the statistics. We can call this effect polarisation of the agents. The quantitative measure of the polarisation of the agent i is the time-averaged difference in the scores of the two strategies

$$v_i = \lim_{t\to\infty} \frac{1}{t} |S_{+,i}(t) - S_{-,i}(t)| \,. \tag{15}$$

In Fig. 4 we can see histogram of the polarisation of the agents. For $\alpha > \alpha_c$ it has two pronounced peaks, showing that there are indeed two types of agents, the first ones switching the strategies all the time, the other having one preferred strategy and they play t all the time. The pressure induced by the difference in scores need

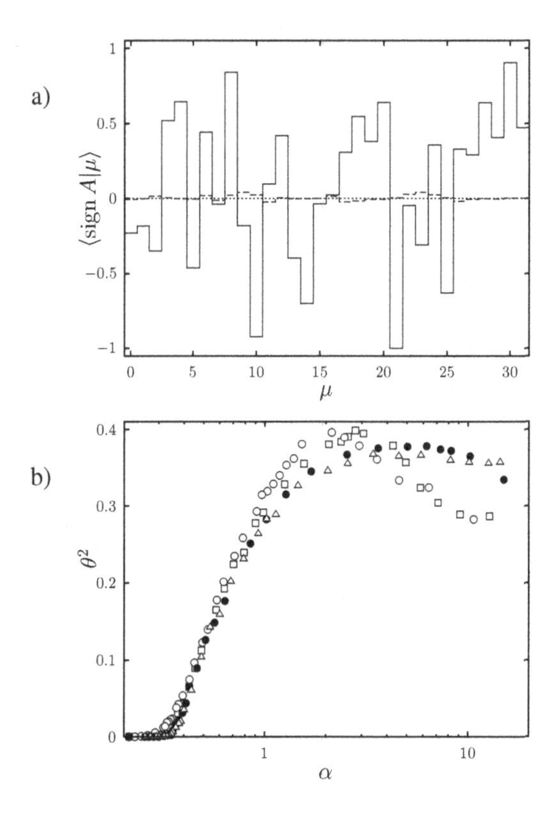

Fig. 3. In the panel a), average sign of the attendance depending on the actual memory pattern μ, for $M = 5$. The number of agents was $N = 51$ (solid line), 81 (dashed line), and 101 (dotted line). In the panel b), the average predictability of the time series, measured from the sign of the attendance, for $M = 5$ (○), 6 (□), 8 (●), 10 (△).

not to be very high, but in the histogram it is clearly visible. The latter agents are called frozen and we should ask how many of them are there. Looking again at Fig. 4 we observe that the higher peak in the histogram vanishes in the efficient phase $\alpha < \alpha_c$, so there are no frozen agents. Quantitative statistics is shown in the right panel of Fig. 4. Surprisingly, the fraction of frozen agents ϕ grows when we decrease α, becomes maximum at $\alpha = \alpha_c$ and then drops discontinuously to zero.

The fraction of frozen agents too can be chosen as an order parameter, as it vanishes in one phase and stays non-zero in the other. And we should not be confused by the fact that the transition looks continuous, i. e. second order, from the point of view of order parameter θ^2, while the second order parameter ϕ indicates a discontinuous, i. e. first-order transition. The classification pertinent to

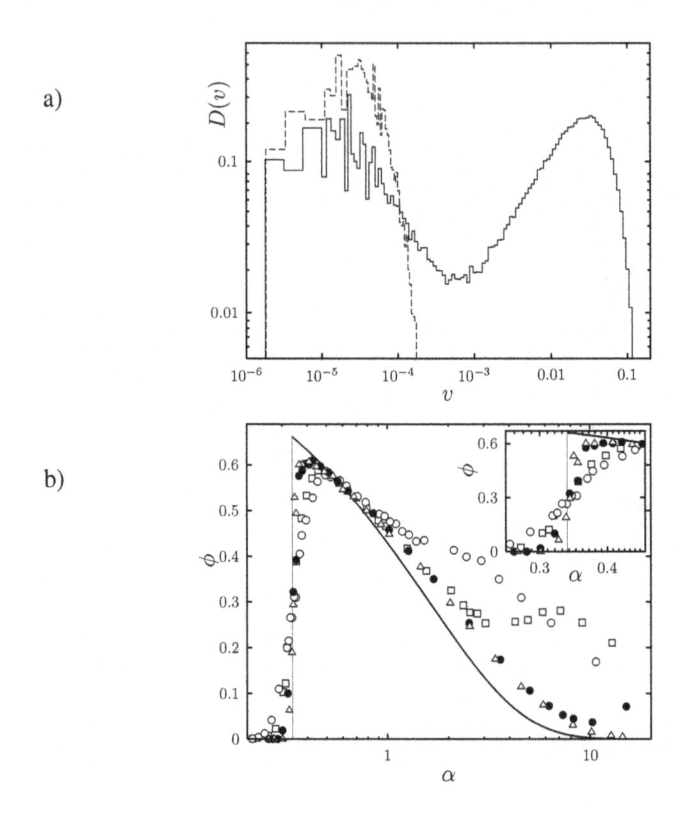

Fig. 4. In the panel a), histogram of the polarisation of the agents, for $M = 8$ and number of agents $N = 251$ (solid line) and $N = 901$ (dashed line). In the panel b), fraction of frozen agents for $M = 5$ ($\circ$), 6 ($\square$), 8 ($\bullet$), 10 ($\triangle$). The line is the analytic result from the replica approach. In the inset, detail of the transition region.

equilibrium order transitions has simply limited applicability when the transition is dynamic one and should not be taken too seriously. For further examples of this phenomenon, the reader can consult e. g.[17,18]

1.4.5. *How many patterns occur?*

There is also a third characteristics distinguishing the efficient and inefficient phases. If we measure the frequency with which certain memory pattern appears, we find, that it is quite homogeneous in the efficient phase, while in the inefficient one it is increasingly uneven.

Denoting $p_\mu = \lim_{T\to\infty} \sum_{t=1}^{T} \delta_{\mu(t)\mu}/T$ the relative frequency with which μ is found in the true dynamics, we can measure the entropy of the distribution established in the course of the dynamics

$$\Sigma = -\sum_{\mu} p_\mu \ln p_\mu \tag{16}$$

and compare it with the entropy $\Sigma_0 = M \ln 2$ of the uniform distribution. We can see in Fig. 5 that the difference $\Sigma_0 - \Sigma$ is indeed negligible in the symmetric phase, $\alpha < \alpha_c$, while in the asymmetric phase it is positive and grows with α. It means that the space of memories is visited inhomogeneously. For very long memory M the effective number of actually occurring μ's can be even rather small, as shown in the inset of Fig. 5. The data suggest that for fixed N the effectively visited volume scales as $e^{\Sigma} \sim 2^{\gamma M}$ with $\gamma \simeq \frac{1}{2}$, so the relative fraction of visited μ's shrinks to zero as $2^{-(1-\gamma)M}$ for $M \to \infty$.

2. Towards Analytical Solution

While numerical experiments with Minority game are readily accessible to anybody who has elementary skills in computer programming, some of the principal questions can be answered only by analytical approaches. Are the efficient and inefficient phases separated by a true phase transition or is it only a crossover phenomenon? Is the efficiency in the low-α phase perfect or just very high? Can we give an exact proof of the scaling property, i. e. that in the thermodynamic limit $M, N \to \infty$ all relevant quantities depend on the memory length and the number of agents only through the parameter $\alpha = 2^M/N$?

It turns out that to tackle the Minority game analytically requires special and rather sophisticated techniques. Despite the big challenges posed by the novelty of the dynamical process underlying MG, large part of the physics behind it is now understood. There are essentially two approaches being used in analytical study of MG. The first one is the replica method,[19] (see Appendix), starting from a mapping of the dynamical process on an effective equilibrium problem. The other one fully accounts for the dynamics using the generating functional method.[20,21] In this chapter, we deal with the replica method only. As for the generating functional, there is entire book[10] devoted to its use in Minority Game.

But before proceeding further, we must introduce some simplifications and modifications of the canonical MG, which do not change the essence of the model, but make it more amenable to solution.

2.1. *A few modifications*

2.1.1. *Quasi-irrelevance of memory*

Historically, the first important step forward was the observation that in the numerical simulations the results remain (nearly) the same if we replace the true dynamics of the memorised outcomes $\mu(t)$ as in (4), by memories μ drawn randomly from the set of all $P = 2^M$ possibilities[22,23]

$$\mathrm{Prob}\{\mu(t) = \mu\} = \frac{1}{P} . \tag{17}$$

We may interpret $\mu(t)$ as an external information, unrelated to the previous results of the game and essentially random. Important point is that all agents are given identical information. We can also consider $\mu(t)$ as a task or requirement the agents have to fulfil. The complexity of the game stems from the fact that all agents must solve the same task simultaneously and individual agents' solutions interfere with each other and there is no general solution available for all. That is the built-in frustration in MG, as we already alluded before.

The independence of $\mu(t)$ on the real history is indeed a crucial simplification, because the processes $q_i(t)$ are now decoupled from the process $\mu(t)$. The original dynamics turns into a Markov process and all memory effects are washed out. When we find it convenient, we shall call this modification Markovian Minority Game.

As the original MG gives nearly the same results as the Markovian version, we can indeed say that the memory in the MG is (nearly) irrelevant. However, the claims about the irrelevance of memory cannot go too far, as the true dynamics of memories does not visit all points in the space of μ's with equal probability, as we have seen in the previous paragraph.

Nevertheless, replacing the true dynamics $\mu(t)$ by randomly drawn μ's we keep all essential complexity of the Minority Game. The Markovian process we get is quantitatively slightly different from the original MG, but we gain much better access to analytical tools.

Besides the technical advantage there is an interesting lesson we can learn immediately. At the beginning we assumed the agents learn inductively how to predict the future attendance. In the Markovian MG there is no future to predict, or, more precisely, there is no past on which such prediction could be based. Instead, the agents simply adapt to each other under various external circumstances, embodied in the binary strings μ. We started with inductively thinking individuals but we find that their behaviour looks very much like they were optimising their position within an environment made up of all the remaining population. Deduc-

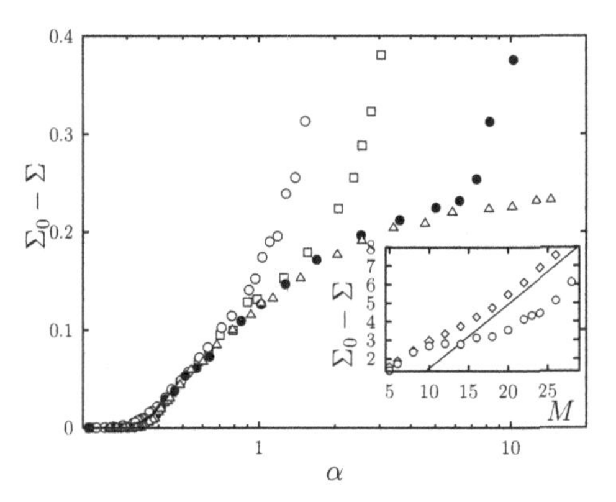

Fig. 5. Entropy of the distribution p_μ of the frequency of the memory patterns μ, relative to the entropy of the uniform distribution Σ_0. In the main figure, the memory length is $M = 5$ (○), 6 (□), 8 (•), 10 (△). In the inset, the same quantity for small number of agents, $N = 3$ (◇) and $N = 5$ (□). The straight line has slope $\ln \sqrt{2}$, indicating behaviour $e^{\Sigma} \sim 2^{M/2}$ for large M.

tive thinking regains a part of its credit.

2.1.2. *Thermal MG with linear payoff*

So far, the strategy was chosen deterministically according to the difference in strategy scores (7). A technically useful modification consists in allowing probabilistic choice of strategy, giving smaller or larger preference to the one with higher score. Similarly as in Monte Carlo simulations of equilibrium systems, we introduce a parameter Γ, analogous to the inverse temperature.[24–26] If the difference of strategies' scores for the agent i is $q_i(t)$, she chooses the strategy s with probability

$$\text{Prob}\{s_i(t) = s\} = \frac{e^{s\,\Gamma\, q_i(t)}}{2\cosh\, \Gamma\, q_i(t)}\ . \tag{18}$$

We recover the original MG prescription in the limit $\Gamma \to \infty$. If the strategy was selected many times with the same score difference, we would get the following average choice

$$\langle s_i(t)\rangle = \tanh \Gamma q_i(t)\ . \tag{19}$$

We shall see soon how and why such average becomes the central quantity the replica approach to MG.

Further simplification can be achieved if we change the payoff rule. Instead of adding or subtracting one point depending on the bare fact that the agent was in the minority or majority, we can provide a payoff proportional to the departure from the ideal equilibrated half-to-half state. Therefore, the payoff will not be proportional to sign $A(t)$ as in (10), but to the attendance $A(t)$ itself. The formula for the update of scores' differences is now linearised and has the form

$$q_i(t+1) - q_i(t) = -\xi_i^{\mu(t)} \left(\Omega^{\mu(t)} + \sum_{j=1}^{N} \xi_j^{\mu(t)} s_j(t) \right) \tag{20}$$

where the Markov processes $\mu(t)$ and $s_i(t)$ are governed by the probabilities (17) and (18).

2.1.3. *Batch MG*

Both in the standard MG and in the modifications described above the scores, i. e. the variables $q_i(t)$ are changed in every step, after each choice of the external information $\mu(t)$. However, we can expect that for large systems the evolution of $q_i(t)$ is rather slow and significant change occurs only on the timescale comparable with the total number of possible μ's, which is $P = 2^M$. This observation suggests a modification which was called the batch Minority Game.

Instead of updating the scores in each step, the variables q_i are changed only after the agents are given all of the P possible μ's (in arbitrary order). The agents react according to the rules of the thermal MG as explained above, but during such round of P steps the choice of the strategy is governed by the same distribution (18). The change in q_i from the round l to $l+1$, i. e. from the time $t = Pl$ to $t' = Pl + P$ is

$$q_i(l+1) - q_i(l) = -\frac{1}{P} \sum_{\mu=0}^{P-1} \xi_i^{\mu} \left(\Omega^{\mu} + \sum_{j=1}^{N} \xi_j^{\mu} s_j(t) \right) . \tag{21}$$

The factor $1/P$ was introduced for further convenience. This expression can be further simplified, because in the thermal MG the choice of $s_j(t)$ is independent of μ and for large P we can write

$$\sum_{\mu=0}^{P-1} \xi_j^{\mu} s_j(t) \simeq \sum_{\mu=0}^{P-1} \xi_j^{\mu} \langle s_j(t) \rangle \tag{22}$$

and the thermal average $\langle s_j(t) \rangle$ is given by (19). So, we conclude that the dynamics is given in a compact form by

$$q_i(l+1) - q_i(l) = -h_i - \sum_{j=1}^{N} J_{ij} \tanh \Gamma\, q_j(l) \tag{23}$$

where we denoted

$$
\begin{aligned}
h_i &= \frac{1}{P}\sum_{\mu=0}^{P-1} \xi_i^\mu \, \Omega^\mu \,, \\
J_{ij} &= \frac{1}{P}\sum_{\mu=0}^{P-1} \xi_i^\mu \, \xi_j^\mu \,.
\end{aligned}
\tag{24}
$$

2.2. *Replica solution*

2.2.1. *Dynamics of magnetisations*

For replica calculations[27,28] it is more convenient to express the dynamics in terms of the magnetisations $m_i = \tanh \Gamma q_i$, instead of the score differences q_i. The time evolution is particularly simple in the limit of small Γ. This is what we shall develop in the following.

In the limit $\Gamma \to 0$ we can expand the change of the magnetisation from one round to the other as

$$
\begin{aligned}
m_i(l+1) - m_i(l) = {}& \Gamma\,(1 - m_i^2(l))(q_i(l+1) - q(l)) \\
& - \Gamma^2\, m_i(l)(1 - m_i^2(l))(q_i(l+1) - q(l))^2 + O(\Gamma^3)
\end{aligned}
\tag{25}
$$

and keeping only the lowest order in Γ we get the following dynamics of the magnetisations

$$
\begin{aligned}
m_i(l+1) - m_i(l) &\simeq -\Gamma(1 - m_i^2(l))\left[h_i + \sum_{j=1}^{N} J_{ij} m_j\right] \\
&= -\Gamma(1 - m_i^2(l))\,\frac{1}{2}\,\frac{\partial}{\partial m_i} H(m_1, m_2, \ldots, m_N)
\end{aligned}
\tag{26}
$$

where we denoted

$$
\begin{aligned}
H(m_1, m_2,\ & \ldots, m_N) \\
&= \frac{1}{P}\sum_{\mu=0}^{P-1} (\Omega^\mu)^2 + 2\sum_{i=1}^{N} h_i m_i + \sum_{i,j=1}^{N} J_{ij} m_i m_j \\
&= \frac{1}{P}\sum_{\mu=0}^{P-1}\left[\sum_{i=1}^{N}\left(\frac{1}{2}(a_{+,i}^\mu + a_{-,i}^\mu) + \frac{1}{2}(a_{+,i}^\mu - a_{-,i}^\mu)\, m_i\right)\right]^2
\end{aligned}
\tag{27}
$$

a function which plays a central role in the subsequent calculation. Indeed, the evolution according to (26) closely follows a gradient descent in a "potential" H. Stationary state corresponds to the minimum of H, which can be considered as a Lyapunov function for the dynamics (26).

Before we proceed to solving the minimisation problem, we should clarify the relation of the Hamiltonian H to observable quantities. In the original MG the agents were rewarded according to the sign of the attendance, while now the gain is proportional to the attendance itself. So, to be consistent, the definition of the predictability according to (14) is less appropriate and we should rather use the quantity

$$\theta_A^2 = \frac{1}{P}\sum_{\mu=0}^{P-1}\langle A|\mu\rangle^2 \tag{28}$$

as a measure of the information content. (For the definition of the symbol $\langle\ldots|\ldots\rangle$ see Eq. (13).) We can see in Fig. 6 that it behaves qualitatively very similar to θ^2, shown above in Fig. 3. Most notably, both θ^2 and θ_A^2 vanish in the symmetric phase $\alpha < \alpha_c$, indicating that all information contained in the signal was used by the agents and thus zeroed.

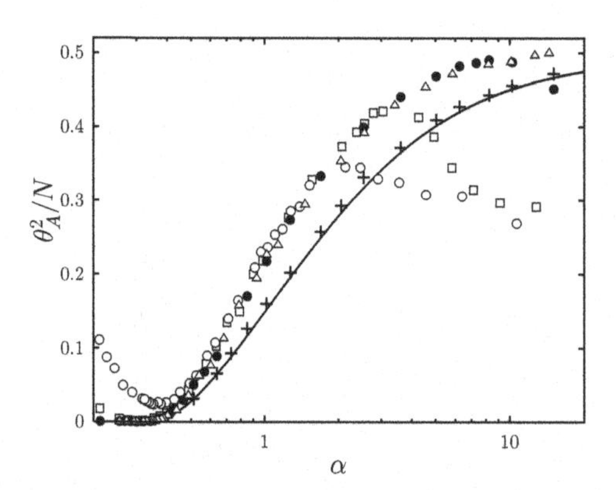

Fig. 6. Average predictability of the time series, measured from the attendance, for $M = 5$ (○), 6 (□), 8 (●), 10 (△). The line is the analytical result from the replica approach, the symbols + denote results of the simulation of the MG with uniformly sampled memories μ, for $M = 8$.

The attendance provided the external information μ is given by (9). Averaging the variables $s_i(t)$ in stationary state is performed according to the probability distribution (18), so $\langle s_i(t)\rangle = m_i$, and inserting the definitions (6) we can see that the expression for the predictability coincides with the Hamiltonian (27). This means that the measured predictability is equal to the minimum of the Hamiltonian because the stationary magnetisations m_i are just those which minimise H.

The relation between the Hamiltonian and the volatility σ^2 is less straightforward. We can write

$$\sigma^2 = \frac{1}{P}\sum_{\mu=0}^{P-1}\langle A^2|\mu\rangle \tag{29}$$

and express the attendance using again (9). The problem arises when we come to averaging the products of $s_i(t)$, because in general they do not factorise, $\langle s_i(t)s_j(t)\rangle \neq \langle s_i(t)\rangle\langle s_j(t)\rangle = m_i m_j$. We obtain

$$\begin{aligned}\sigma^2 &= \frac{1}{P}\sum_{\mu=0}^{P-1}\left(\sum_{i=1}^{N}(\omega_i^\mu + \xi_i^\mu m_i)\right)^2 + \frac{1}{P}\sum_{\mu=0}^{P-1}\sum_{i=1}^{N}(\xi_i^\mu)^2(1-m_i^2)\\ &+\frac{1}{P}\sum_{\mu=0}^{P-1}\sum_{i,j=1}^{N}(1-\delta_{ij})\xi_i^\mu\xi_j^\mu\langle(s_i(t)-m_i)(s_j(t)-m_j)\rangle\,. \end{aligned}\tag{30}$$

In the first term we recognise the Hamiltonian (27) and the second term can be easily computed once we obtain the magnetisations by minimisation of H. However, the third term cannot be obtained only from the dynamics of magnetisations according to (26). More detailed studies[29] show that in the ergodic phase $\alpha > \alpha_c$ the product averages do factorise and therefore the third term in (30) can safely be neglected. On the other hand, in the non-ergodic phase the factorisation breaks down and the value of the third term depends on the thermal parameter Γ. However, in the limit $\Gamma \to 0$ it vanishes again, so for infinitesimally small Γ we can calculate σ^2 for all values of the parameter α by studying only the static properties of the Hamiltonian H. The price to pay will be the quantitative disagreement between σ^2 as calculated in the non-ergodic regime and numerical simulations of the original MG, because the latter corresponds to the opposite limit $\Gamma \to \infty$. We shall see later how large that disagreement is and to placate the impatient reader we can say in advance that qualitatively the analytic results from minimisation the Hamiltonian H give a reasonably true picture of what is seen in the simulations.

Finally, we can also learn an important general feature of the behaviour of agents in MG. Each of them tries hard to maximise her individual profit and that is why she assesses the quality of the strategies and chooses the best one. However, we have just seen that it is not the overall loss, measured by the volatility σ^2, but the information content, or H, that is minimised in the dynamics. In short, the agents are not fully aware of the collective effect of their actions. Instead of optimising the global performance, they merely devour as much information as they can.

2.2.2. *Effective spin model*

We have seen that we can formulate the problem as finding the ground state for the system of soft spins $m_i \in [-1,1]$ described by the Hamiltonian H. To this end we shall first investigate its behaviour at finite temperature β and eventually determine the ground-state properties by sending $\beta \to \infty$. To pursue such program we need to overcome a serious hindrance. The Hamiltonian depends on the strategies $a^{\mu}_{s,i}$, which are selected randomly at the beginning and introduce quenched disorder into the Hamiltonian.

In fact, such problem is nothing much new in statistical physics. Spin glasses and neural networks [19] are very well described by very similar Hamiltonians. The presence of quenched randomness is the feature all these models have in common and it can be very effectively tackled using the replica method. (See the Appendix.) The quantity of prime interest is

$$Z(n) = \overline{\left(\int_{-1}^{1} \mathrm{d}[m_i]\, \mathrm{e}^{-\beta H}\right)^n} \tag{31}$$

where the overbar denotes the disorder average, i. e. the average over realisations of the strategies $a^{\mu}_{s,i}$ and we introduced a shortcut notation for multiple integrations $\int_{-1}^{1} \mathrm{d}[m_i] \equiv \int_{-1}^{1} \mathrm{d}m_1 \int_{-1}^{1} \mathrm{d}m_2 \ldots \int_{-1}^{1} \mathrm{d}m_N$ which will be used in various modifications throughout the calculation.

To perform the disorder average in (31) for positive integer n we formally introduce n replicas of the system, with state variables m_i^a, $a = 1, 2, \ldots, n$. Then, we can see that the Hamiltonian (27) consists of $P = 2^M$ terms, each of them being the sum of n squares. The squares in exponents are conveniently simplified using the Hubbard-Stratonovich transform (see Appendix), introducing in revanche nP new auxiliary fields z_a^{μ}. But it turns out that all contributions for different μ are disorder-averaged independently of the others and we end with a product of P identical factors, each of them containing only n auxiliary fields z_a. Explicitly, we find

$$\begin{aligned} Z(n) = \int_{-1}^{1} \mathrm{d}[m_i^a] \Bigg\{ \prod_{i=1}^{N} \int_{-\infty}^{\infty} \mathrm{d}\left[\frac{z_a}{\sqrt{2\pi}}\right] \mathrm{e}^{-\frac{1}{2}\sum_{a=1}^{n} z_a^2} \\ \overline{\times \exp\left(\mathrm{i}\sqrt{\frac{\beta}{2P}} \sum_{a=1}^{n} z_a\,(1+m_i^a)\, z_a\, a_{+,i}\right)} \\ \overline{\times \exp\left(\mathrm{i}\sqrt{\frac{\beta}{2P}} \sum_{a=1}^{n} z_a\,(1-m_i^a)\, z_a\, a_{-,i}\right)} \Bigg\}^P . \end{aligned} \tag{32}$$

The averages over the variables $a_{s,i}$ are performed using the following simple trick, which works for $P \to \infty$

$$\overline{\exp(\mathrm{i}\frac{C}{\sqrt{P}}a_{s,i})} = \frac{1}{2}\sum_{a=\pm 1} \mathrm{e}^{\mathrm{i}Ca/\sqrt{P}}$$

$$= \cos(\frac{C}{\sqrt{P}}) \simeq 1 - \frac{1}{2P}C^2 \simeq \exp(-\frac{C^2}{2P}) \,. \tag{33}$$

This way we get in the exponent terms of the type $m_i^a m_i^b$ indicating interaction between replicas. A short algebra yields

$$Z(n) = \int_{-1}^{1} \mathrm{d}[m_i^a] \Big\{ \int_{-\infty}^{\infty} \mathrm{d}\Big[\frac{z_a}{\sqrt{2\pi}}\Big] \mathrm{e}^{-\frac{1}{2}\sum_{a=1}^{n} z_a^2}$$

$$\times \exp\Big(-\frac{1}{2}\frac{\beta}{\alpha}\sum_{a,b=1}^{n} z_a z_b (1 + \frac{1}{N}\sum_{i=1}^{N} m_i^a m_i^b)\Big)\Big\}^P \,. \tag{34}$$

The pivotal parameter $\alpha = P/N$ appears for the first time here. For the integration over magnetisations m_i^a it is very inconvenient that the term $\sum_{i=1}^{N} m_i^a m_i^b$ appears inside the bracket $\{\ldots\}^P$. To take it out, we introduce another set of $n(n+1)/2$ variables q_{ab}, $a \le b$, with δ-functions guaranteeing that $q_{ab} = \frac{1}{N}\sum_{i=1}^{N} m_i^a m_i^b$. Thus

$$Z(n) = \int_{-1}^{1} \mathrm{d}[m_i^a] \int_{-\infty}^{\infty} \mathrm{d}[q_{ab}] \Big(\prod_{a\le b}^{n} \delta\big(q_{ab} - \frac{1}{N}\sum_{i=1}^{N} m_i^a m_i^b\big)\Big)$$

$$\times \Big\{ \int_{-\infty}^{\infty} \mathrm{d}\Big[\frac{z_a}{\sqrt{2\pi}}\Big] \mathrm{e}^{-\frac{1}{2}\sum_{a=1}^{n} z_a^2} \exp\Big(-\frac{1}{2}\frac{\beta}{\alpha}\sum_{a,b=1}^{n} z_a z_b (1 + q_{ab})\Big)\Big\}^P \,. \tag{35}$$

The δ-functions are then expressed by integral representation $\delta(x) = \int_{-\infty}^{\infty} \frac{\mathrm{d}y}{2\pi} \mathrm{e}^{\mathrm{i}xy}$. Furthermore, in the exponent we recognise a quadratic form in the variables z_a, so the integration over these variables is straightforward in principle. We complete the definition of quantities q_{ab} by symmetrisation, $q_{ab} = q_{ba}$, and define a $n \times n$ matrix M with elements $M_{ab} = \delta_{ab} + \frac{\beta}{\alpha}(1 + q_{ab})$. The integration over all z's gives $(\det M)^{-1/2}$. Therefore

$$Z(n) = \int_{-\infty}^{\infty} \mathrm{d}[q_{ab}] \int_{-\mathrm{i}\infty}^{\mathrm{i}\infty} \mathrm{d}\Big[\frac{-\mathrm{i}N\, r_{ab}}{2\pi}\Big] \exp\Big(- N\sum_{a\le b}^{n} r_{ab}\, q_{ab}\Big)$$

$$\times (\det M)^{-P/2} \Big[\int_{-1}^{1} \mathrm{d}[m^a] \exp\Big(\sum_{a\le b} r_{ab}\, m^a m^b\Big)\Big]^N$$

$$= \int_{-\infty}^{\infty} \mathrm{d}[q_{ab}] \int_{-\mathrm{i}\infty}^{\mathrm{i}\infty} \mathrm{d}\Big[\frac{-\mathrm{i}N\, r_{ab}}{2\pi}\Big] \exp\big(- N\beta n \mathcal{F}\big) \tag{36}$$

with effective replicated free energy

$$\mathcal{F}(q_{ab}, r_{ab}) = \frac{1}{n\beta} \sum_{a \leq b}^{n} r_{ab}\, q_{ab} + \frac{\alpha}{2n\beta} \ln \det M$$
$$- \frac{1}{n\beta} \ln \int_{-1}^{1} \mathrm{d}[m^a] \exp\left(\sum_{a \leq b} r_{ab}\, m^a m^b \right) . \qquad (37)$$

The last integral in (36) can be taken by the saddle-point method, as in the thermodynamic limit $N \to \infty$ we have

$$Z(n) \simeq \exp\big(- N\beta n \mathcal{F}(q^*_{ab}, r^*_{ab})\big) \qquad (38)$$

and the calculation reduces to finding the position q^*_{ab}, r^*_{ab} of the minimum of $\mathcal{F}$, i. e. solving the set of $n(n+1)$ equations

$$\frac{\partial}{\partial q_{ab}} \mathcal{F}(q^*_{ab}, r^*_{ab}) = \frac{\partial}{\partial r_{ab}} \mathcal{F}(q^*_{ab}, r^*_{ab}) = 0 . \qquad (39)$$

Fortunately, we can look for the solution in a very simple replica-symmetric form

$$\begin{aligned} q^*_{ab} &= q + (Q - q)\delta_{ab} \\ r^*_{ab} &= r + (R - r)\delta_{ab} \end{aligned} \qquad (40)$$

reducing the number of free parameters to only four. It is possible to prove that this solution is thermodynamically stable, which justifies the replica symmetry.[30] The effective free energy becomes

$$\frac{1}{n}\mathcal{F} = \frac{1}{\beta}\left(RQ - \frac{1}{2} rq \right)$$
$$+ \frac{\alpha}{2\beta} \ln\left(1 + \frac{\beta(Q-q)}{\alpha} \right) + \frac{\alpha(1+q)}{2(\alpha + \beta(Q-q))}$$
$$- \frac{1}{\beta} \int_{-\infty}^{\infty} \frac{\mathrm{d}z}{\sqrt{2\pi}}\, \mathrm{e}^{-\frac{1}{2}z^2} \ln \int_{-1}^{1} \mathrm{e}^{\left(R - \frac{1}{2}r\right) m^2 + z\sqrt{r}\, m} \mathrm{d}m + O(n) . \qquad (41)$$

The last term, containing the integrals over m and z, looks like a free energy of a particle in potential $V_z(m) = -\left(R - \frac{1}{2}r\right) m^2 - z\sqrt{r}\, m$, averaged over Gaussian-distributed random external field z. We shall use that analogy in practical calculations, introducing a z-dependent averages with respect to the potential $V_z(m)$ as $\langle \ldots \rangle_z = \int_{-1}^{1} \ldots \mathrm{e}^{-V_z(m)} \mathrm{d}m / \int_{-1}^{1} \mathrm{e}^{-V_z(m)} \mathrm{d}m$. Note also that the dependence on n occurs only in a term of higher order $O(n)$, so we can safely perform the limit $n \to 0$ now. The minimum of the effective free energy is found by differentiating $\mathcal{F}$ with respect to the four free parameters. After a short calculation we get four coupled transcendental equations

$$
\begin{aligned}
R - \frac{r}{2} &= -\frac{\alpha\beta}{2} \frac{1}{\alpha + \beta(Q-q)} , \\
r &= \alpha\beta^2 \frac{1+q}{(\alpha + \beta(Q-q))^2} , \\
Q - q &= \frac{1}{\sqrt{r}} \int_{-\infty}^{\infty} \frac{\mathrm{d}z}{\sqrt{2\pi}} \, \mathrm{e}^{-\frac{1}{2}z^2} \, z \langle m \rangle_z , \\
Q &= \int_{-\infty}^{\infty} \frac{\mathrm{d}z}{\sqrt{2\pi}} \, \mathrm{e}^{-\frac{1}{2}z^2} \, \langle m^2 \rangle_z .
\end{aligned}
\tag{42}
$$

Let us recall that we are looking for the minimum of the Hamiltonian (27), so we need the solution of (42) in the limit $\beta \to \infty$. We should distinguish two cases. First, the difference $Q - q$ can approach a finite limit, so that the quantity

$$
\chi \equiv \beta(Q - q) \tag{43}
$$

diverges. We shall see later that this is the case of the symmetric, non-ergodic phase with $\alpha < \alpha_c$. Indeed, χ can be interpreted as susceptibility, measuring sensitivity to initial conditions. Diverging susceptibility implies that arbitrarily small change in initial conditions persists infinitely long, marking the non-ergodic behaviour.

Second solution is characterised by finite susceptibility, so $Q - q \to 0$ for $\beta \to \infty$. This is the asymmetric phase, $\alpha > \alpha_c$ and we shall spend some time now analysing the results we can infer from the solution of (42).

2.2.3. *Properties of the ergodic phase*

The most technically difficult part of the set (42) is calculation of the averages $\langle m \rangle_z$ and $\langle m^2 \rangle_z$ and then integrating over z. However, in the limit $\beta \to \infty$ the algebra simplifies significantly, as we can see from the first two equations in (42) that $r \sim \beta^2$ and $R - \frac{1}{2}r \sim \beta$, so the potential $V_z(m) \sim \beta$ for $\beta \to \infty$. The averages with respect to such potential are dominated by its single minimum at a point $m = m^*$, which can be readily obtained. To simplify the notation we write $V_z(m) = \sqrt{r}\big(\frac{1}{2}\zeta\, m^2 - z\, m\big)$, where $\zeta = -\frac{2}{\sqrt{r}}\big(R - \frac{1}{2}r\big)$. For $\beta \to \infty$ the parameter ζ approaches a finite limit,

$$
\zeta = \sqrt{\frac{\alpha}{1+Q}} \, . \tag{44}
$$

According to the value of z the minimum lies either inside the interval $(-1, 1)$ or at the endpoints -1 or 1. We list the results in the Table 1, showing the values

Table 1. Various pieces needed to calculate Q and χ.

$z \in$	$(-\zeta,\zeta)$	$(-\infty,-\zeta]$	$[\zeta,\infty)$
m^*	z/ζ	-1	1
$\langle m\rangle_z$	z/ζ	-1	1
$\langle m^2\rangle_z$	z^2/ζ^2	1	1
$\int \frac{dz\, e^{-z^2/2}}{\sqrt{2\pi}}\, z\langle m\rangle_z$	$\frac{1}{\zeta}\,\mathrm{erf}\big(\frac{\zeta}{\sqrt{2}}\big) - \sqrt{\frac{2}{\pi}}\, e^{-\frac{\zeta^2}{2}}$	$\sqrt{\frac{2}{\pi}}\, e^{-\frac{\zeta^2}{2}}$	
$\int \frac{dz\, e^{-z^2/2}}{\sqrt{2\pi}}\, \langle m^2\rangle_z$	$\frac{1}{\zeta^2}\,\mathrm{erf}\big(\frac{\zeta}{\sqrt{2}}\big) - \sqrt{\frac{2}{\pi}}\,\frac{1}{\zeta}\, e^{-\frac{\zeta^2}{2}}$	$1-\mathrm{erf}\big(\frac{\zeta}{\sqrt{2}}\big)$	

of the averages in each of the three ranges of z, together with the results of the integration over z in the corresponding intervals.

Putting together the pieces contained in Table 1 and the relation between ζ, α and Q given by (44), we obtain the following expressions for Q and the susceptibility χ, parametrised by ζ.

$$
\begin{aligned}
\alpha &= 2\zeta^2 + (1-\zeta^2)\,\mathrm{erf}\Big(\frac{\zeta}{\sqrt{2}}\Big) - \sqrt{\frac{2}{\pi}}\,\zeta\, e^{-\frac{\zeta^2}{2}}\,, \\
Q &= 1 - \mathrm{erf}\Big(\frac{\zeta}{\sqrt{2}}\Big) + \frac{1}{\zeta^2}\,\mathrm{erf}\Big(\frac{\zeta}{\sqrt{2}}\Big) - \sqrt{\frac{2}{\pi}}\,\frac{1}{\zeta}\, e^{-\frac{\zeta^2}{2}}\,, \\
\chi &= \frac{\alpha\,\mathrm{erf}\big(\frac{\zeta}{\sqrt{2}}\big)}{\alpha - \mathrm{erf}\big(\frac{\zeta}{\sqrt{2}}\big)}\,.
\end{aligned}
\tag{45}
$$

This solution breaks down when the susceptibility diverges, i. e. for $\alpha \to \mathrm{erf}\big(\frac{\zeta}{\sqrt{2}}\big)$. This condition fixes the critical value of the control parameter α through the pair of equations

$$
\begin{aligned}
2\zeta_c &= \zeta_c\,\mathrm{erf}\Big(\frac{\zeta_c}{\sqrt{2}}\Big) + \sqrt{\frac{2}{\pi}}\, e^{-\frac{\zeta_c^2}{2}}\,. \\
\alpha_c &= \mathrm{erf}\Big(\frac{\zeta_c}{\sqrt{2}}\Big)\,.
\end{aligned}
\tag{46}
$$

Numerical solution gives

$$
\zeta_c = 0.43632656\ldots,\quad \alpha_c = 0.33740018\ldots \tag{47}
$$

which agrees very well with the value observed in simulations, $\alpha_c \simeq 0.34$.

Now we are ready to compare the analytical results with simulations in the ergodic regime, $\alpha > \alpha_c$. The predictability (28), corresponds to the minimum of the Hamiltonian (27) and in our calculation it is just the minimum of the free energy (41) in the limit $\beta \to \infty$. We find

$$
\theta_A^2 = \frac{\alpha^2}{2}\,\frac{1+Q}{(\alpha+\chi)^2}\,. \tag{48}
$$

From (30) it follows that the volatility, averaged over the realisations of the strategies, is related to the predictability as

$$\sigma^2 = \theta_A^2 + \frac{1-Q}{2} \,. \tag{49}$$

As a brief comparison of the analytical results with numerical data, look at Fig. 6. Clearly, the original MG is slightly off the analytical prediction, which is due to the fact that we assumed all memory patterns to be equally probable. If we implement this assumption in simulations, the agreement is excellent.

We can also look at the volatility in Fig. 2. The agreement is rather satisfactory and even more, we can see that it is the better the closer we are to the critical point. Indeed, this is due to the fact that the distribution of memory patterns becomes uniform at α_c. This leads us to conjecture that the analytically found critical value (47) of α_c is in fact an exact result.

3. Is the Minority Game Useful?

No doubt, if the econophysicists contributed to the progress of science only by inventing Minority Game, they would already deserve fame. Indeed, MG is a beautiful example of a simple, clear and difficult, yet soluble kodel of statistical physics. This makes some call the Minority Game the "Ising model of econophysics", although it is certainly exaggerated.

On the other hand, we may ask if it fulfilled the initial expectations, i. e. explaining what happens when people trade shares on a stock market. At this point, there is no direct success. It is a well established fact that the distribution of price changes at all markets has a power-law tail. This is the feature which makes stock market complex but this feature is, unfortunately, not reproduced in the Minority Game. However, there are tricks how to alleviate this failure. We shall not go to details, but just briefly describe the ideas.

3.1. *Producers and speculators*

In the standard MG all agents are essentially equal. They differ only in their randomly chosen strategies. An important step towards reality is to include two types of behaviour, which we call producers and speculators.[31] Suppose there are N_p and N_s of them, respectively. In fact, it is now unnecessary to keep more than one strategy for an agent, so each agent has unique prescription what to do if certain memory pattern μ arises.

The producers are more simple. They just always do what the strategy says. The speculators are a bit more complicated, as they may decide whether they want to play or not. If they play, they do what the strategy prescribes, but if they abstain from the game, their contribution to the attendance is zero.

To specify whether a given agent should play or not, we look at the score of her strategy, which is updated even if the agent does not play. If the score exceeds a crucial control parameter ϵ, the agent participates, otherwise she abstains.

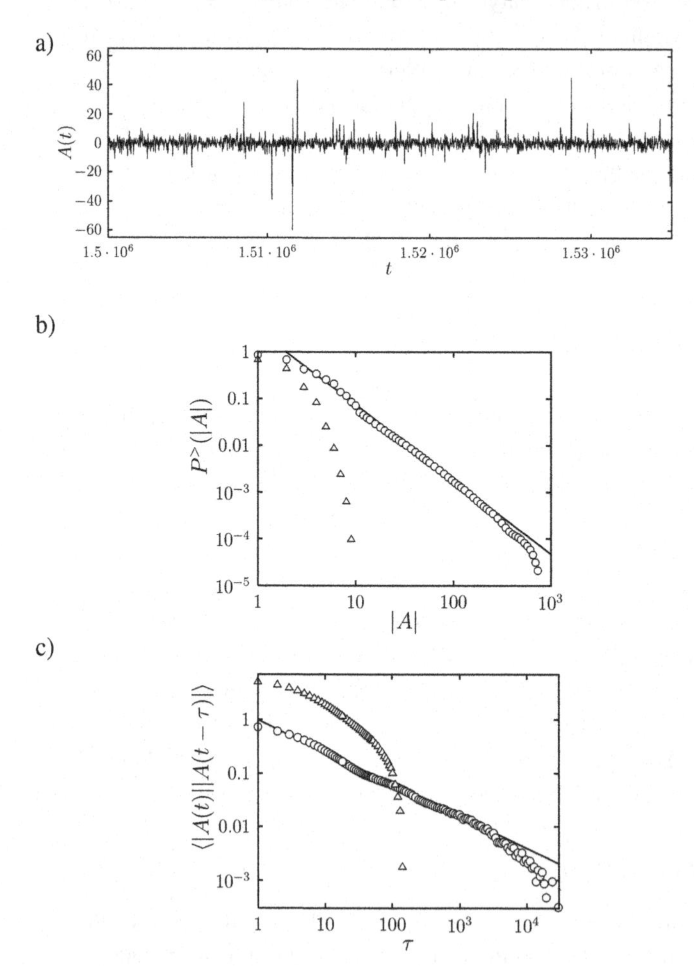

Fig. 7. Properties of the time series of attendance in the Grand-canonical Minority Game. In the panel a), a typical realisation of the time series. In the panel b), cumulative distribution of the absolute attendance. The symbols distinguish two typical situations: the realisation with high kurtosis (○) and with low kurtosis (△). The parameters were $N_p = 16$, $N_s = 1601$, $M = 4$ and $\epsilon = 0.01$. The line is the power law $|A|^{-1.6}$. In the panel c), the aurocorrelation of absolute attendance is shown. Again, we show typical realisation with strong volatility clustering (○) and with weak one (△). The line is the power $\sim \tau^{-1/2}$.

3.2. *Provides MG power-law distributions?*

In analogy with the grand-canonical ensemble, which permits fluctuations in the number of particles, the above explained variant of the MG is called Grand-canonical Minority Game.[32–37] In Fig. 7 we can see the typical fluctuations of the attendance. Due to speculators coming in and out, the attendance fluctuates much more than in the classical MG. We shall look at its distribution

$$P^{>}(|A|) = \lim_{T\to\infty} \frac{1}{T} \sum_{t=1}^{T} \theta\Big(|\sum_{i=1}^{N} a_i(t)| - |A| \Big) \tag{50}$$

where in this definition $\theta(x) = 1$ for $x \geq 0$ and $\theta(x) = 0$ for $x < 0$. To see the properties of the time series quantitatively, we show, again in Fig. 7, the distribution of absolute values of the attendance, $P^{>}(|A|)$ which should be compared to the return in price series. We also show the autocorrelation of the absolute attendance.

There is a surprise in the data. In fact, we observe two types of behaviour, depending on the initial conditions. Either the attendance is distributed as a power law, $P^{>}(|A|) \sim |A|^{1-\gamma}$ with exponent $\gamma \simeq 2.6$, or the distribution is exponential. Similarly, in some realisations the autocorrelation falls off exponentially and in others it decreases slowly as a power law with exponent close to $\simeq 0.5$. So, what is the generic behaviour of the game?

As a partial but well-defined quantitative measure, we can calculate the kurtosis of the distribution of the attendances. If it were Gaussian, the kurtosis would be $\kappa = 3$, so the excess kurtosis, defined as $|\kappa - 3|$, tells us how much the distribution deviates from the Gaussian shape. We expect that power-law distribution will have very large excess kurtosis, while slight modifications of the Gaussian will exhibit much smaller value.

It is possible to plot the histogram of the excess kurtosis found in many realisations of the Grand-canonical MG. If we do that, we observe two distinct peaks, proving that there are indeed two qualitatively different types of behaviour. The key parameter for the appearance of the peak with high kurtosis is the quantity $n_s = N_s/P$, where $P = 2^M$ is again the number of distinct memory patterns. It comes up at about $n_s \simeq 10$ and for larger n_s the kurtosis grows to very high values, typically $\kappa \simeq 10^3$ for $n_s \simeq 100$. Also the overall weight of the high-kurtosis peak is higher for higher number of speculators.

We can conclude this paragraph by positive answer to the question raised in its title. Yes, there is quite natural way how to reproduce the behaviour of stock-market prices, namely power-law tail in the distribution of price changes. There is a variant of the Minority Game to provide that. Although some doubts still persist, for example how comes that some realisations are in accord with stylised facts and some not, MG seems to be one of the most fertile grounds for econophysics modelling.

4. Further Reading

Minority Game was elaborated by various ways. To learn more, the reader may resort to books[8–10] and reviews.[38,39] Let us briefly note several important directions we have not covered here. First, the replica method is not the only analytical approach applicable. You may use the more intuitive crowd-anticrowd theory of Johnson et al.[40–43] On the other side, much more sophisticated approach, which provides deeper and wider results is the generating functional method of Coolen et al.[44–50]

The Minority Game was also generalised in various other ways, besides the already mentioned Grand-canonical MG. For example, including the market impact into the strategies of agents leads to a model with replica-symmetry broken state.[30] The reward may be distributed in various more sophisticated ways, resulting in new phenomena[51,52]

One may also ask, what if the rule is not the minority, but the majority one.[53–55] The agents may be allowed to evolve by selection.[56–58] Local versions of MG, where the agents interact only with nearest neighbours were investigated.[59–62] If the agents accept leadership of their more successful neighbours, complex imitation structures, resulting in complex scale-free networks emerge.[63–66] The MG was also played on-line with human players.[67,68] And finally, there are also quantum versions of the Minority Game.[69,70] Therefore, we believe there is enough space for an ambitious researcher to push the frontier further on.

Appendix

Frustration

Let us explain the concept of frustration on a simple example of a spin system, as shown in the following figure.

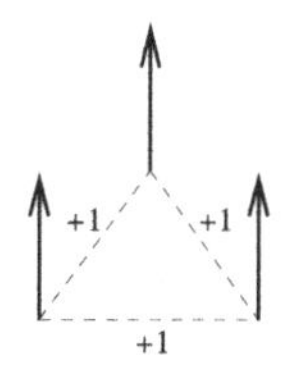

Ferromagnetic case: $1 \times 2 = 2$ ground states. All bonds satisfied.

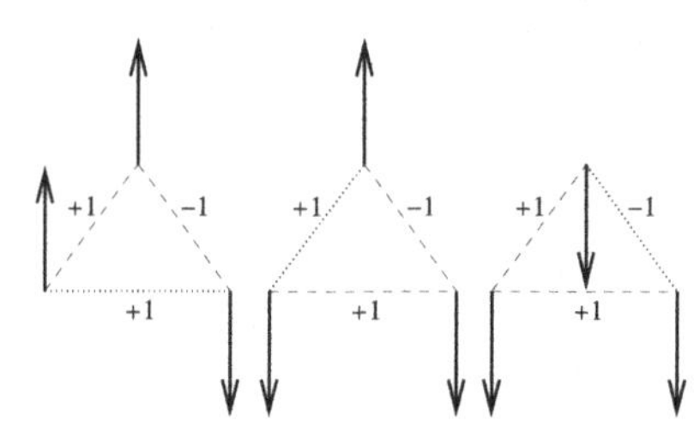

Frustrated case: $3 \times 2 = 6$ ground states. Unsatisfied bond is shown as a dotted line.

If three Ising spins are bound together by ferromagnetic interaction, so the Hamiltonian is $H = s_1s_2 + s_2s_3 + s_3s_1$, the ground state is obviously the uniform configuration $s_1 = s_2 = s_3$. All three bonds can be considered "satisfied", as all three pairs of spins individually are in their lowest-energy configuration. The ground state is twice degenerate due to the global symmetry; the energy does not change if we flip all spins simultaneously. On the other hand, if one of the bonds is antiferromagnetic, so $H = s_1s_2 + s_2s_3 - s_3s_1$, there is no configuration in which all three bonds were satisfied. Such situation is called frustration. Since there are three equivalent choices of the unsatisfied bond, there are three different ground state spin configurations (taking into account the spin-flip symmetry, we have altogether six ground states).

Frustration is the source of complex behaviour of many systems, especially disordered ones. As seen already in our example, frustration leads to proliferation of equilibrium states. For example in spin glasses the number of equilibria increases exponentially with the number of spins.

Replica trick

Replica trick is a technique for computing averaged properties of a random system in thermodynamic equilibrium. Let us have a Hamiltonian $H(s, a)$ depending on state variables s (e. g. spins) and moreover on random parameters a (e. g. randomly placed impurities). As a are random variables, so is also the Hamiltonian, the partition function $Z = \sum_s \mathrm{e}^{-H}$, the free energy $F = -\ln Z$ and all other

thermodynamic quantities (for brevity we set the inverse temperature $\beta = 1$). Physically relevant results are the averages over the disorder, most notably the mean free energy $\overline{F} \equiv \int P(a) F \mathrm{d}a$. From the probability theory we know that all information on the random variable F is contained in its characteristic function $Z(n) = \overline{\mathrm{e}^{-nF}}$ depending on the complex variable $n \in \mathbb{C}$. For example the average $\overline{F} = -\lim_{n\to 0} \frac{\mathrm{d}}{\mathrm{d}n} Z(n)$ and the fluctuations related to the second moment $\overline{F^2} = \lim_{n\to 0} \frac{\mathrm{d}^2}{\mathrm{d}n^2} Z(n)$ can be readily obtained.
The difficulty arises when we want to actually compute $Z(n)$ for a given model. It comes out that we are unable to get it for general n, except for positive integer values $n \in \mathbb{N}_+$. We are saying that we have n "replicas" of the original system. To make the derivatives and the limit $n \to 0$ we must first make the analytic continuation of the function $Z(n)$ from $\mathbb{N}_+$ to the rest of the complex plane, or at least to certain neighbourhood of the point $n = 0$. The existence and uniqueness of this continuation is one of the hardest open problems of contemporary mathematical physics. The best we can do here is to assume that the continuation does exist and that it is unique. Some general features can be stated, though. The continuation must be always done around an accumulation point of the set on which $Z(n)$ is known. Here the only accumulation point is $n = \infty$, so we should effectively work with infinitely many replicas. The continuation in fact goes from the neighbourhood of $n = \infty$ to the neighbourhood of $n = 0$.

Hubbard-Stratonovich transform

is based on the identity

$$\int_{-\infty}^{\infty} \exp\left(-\tfrac{1}{2}x^2 + Ax\right) \frac{\mathrm{d}x}{\sqrt{2\pi}} = \exp\left(\tfrac{1}{2}A^2\right).$$

It is used very often to convert four-spin interactions to two-spin ones and terms which are quadratic in state variables into linear terms, on the expense of introducing a new auxiliary continuous variable. A posteriori, such variable is usually identified with certain equilibrium characteristics of the model.

Originality of the results

For the purpose of writing this chapter we developed our own codes implementing the canonical MG, as well as its variants, namely the grand-canonical MG. All simulation results shown in this chapter come from our own data resulting from our own codes.

Acknowledgement

This work was partially supported by the MŠMT of the Czech Republic, grant no. OC09078.

References

1. R. N. Mantegna and H. E. Stanley, *Introduction to Econophysics: Correlations and Complexity in Finance* (Cambridge University Press, Cambridge, 2000).
2. M. Levy, H. Levy, and S. Solomon, *Microscopic Simulation of Financial Markets* (Academic Press, San Diego, 2000).
3. B. M. Roehner, *Patterns of Speculation: A Study in Observational Econophysics* (Cambridge University Press, Cambridge, 2002).
4. J.-P. Bouchaud and M. Potters, *Theory of Financial Risk and Derivative Pricing* (Cambridge University Press, Cambridge, 2003).
5. J. Voit, *The Statistical Mechanics of Financial Markets*, (Springer, Berlin, 2003).
6. D. Sornette, *Why Stock Markets Crash: Critical Events in Complex Financial Systems* (Princeton University Press, Princeton, 2003).
7. F. Slanina, *Essentials of Econophysics Modelling* (Oxford University Press, to appear).
8. N. F. Johnson, P. Jefferies, and P. M. Hui, *Financial Market Complexity*, (Oxford University Press, Oxford, 2003).
9. D. Challet, M. Marsili, and Y.-C. Zhang, *Minority Games*, (Oxford University Press, Oxford, 2005).
10. A. C. C. Coolen, *The Mathematical Theory of Minority Games*, (Oxford University Press, Oxford, 2005).
11. W. B. Arthur, Amer. Econ. Review (Papers and Proceedings) **84**, 406 (1994).
12. D. Challet and Y.-C. Zhang, Physica A **246**, 407 (1997).
13. D. Challet and Y.-C. Zhang, Physica A **256**, 514 (1998).
14. R. Savit, R. Manuca and R. Riolo, adap-org/9712006.
15. R. Savit, R. Manuca, and R. Riolo, Phys. Rev. Lett. **82**, 2203 (1999).
16. D. Challet and M. Marsili, Phys. Rev. E **60**, R6271 (1999).
17. F. Slanina and P. Chvosta, Phys. Rev. E **69**, 041502 (2004).
18. F. Slanina, Eur. Phys. J. B **79**, 99 (2011).
19. M. Mézard, G. Parisi, and M. A. Virasoro, *Spin Glass Theory and Beyond* (World Scientific, Singapore 1987).
20. P. C. Martin, E. D. Siggia, and H. A. Rose, Phys. Rev. A **8**, 423 (1973).
21. C. De Dominicis, Phys. Rev. B **18**, 4913 (1978).
22. A. Cavagna, Phys. Rev. E **59**, R3783 (1999).
23. D. Challet and M. Marsili, Phys. Rev. E **62**, 1862 (2000).
24. A. Cavagna, J. P. Garrahan, I. Giardina, and D. Sherrington, Phys. Rev. Lett. **83**, 4429 (1999).
25. D. Challet, M. Marsili, and R. Zecchina, Phys. Rev. Lett. **85**, 5008 (2000).
26. A. Cavagna, J. P. Garrahan, I. Giardina and D. Sherrington, Phys. Rev. Lett. **85**, 5009 (2000).
27. D. Challet, M. Marsili, and R. Zecchina, Phys. Rev. Lett. **84**, 1824 (2000).
28. M. Marsili, D. Challet, and R. Zecchina, Physica A **280**, 522 (2000).
29. M. Marsili and D. Challet, Phys. Rev. E **64**, 056138 (2001).
30. A. De Martino and M. Marsili, J. Phys. A: Math. Gen. **34**, 2525 (2001).
31. F. Slanina and Y.-C. Zhang, Physica A **272**, 257 (1999).
32. Y.-C. Zhang, Europhysics News **29**, 51 (1998).

33. D. Challet, A. De Martino, M. Marsili, and I. Pérez Castillo, J. Stat. Mech. P03004 (2006).
34. I. Giardina, J.P. Bouchaud, and M. Mézard, Physica A **299**, 28 (2001).
35. D. Challet and M. Marsili, Phys. Rev. E **68**, 036132 (2003).
36. M. Marsili, Physica A **324**, 17 (2003).
37. D. Challet, M. Marsili, and Y.-C. Zhang, Physica A **299**, 228 (2001).
38. A. De Martino and M. Marsili, J. Phys. A: Math. Gen. **39**, R465 (2006).
39. T. Galla, G. Mosetti, and Y.-C. Zhang, physics/0608091.
40. N. F. Johnson, P. M. Hui, D. Zheng, and M. Hart, J. Phys. A: Math. Gen. **32**, L427 (1999).
41. N. F. Johnson, P. M. Hui, R. Jonson, and T. S. Lo, Phys. Rev. Lett. **82**, 3360 (1999).
42. M. Hart, P. Jefferies, N. F. Johnson, and P. M. Hui, Physica A **298**, 537 (2001).
43. M. Hart, P. Jefferies, P. M. Hui, and N. F. Johnson, Eur. Phys. J. B **20**, 547 (2001).
44. J. A. F. Heimel and A. C. C. Coolen, Phys. Rev. E **63**, 056121 (2001).
45. A. C. C. Coolen, J. A. F. Heimel, and D. Sherrington, Phys. Rev. E **65**, 016126 (2002).
46. A. C. C. Coolen and J. A. F. Heimel, J. Phys. A: Math. Gen. **34**, 10783 (2001).
47. T. Galla, A. C. C. Coolen, and D. Sherrington, J. Phys. A: Math. Gen. **36**, 11159 (2003).
48. A. C. C. Coolen, J. Phys. A: Math. Gen. **38**, 2311 (2005).
49. A. C. C. Coolen, cond-mat/0205262.
50. J. A. F. Heimel and A. De Martino, J. Phys. A: Math. Gen. **34**, L539 (2001).
51. F. Slanina and Y.-C. Zhang, Physica A **289**, 290 (2001).
52. J. V. Andersen and D. Sornette, Eur. Phys. J. B **31**, 141 (2003).
53. M. Marsili, Physica A **299**, 93 (2001).
54. P. Kozłowski and M. Marsili, J. Phys. A: Math. Gen. **36**, 11725 (2003).
55. A. De Martino, I. Giardina, and G. Mosetti, J. Phys. A: Math. Gen. **36**, 8935 (2003).
56. R. Manuca, Y. Li, R. Riolo, and R. Savit, adap-org/9811005.
57. Y. Li, R. Riolo, and R. Savit, Physica A **276**, 234 (2000).
58. Y. Li, R. Riolo, and R. Savit, Physica A **276**, 265 (2000).
59. S. Moelbert and P. De Los Rios, Physica A **303**, 217 (2002).
60. E. Burgos, H. Ceva, and R. P. J. Perazzo, Physica A **337**, 635 (2004).
61. E. Burgos, H. Ceva, and R. P. J. Perazzo, Physica A **354**, 518 (2005).
62. G. Fagiolo and M. Valente, Computational Economics **25**, 41 (2005).
63. F. Slanina, Physica A **286**, 367 (2000).
64. F. Slanina, Physica A **299**, 334 (2001).
65. H. Lavička and F. Slanina, Eur. Phys. J. B **56**, 53 (2007).
66. M. Anghel, Z. Toroczkai, K. E. Bassler, and G. Korniss, Phys. Rev. Lett. **92**, 058701 (2004).
67. P. Ruch, J. Wakeling, and Y.-C. Zhang, cond-mat/0208310.
68. P. Laureti, P. Ruch, J. Wakeling, and Y.-C. Zhang, Physica A **331**, 651 (2004).
69. R. Kay, N. F. Johnson, and S. C. Benjamin, J. Phys. A: Math. Gen. **34**, L547 (2001).
70. N. F. Johnson, Phys. Rev. A **63**, 020302 (2001).

Subject Index